10th Anniversary of Nanomaterials—Recent Advances in Nanocomposite Thin Films and 2D Materials

10th Anniversary of Nanomaterials—Recent Advances in Nanocomposite Thin Films and 2D Materials

Editors

Jordi Sort
Gemma Rius

MDPI • Basel • Beijing • Wuhan • Barcelona • Belgrade • Manchester • Tokyo • Cluj • Tianjin

Editors
Jordi Sort
ICREA
Universitat Autònoma de
Barcelona
Spain

Gemma Rius
Institut de Microelectrònica de
Barcelona, IMB-CNM-CSIC
Spain

Editorial Office
MDPI
St. Alban-Anlage 66
4052 Basel, Switzerland

This is a reprint of articles from the Special Issue published online in the open access journal *Nanomaterials* (ISSN 2079-4991) (available at: http://www.mdpi.com).

For citation purposes, cite each article independently as indicated on the article page online and as indicated below:

LastName, A.A.; LastName, B.B.; LastName, C.C. Article Title. *Journal Name* **Year**, *Volume Number*, Page Range.

ISBN 978-3-0365-2159-6 (Hbk)
ISBN 978-3-0365-2160-2 (PDF)

Contents

About the Editors

Prof. Jordi Sort received his PhD Degree in Materials Science from Universitat Autònoma de Barcelona (UAB) in 2002 (Extraordinary Award). The topic of his PhD dissertation was the study of magnetic exchange interactions in ferromagnetic–antiferromagnetic systems. He worked for two years as a Postdoctoral Researcher at SPINTEC (Grenoble) and almost one year at the Argonne National Laboratory. Prof. Sort now leads the "Group of Smart Nanoengineered Materials, Nanomechanics and Nanomagnetism", which focuses its activities on the synthesis of a wide variety of functional materials and the study of their structural, mechanical and magnetic properties. Prof. Sort's work has received awards from the Catalan Physical Society (2000), the Spanish Royal Physical Society (Young Researcher Award in Experimental Physics, 2003) and the Federation of Materials Societies (FEMS Prize in Materials Science & Technology, 2015). In 2014, Prof. Sort was awarded a Consolidator Grant from the European Research Council.

Dr. Gemma Rius is a Ramon y Cajal Scientist at the Institute of Microelectronics of Barcelona, IMB-CNM-CSIC (2018-present). She graduated in Physics from the Universitat Autònoma de Barcelona (UAB) in 2002 and has since been working in nanotechnology, applying nano- and microfabrication methods. She obtained her PhD with her thesis "Electron Beam Lithography for Nanofabrication" from the UAB (2008). Her postdoctoral experience of 7 years, in Japan, includes a Tenure Track position at the Nagoya Institute of Technology, while she returned to the IMB-CNM-CSIC as a Marie Curie Fellow in 2015. She has realized pioneering work in nanodevice fabrication, developed internationally competitive work in graphene synthesis and nanoelectronics, and achieved other developments such as CNT and carbon nanofiber functionalized probes for atomic force microscopy (AFM) or graphene-based supercapacitors and batteries or the application of original nanopatterning methods based on either AFM advanced modes or fine ion beams.

nanomaterials

Editorial

Editorial for the Special Issue on "10th Anniversary of Nanomaterials—Recent Advances in Nanocomposite Thin Films and 2D Materials"

Jordi Sort [1,2,*] **and Gemma Rius** [3]

1   Departament de Física, Universitat Autònoma de Barcelona (UAB), E-08193 Cerdanyola Del Vallès, Spain
2   Institució Catalana de Recerca i Estudis Avançats (ICREA), Pg. Lluís Companys 23, E-08010 Barcelona, Spain
3   Instituto de Microelectrónica de Barcelona (IMB-CNM, CSIC), Campus UAB, E-08193 Cerdanyola del Vallès, Spain; gemma.rius@csic.es
*   Correspondence: Jordi.Sort@uab.cat

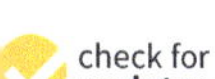

Citation: Sort, J.; Rius, G. Editorial for the Special Issue on "10th Anniversary of Nanomaterials— Recent Advances in Nanocomposite Thin Films and 2D Materials". *Nanomaterials* **2021**, *11*, 2069. https://doi.org/10.3390/nano11082069

Received: 4 August 2021
Accepted: 11 August 2021
Published: 15 August 2021

Publisher's Note: MDPI stays neutral with regard to jurisdictional claims in published maps and institutional affiliations.

As a way to celebrate the 10th anniversary of the journal Nanomaterials, this Special Issue within the section 'Nanocomposite thin film and 2D materials' provides an overview of the wide spectrum of research challenges and applications in the field, represented by a collection of 12 contributions, including three up-to-date review articles plus nine original works, in different targeted topics as described below.

The importance of synthesis and processing in the properties of compound or alloy thin films and their applications as functional coatings is reported in three of the contributed articles. Such is the case for perovskite $BiFeO_3$ films presented by Micard et al. [1]). They use metal organic chemical vapor deposition (MOCVD) to optimize polycrystalline, pure phase thin films, as understood from XRD, EDX, and FE-SEM characterizations. Films piezoelectricity and ferroelectric property is confirmed by piezo force microscopy and spectroscopy, which envisions their use as lead-free hybrid energy harvesters. Another synthesis work is presented by Sudiyarmanto and Kondoh in [2]. In this case, supercritical fluid chemical deposition is applied to the production of Ni-Pt alloy thin films. By tuning the deposition rate with the precursors ratio, they obtain, as well, single-phase, polycrystalline material of the Ni-Pt alloy. The films are intended to be used as model catalyst surfaces due to their high activity and stability. In [3], the relevance of amorphous films is exemplified in contribution of Yan et al., in which a processing development on Mg-based metallic glasses for improving the corrosion resistance is presented. Particularly, first, MgO nanoplate arrays are coated by cyclic voltammetry treatments, and then, stearic acid is efficiently adhered to their surface, so that the corrosion resistance to NaCl solution is increased.

The versatility and flexibility of the growing and continuously evolving family of 2D materials is also well represented in this Special Issue. The advances in graphene materials are still generating new knowledge while still requiring significant technological progress. Novel methods to obtain graphene–polymer composite films with multiple functionalities, or their use for sensing platforms are two examples.

Van der Schueren et al. [4] show how co-mixing aqueous colloids plus casting of PVA and few-layer graphene can be used to obtain composite films with multiple functional properties. As their exfoliation method provided relatively large graphene flakes, the PVA-FLG composite exhibits good mechanical and electrical conductivity characteristics, as well as potential as an O_2 barrier membrane based on a transmission rate reduction of 60%. In another contributed paper [5], graphene is used to realize interference-enhanced Raman scattering. Here, the key is to combine single-layer graphene with an ultra-thin alumina film on top of a metallic aluminum support. Correlating both experimental and theoretical results, their interference amplification can also be implemented. With this platform, on the basis of more conventional SERS, good results were obtained by simply adding ultra-small silver particles.

1

Closely related to the relevant and relatively new graphene and composites investigations domain, families of emerging materials such as 2D carbides and nitrides (MXene) and transistion metal dichalcogenides (TMD) are also receiving lots of attention and are thus represented in this Special Issue. The paper by Raagulan et al. [6] reviews reported works on MXene–graphene aerogel composites, with the focus on their use as electromagnetic interference shielding materials. Their efficiency is correlated with the obtained morpho structural characteristics, while they compile relevant information in terms of processing techniques. In addition, centered on the specific functionalities of MXene materials, another review is provided by Ibrahim et al. [7] on their application for supercapacitors. In partic ular, they summarize the current knowledge and assess the progress in the self-standing MXenes as understood from their mechanical properties analysis and comparison to hybrid MXenes or other 2D materials.

Finding energy solutions in nanomaterials and, particularly, exploiting higher effi ciencies derived from intrinsically high and, for instance, electrochemically-active, surface area of 2D materials has been an intense quest in the field. The original work provided by Hussain et al. [8] is a good example of the opportunities to investigate combinations of emerging materials and new synthetic processes. They develop a simple, one-pot chemical reaction to produce hybrid W_2C/WS_2 nanostructured electrodes, which show good electrochemical performances for energy applications, such as hydrogen evolution and supercapacitors.

Another demonstration of the new developments needed for nanostructured, emer gent 2D materials is given by Li et al. in [9]. In this case, they aim to facilitate integration toward industry by providing automated software methods and hyperspectral imaging hardware for characterizing MoS_2 materials. Their convolutional neural-network-based al gorithms demonstrated identification capabilities with a special resolution down to 100 nm and reasonable acquisition times in relatively large images.

Concluding, nanomaterials combinations including structural carbon films have a long run and will be the object of many research studies and developments in view of their commercial adoption. Versatile, abundant carbon, engineered at the nanoscale, still brings advances such as reflected in the following three papers, two of them about original developments of functional films or coatings and a concise review on RRAM devices.

Zhang et al. [10] combine carbon nitride films with TiO_2 to obtain visible light en hanced photocatalysis, where the key of its efficiency comes from the optimization of plasma processing gases, while Sharma et al. [11] also use ion-based methods and study the synthesis of Li-C nanocomposites to evaluate their material potential for alternative batteries, such as based on Li–air interfaces. A comprehensive review on RRAM resis tive switching mechanisms, materials, and bionic synaptic application is provided by Shen et al. [12] as a good example of the central role of nanomaterials in enabling sustain able and efficient solutions to current and future demands of our society.

Funding: This research was funded by the *Agencia Estatal de Investigación* of the Spanish Government (Grant Nos. MAT2017-86357-C3-1-R, PID2019-104670GB-I00 and associated FEDER, and RYC-2016 21412), and the Generalitat de Catalunya (Grant Nos. 2017-SGR-292 and 2017-SGR-105).

Conflicts of Interest: The authors declare no conflict of interest.

References

1. Micard, Q.; Condorelli, G.G.; Malandrino, G. Piezoelectric BiFeO3 thin films: Optimization of MOCVD process on Si. *Nanomaterials* **2020**, *10*, 603. [CrossRef] [PubMed]
2. Sudiyarmanto; Kondoh, E. Synthesis and characterization of Ni-Pt alloy thin films prepared by supercritical fluid chemica deposition technique. *Nanomaterials* **2021**, *11*, 151. [CrossRef] [PubMed]
3. Yan, Y.; Liu, X.; Xiong, H.; Zhou, J.; Yu, H.; Qin, C.; Wang, Z. Stearic Acid Coated MgO Nanoplate Arrays as Effective Hydrophobic Films for Improving Corrosion Resistance of Mg-Based Metallic Glasses. *Nanomaterials* **2020**, *10*, 947. [CrossRef] [PubMed]
4. Van der Schueren, B.; El Marouazi, H.; Mohanty, A.; Lévêque, P.; Sutter, C.; Romero, T.; Janowska, I. Polyvinyl alcohol-few layer graphene composite films prepared from aqueous colloids. Investigations of mechanical, conductive and gas barrier properties *Nanomaterials* **2020**, *10*, 858. [CrossRef] [PubMed]

Aguilar-Pujol, M.; Ramírez-Jiménez, R.; Xifre-Perez, E.; Cortijo-Campos, S.; Bartolomé, J.; Marsal, L.; de Andrés, A. Supported ultra-thin alumina membranes with graphene as efficient interference enhanced raman scattering platforms for sensing. *Nanomaterials* **2020**, *10*, 830. [CrossRef] [PubMed]

Raagulan, K.; Kim, B.M.; Chai, K.Y. Recent advancement of electromagnetic interference (EMI) shielding of two dimensional (2D) MXene and graphene aerogel composites. *Nanomaterials* **2020**, *10*, 702. [CrossRef] [PubMed]

Ibrahim, Y.; Mohamed, A.; AbdelGawad, A.M.; Eid, K.; Abdullah, A.M.; Elzatahry, A. The recent advances in the mechanical properties of self-standing two-dimensional MXene-based nanostructures: Deep insights into the supercapacitor. *Nanomaterials* **2020**, *10*, 1916. [CrossRef] [PubMed]

Hussain, S.; Rabani, I.; Vikraman, D.; Feroze, A.; Ali, M.; Seo, Y.-S.; Kim, H.-S.; Chun, S.-H.; Jung, J. One-pot synthesis of W_2C/WS_2 hybrid nanostructures for improved hydrogen evolution reactions and supercapacitors. *Nanomaterials* **2020**, *10*, 1597. [CrossRef] [PubMed]

Li, K.-C.; Lu, M.-Y.; Nguyen, H.T.; Feng, S.-W.; Artemkina, S.B.; Fedorov, V.E.; Wang, H.-C. Intelligent identification of MoS2 nanostructures with hyperspectral imaging by 3D-CNN. *Nanomaterials* **2020**, *10*, 1161. [CrossRef] [PubMed]

10. Zhang, B.; Peng, X.; Wang, Z. Noble metal-free TiO2-coated carbon nitride layers for enhanced visible light-driven photocatalysis. *Nanomaterials* **2020**, *10*, 805. [CrossRef] [PubMed]

11. Sharma, S.; Osugi, T.; Elnobi, S.; Ozeki, S.; Jaisi, B.P.; Kalita, G.; Capiglia, C.; Tanemura, M. Synthesis and characterization of Li-C nanocomposite for easy and safe handling. *Nanomaterials* **2020**, *10*, 1483. [CrossRef] [PubMed]

12. Shen, Z.; Zhao, C.; Qi, Y.; Xu, W.; Liu, Y.; Mitrovic, I.Z.; Yang, L.; Zhao, C. Advances of RRAM devices: Resistive switching mechanisms, materials and bionic synaptic application. *Nanomaterials* **2020**, *10*, 1437. [CrossRef] [PubMed]

 nanomaterials

Article

Synthesis and Characterization of Ni-Pt Alloy Thin Films Prepared by Supercritical Fluid Chemical Deposition Technique

Sudiyarmanto [1,2] and Eiichi Kondoh [1,*]

[1] Integrated Graduate School of Medicine, Engineering, and Agricultural Sciences, University of Yamanashi, Kofu 400-8511, Japan; g18dtea3@yamanashi.ac.jp or sudi012@lipi.go.id
[2] Research Center for Chemistry, Indonesian Institute of Sciences (LIPI), Kawasan PUSPIPTEK Serpong, Tangerang Selatan, Banten 15310, Indonesia
* Correspondence: kondoh@yamanashi.ac.jp; Tel.: +81-55-220-8472

Abstract: Ni-Pt alloy thin films have been successfully synthesized and characterized; the films were prepared by the supercritical fluid chemical deposition (SFCD) technique from $Ni(hfac)_2 \cdot 3H_2O$ and $Pt(hfac)_2$ precursors by hydrogen reduction. The results indicated that the deposition rate of the Ni-Pt alloy thin films decreased with increasing Ni content and gradually increased as the precursor concentration was increased. The film peaks determined by X-ray diffraction shifted to lower diffraction angles with decreasing Ni content. The deposited films were single-phase polycrystalline Ni-Pt solid solution and it exhibited smooth, continuous, and uniform distribution on the substrate for all elemental compositions as determined by scanning electron microscopy and scanning transmission electron microscopy analyses. In the X-ray photoelectron spectroscopy (XPS) analysis, the intensity of the Pt 4f peaks of the films decreased as the Ni content increased, and vice versa for the Ni 2p peak intensities. Furthermore, based on the depth profiles determined by XPS, there was no evidence of atomic diffusion between Pt and Ni, which indicated alloy formation in the film. Therefore, Ni-Pt alloy films deposited by the SFCD technique can be used as a suitable model for catalytic reactions due to their high activity and good stability for various reactions.

Keywords: supercritical fluids; deposition; Ni-Pt alloy films; thin films

Citation: Sudiyarmanto; Kondoh, E. Synthesis and Characterization of Ni-Pt Alloy Thin Films Prepared by Supercritical Fluid Chemical Deposition Technique. *Nanomaterials* **2021**, *11*, 151. https://doi.org/10.3390/nano11010151

Received: 14 December 2020
Accepted: 7 January 2021
Published: 9 January 2021

Publisher's Note: MDPI stays neutral with regard to jurisdictional claims in published maps and institutional affiliations.

1. Introduction

Currently, the development of novel functional micro- and nanomaterials such as thin films [1,2], nanoparticles [3,4], nanowires [5,6], and nanorods [7,8] has attracted the interest of both scientists and industrial societies around the world. Such materials have been the subject of intense research in the field of inorganic or organic–inorganic hybrid materials aiming to improve the physicochemical properties compared to existing materials. In addition, the materials' characteristics and novel application opportunities depend on the fabrication method, creating the need to design and optimize new synthetic approaches [9,10]. For these reasons, a variety of materials processing technologies are utilized in a wide range of applications areas, including microelectronic devices, optics, energy conversion/storage, chemistry, and catalysis. Among these, the use of supercritical fluids (SCF) could be considered as a versatile approach for designing micro and nanomaterials due to their thermophysical properties. A SCF is a hybrid phase of both liquid and gas that is easily tunable with small variations in temperature or pressure. Under typical processing conditions, SCF are characterized by liquid-like density, gas-like viscosity, higher diffusivity than liquids, and zero surface tension [9–20].

The most frequently applied solvent in SCF is carbon dioxide, which is called supercritical carbon dioxide ($scCO_2$). It has relatively mild critical temperature ($T_c = 31\ °C$) and critical pressure ($P_c = 7.38$ MPa) conditions, which make CO_2 a useful medium for a variety of applications, especially for thermochemical reactions. Moreover, by the addition of co-solvents (ethanol, acetone, hexane, isopropyl alcohol, etc.) in CO_2, the solubility of polar precursors can be increased due to changes in polarity and density of the solvent, which makes it more attractive as a solvent medium for controlling the solubility of precursors in the processing of micro/nanomaterials [9,10]. One of the SCF-based technologies that has attracted interest in the development of functional materials is supercritical fluid chemical deposition (SFCD). This promising technique basically consists of the following process steps: Dissolution of the precursor in the SCF, adsorption of the precursor and a reducing agent onto the substrate surface, reduction of the adsorbed precursor to its metal form on the substrate surface by the reducing agent, and desorption of hydrogenated ligands from the substrate into the SCF phase [15,19].

The SFCD technique can be employed to deposit metals, metal oxides, alloys, and ceramic materials on crystalline substrates, supported materials, and polymers in the form of thin films or nanoparticles. Using this technique, many metals or metal oxides (Cu, Ni, Ru, Co, Pt, Pd, Au, Ag, etc.) have been successfully deposited onto a wide variety of substrates, with good conformal step coverage and high-aspect-ratio features [12,21–27]. However, for synthesized alloy films, the SFCD technique has been employed in relatively few studies, and only a few metal alloys have been reported such as Cu-Ag, Cu-Mn, and Cu-Ni. Zhao et al. reported that single-phase alloys (solid solutions) were not formed during deposition of Cu-Ag and Cu-Mn films, while Rasadujjaman et al. reported that the formation of a Cu-rich solid solution occurred after annealing the deposited films at a higher temperature [28–30]. Such alloy systems are typically known to involve complex intermetallic, phase-separated, and solid solution structures, which makes it difficult to obtain a better understanding of metal alloy deposition. For this reason, knowledge of alloy systems attainable by SFCD synthesis is still needed, especially study of the characteristics of metal alloy films.

Currently, one of the metal alloy films with the greatest development interest is Ni-Pt alloy thin films, which are widely used in many applications such as magnetic micro electromechanical system (MEMS) devices [31], polymer electrolyte fuel cell (PEFC) cath odes [32], and especially for catalytic reactions as diverse as the oxygen reduction reaction (ORR) [32,33], hydrogen evolution reaction (HER) [34,35], and methanol/ethanol oxida tion [9,36–38]. This is because Ni is a homologous element with Pt over a range of mass ratios and shows high catalytic activity and good stability for various reactions [31,32,39]. In addition, alloying Pt with Ni is expected to greatly reduce the overall cost by decreasing the use of noble metal in the prepared catalyst materials. Generally, the formation of a single-phase Ni-Pt alloy (solid solution) depends on the thermodynamics of the fluid system such as the phase diagram and the mixing enthalpy of the element pair. The Ni-Pt phase diagram exhibits full miscibility with the existence of ordered phases, and an approx imately linear relationship of the variation Curie temperature (T_C) on composition in the Ni-rich solid solution, yielding 100 °C at 27 at.% Pt [40]; whereas, the Ni-Pt system indicates a negative mixing enthalpy over the entire range of composition, favoring the formation of solid solution structures [41]. Moreover, the formation of single-phase Ni-Pt alloy that shows full miscibility thanks to their identical face-centered cubic lattice structure (fcc) and their equivalent atomic radii [42]. To date, Ni-Pt alloy thin films with varying compositions have been deposited by sputtering, electrodeposition, electron beam evaporation, and chemical vapor deposition (CVD) methods [9,31–34,43,44].

Herein, in this study, we demonstrate the feasibility of Ni-Pt alloy thin film deposition on TiN/SiO$_2$/Si substrates through hydrogen reduction of nickel(II) hexafluoroacetylaceto nate hydrate and platinum(II) hexafluoroacetylacetonate precursors from scCO$_2$ solutions with various element compositions. In this work, we employed an SFCD flow-type reactor system, which is a single-step process enabling a simple integration procedure for alloy film deposition. Furthermore, we characterized the deposited Ni-Pt alloy thin films by using various analytical techniques.

2. Materials and Methods

2.1. Materials

The substrates used in this research were Si(100) wafers with a TiN/SiO$_2$ layer on top. The other raw materials were liquid CO$_2$, nickel(II) hexafluoroacetylacetonate hydrate [Ni(hfac)$_2$·3H$_2$O] and platinum(II) hexafluoroacetylacetonate [Pt(hfac)$_2$] as precursors, H$_2$ gas as the reducing agent, and acetone as co-solvent.

2.2. Synthesis of Ni-Pt Alloy Thin Films

In this study, Ni-Pt alloy thin films were prepared by a flow-type SFCD reaction system, whose detailed schematic diagram is shown in Figure 1 [12,17].

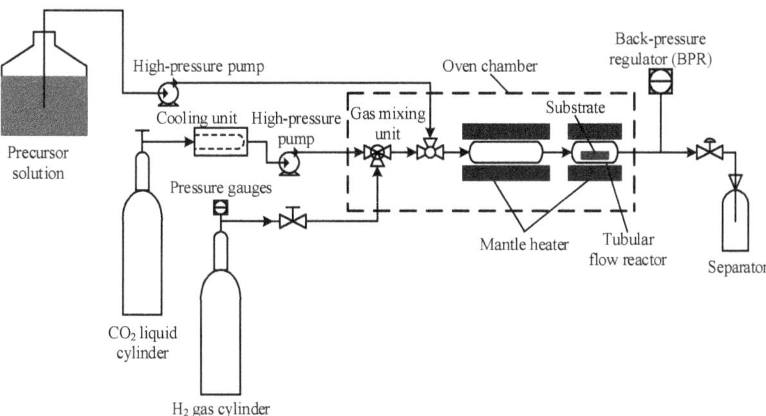

Figure 1. Schematic diagram of flow-type supercritical fluid chemical deposition (SFCD) reaction system.

Ni(hfac)$_2$·3H$_2$O and Pt(hfac)$_2$ precursors were mixed and stirred with acetone solvent listing these concentrations in Table 1 in a glass beaker, and hereafter this solution is referred to as the precursor solution. Meanwhile, the scCO$_2$ solution was mixed with H$_2$ gas by using a gas mixing unit. Then, the scCO$_2$ solution (flow rate: 3.0 mL/min) was blended with the precursor solution (flow rate: 0.5 mL/min) by using a triangular pipe, and the mixture was continuously flowed into a tubular flow reactor. Before entering the reactor, the solution was preheated by the electric heater at 150 °C. Subsequently, the tubular flow reactor was heated by the electric heater (type: P-21, 100 V, 100 W) at 300–330 °C with a heating rate of 5 °C/min. All temperatures in the reactor were controlled by a type K calibrated thermocouple. Prior to the deposition process, the substrate was placed inside the tubular flow reactor, which has dimensions of 60 mm in length, 10 mm in inner diameter, and 1 mm in wall thickness. All of the equipment including the tubular flow reactor, gas mixing unit, and related valves and piping, were placed in an oven chamber maintained at 40 °C. The pressure of this system was controlled by a back-pressure regulator (BPR) at 10 MPa located downstream of the reactor. The deposition conditions of the Ni-Pt alloy thin films used in this work are summarized in Table 1. Furthermore, the thickness of the deposited Ni-Pt alloy thin films was measured using a Veeco-Dektak 150 profilometer.

Table 1. Deposition conditions for Ni-Pt alloy thin films.

Parameters	Value
Precursor concentration	$(7.7-15.3) \times 10^{-3}$ (mole%)
Deposition temperature	300–330 °C
H_2 concentration	0.94 (mole%)
Total pressure	10 MPa
Deposition time	30 min
Flow rate of CO_2 solution	3.1 mL/min
Flow rate of precursor solution	0.4 mL/min

2.3. Ni-Pt Alloy Thin Film Characterization

The deposited Ni-Pt alloy thin films were characterized in order to obtain their chemical composition, crystallinity, morphology, and topography. X-ray diffraction (XRD) analysis was carried out for crystallographic phase identification of the deposited Ni-Pt alloy thin films using a Shimadzu XRD-6000 (Shimadzu Corp., Kyoto, Japan) diffractometer with Cu Kα radiation at 40 kV and 30 mA in the 2θ range of 10–80°. Elemental analysis of the Ni-Pt alloy thin films was carried out by X-ray fluorescence with a Rigaku ZSX Primus wavelength dispersive X-ray fluorescence spectrometer. For surface morphology analysis, the Ni-Pt alloy thin film samples were observed by field-emission scanning electron microscopy coupled with energy-dispersive X-ray spectroscopy (FE SEM-EDX; JSM-6500 JEOL Ltd., Tokyo, Japan) at an acceleration voltage of 15 kV. Moreover, the chemical state of elements on the surface of the Ni-Pt alloy thin films and the depth profile were analyzed by X-ray photoelectron spectroscopy (XPS; JPS-9200, JEOL Ltd., Tokyo, Japan) with Mg Kα radiation operated at 15 kV and 10 mA accelerating voltage and emission current respectively. Then, the Ni-Pt alloy thin film was observed by a scanning transmission electron microscope coupled with energy-dispersive X-ray spectroscopy (STEM-EDX; Tecnai Osiris, FEI, Eindhoven, Netherlands) with an acceleration voltage of 200 kV to gain an understanding of the crystalline topography and morphological structure.

3. Results and Discussion

3.1. Deposition of Ni-Pt Alloy Thin Films

Ni-Pt alloy thin films were synthesized by the SFCD technique from a precursor mixture of $Ni(hfac)_2 \cdot 3H_2O$ and $Pt(hfac)_2$ with various elemental compositions via hydrogen reduction at different deposition temperatures, obtaining shiny, continuous, and reflective films. After the deposition of Ni-Pt alloy thin films, the film thickness was analyzed by Dektak-150 measurement (Table S1) and the deposition rates were calculated by dividing the total thickness by the deposition time.

Figure 2 shows the deposition rates for Ni-Pt alloy thin films with various elemental compositions as a function of temperature, which was varied over the temperature range 300 °C to 330 °C. At all temperatures, the deposition rate for Ni-Pt alloy thin films decreased with increasing Ni content from 20 to 80 (at.%) in the alloying systems. Decreasing the deposition rate of the Ni-Pt alloy thin films was due to the use of hydrate-based of Ni precursor, which this precursor species typically has low solubility in scCO$_2$ owing to the presence of water molecules bound to Ni in the precursor $Ni(hfac)_2 \cdot 3H_2O$ [16]. Other authors have previously reported that the lower solubility values in scCO$_2$ for hydrated precursor in the copper (II) hexafluoroacetylacetonate [45] and acetylacetonate complexes of cobalt and manganese [46]. In addition, increasing the deposition temperature increased the deposition rate of pure Pt films, whereas this phenomenon occurs in reverse for pure Ni films. Lee et al. reported that the highest growth rate for Pt thin films occurred in the temperature range 280 °C to 340 °C. At higher temperatures, the Pt precursor typically is more exothermic, which is enhanced the chemisorption between precursor and converting agent on surface-active sites of the substrate, resulting in a fast growth rate [47]. Battiston et al. reported that the starting temperatures for Pt film deposition should be at least 240 °C and they divided the deposition temperatures range 240–300 °C in the presence of oxygen and 280–440 °C in the water vapor. The Pt thin film deposition is an autocatalytic process in which Pt has an "induction period"; the time required to build up enough catalytic platinum to cause rapid acceleration of the deposition rate. The induction period for platinum can be achieved by increasing the temperature and the percentage of oxygen [48]. While Ni thin films have a tendency to show a decrease in growth rate at temperatures above 270 °C due to the thermal decomposition of Ni precursor, the resulting Ni precursor available for hydrogen reduction decreases in concentration with increasing reaction temperature above 270 °C [49]. For these reasons, the increase in the Ni content in the Ni-Pt alloy thin films caused a decrease in the deposition rate, which might be related to the grain/crystallite sizes of Ni and Pt elements on the surface of the substrate.

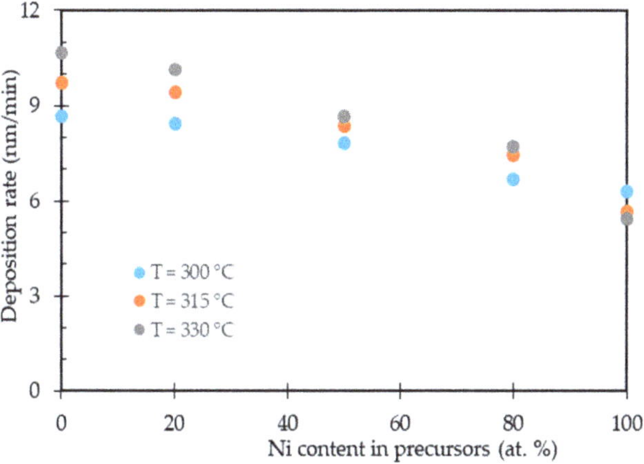

Figure 2. Deposition rate of Ni-Pt alloy thin films as a function of Ni content in the precursor solution.

The effect of precursor concentration with the Ni/Pt ratio of 50:50 (at.%) on the deposition rate was observed at deposition temperatures of 300 °C and 315 °C, as shown in Figure 3. The deposition rate of Ni-Pt thin alloy films gradually increased as the precursor concentration was increased from 7.7×10^{-3} to 15.3×10^{-3} (mole%). Increasing the precursor concentration will lead to more collisions of that precursor, which may favor the precursor available for hydrogen chemisorption increases on surface-active sites of the substrate; this consequently increases the deposition rate of the Ni-Pt thin alloy films. In addition, increasing the precursor concentration results in more frequent molecular interactions between the molecules of precursors in a $scCO_2$/acetone solvent that tend to form complexes such as hydrogen bond or van der Waals forces, which caused increasing the precursor solubility in the fluid system. Teoh et al. reported that the presence of inter- or intra-molecular hydrogen bonds in a $scCO_2$/co-solvent can enhance the solubility of a number of polar solutes due to the increase of polarization of one of the C=O bonds of CO_2 [50]. Regarding the temperature dependence, this figure also shows that the deposition rate increased with the increase in temperature. This is because the solubility of the precursors increases with temperature at higher concentrations, owing to the effect of solute vapor pressure being more dominant compared to the effect of a rapid decrease in density of the fluid, which yielded a higher growth rate for the Ni-Pt thin alloy films. The effect of temperature on solubility usually is more intricate due to the influence of both the density of the $scCO_2$ solvent and the vapor pressure of the solute [16,50]. The highest deposition rate of Ni-Pt alloy thin films obtained was 11.25 nm/min at a precursor concentration of 15.3×10^{-3} (mole%) and a temperature of 315 °C.

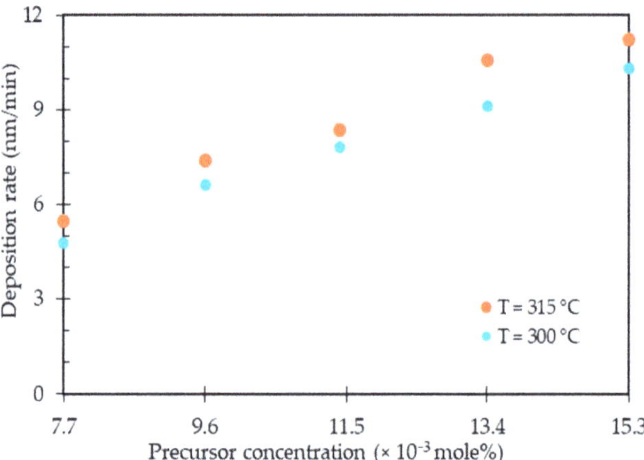

Figure 3. Deposition rate of Ni-Pt alloy thin films as a function of precursor concentration.

To determine the exact elemental composition of the Ni-Pt alloy thin films, the final chemical composition of the deposited Ni-Pt films was measured by XRF as shown in the table below.

The relationship between the Ni-Pt ratio in the precursor solution and in the deposited film at temperatures of 300 °C, 315 °C, and 330 °C for different elemental compositions is tabulated in Table 2. The Ni-Pt ratio in the precursor solution did not agree with the final film composition. This may be related to differences in the crystallographic structure and atomic radii of the metals in the Ni-Pt alloy films [31,51].

Table 2. Elemental compositions of Ni-Pt in the precursors and in the deposited films.

Ni-Pt in the Precursors (at.%)	Ni-Pt in the Deposited Films (at.%)		
	T = 300 °C	T = 315 °C	T = 330 °C
Ni_{100}	Ni_{100}	Ni_{100}	Ni_{100}
$Ni_{80}Pt_{20}$	$Ni_{94}Pt_{6}$	$Ni_{91}Pt_{9}$	$Ni_{90}Pt_{10}$
$Ni_{50}Pt_{50}$	$Ni_{79}Pt_{21}$	$Ni_{72}Pt_{28}$	$Ni_{68}Pt_{32}$
$Ni_{20}Pt_{80}$	$Ni_{38}Pt_{62}$	$Ni_{3}Pt_{97}$	$Ni_{7}Pt_{93}$
Pt_{100}	Pt_{100}	Pt_{100}	Pt_{100}

Furthermore, in this study, we analyzed the Ni-Pt atomic ratio in the precursor and in the film as determined by XRF and EDX measurements. From these results, the measurement of the Ni-Pt atomic ratio obtained by EDX was in fairly good agreement with that determined by XRF (see Figure 4a). In addition, Figure 4b shows the correlation between the Ni-Pt atomic ratio in the precursor solution used for deposition and that of the final films obtained at temperatures of 300 °C, 315 °C, and 330 °C. A good linear relation was observed between the Ni-Pt atomic ratio in the precursor solution and that in the films. However, there was a deviation in the values obtained from these correlations, which may be due to the different atomic radii of Ni and Pt [31].

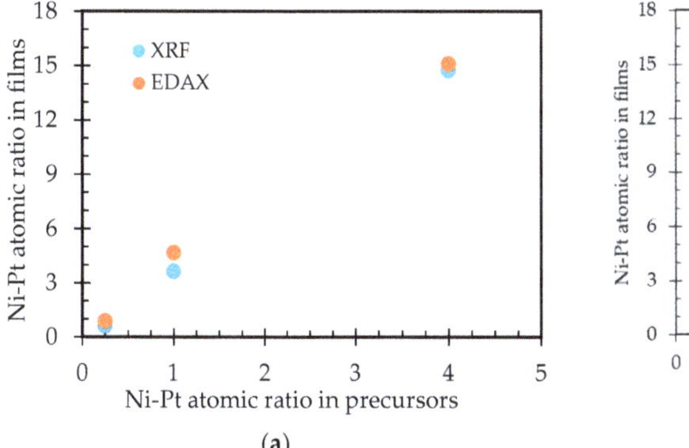

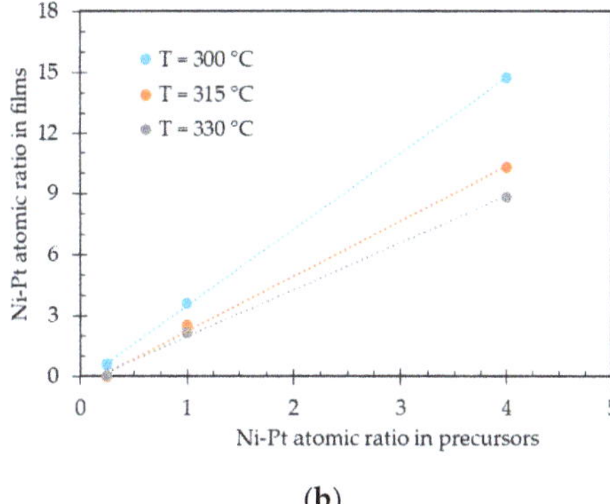

Figure 4. Ni-Pt atomic ratios in precursors and in films: (**a**) Determined by XRF and SEM-EDX; (**b**) at different deposition temperatures.

3.2. Characterization of Ni-Pt Alloy Thin Films

The synthesized Ni-Pt alloy thin films were characterized by using various instrumental analysis techniques such as XRD, SEM-EDX, TEM-EDX, and XPS.

The XRD patterns for Ni, Ni-Pt alloy, and Pt films with different elemental compositions deposited at 315 °C, 300 °C, and 330 °C are shown in Figure 5, Figure S1a,b respectively. The XRD spectra indicate that the yielded Ni and Pt films are purely in the metallic state, and no oxide peaks (Ni oxides and Pt oxides) are apparent. Sharp diffraction peaks for Ni thin films occurred at 2θ: 44.4° and 51.7° for Ni(111) and Ni(200), respectively (Ni JCPDS No. 040850), while for the Pt thin films, sharp diffraction peaks were observed at 2θ: 39.5°, 46.0°, and 67.2° for Pt(111), Pt(200), and Pt(220), respectively (Pt JCPDS No. 040802). Subsequently, the XRD peaks of the deposited Ni-Pt alloy thin films were located at angles between those of the Ni and Pt film peaks. These peaks shifted to lower diffraction angles with decreasing Ni content, which might indicate the solubility of nickel in the platinum lattice and a single-phase Ni-Pt alloy formation [43]. At all temperatures, the XRD patterns consistently exhibited the presence of a single phase Ni-Pt alloy formation. Such shifted peaks of the Ni-Pt alloy films in this work agreed with the reports of Takahashi et al. [32], Eiler et al. [31,35], and Park et al. [38].

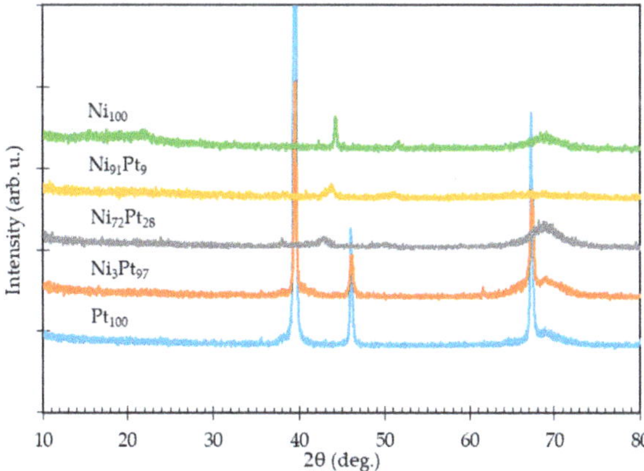

Figure 5. XRD patterns for deposited Ni-Pt alloy thin films at different elemental compositions.

Figure 6 shows that the correlation between lattice parameters for face-centered cubic (fcc) Ni and fcc Pt with Ni content in the Ni-Pt alloy thin films followed Vegard's Law.

A plot of the cell parameter of the Ni-Pt phase against the alloy elemental composition showed good linearity, which is similar to results in the literature [31,32]. In addition, the calculated values of the lattice parameter (a) decrease with increasing Ni content (at.% as determined in this work, as well as in the works of Takahashi et al., Eiler et al., and Zhang et al. [31,32,52]. According to Zhang et al., decreasing the Pt concentration in Ni-Pt alloys reduced the value of the lattice constant due to the smaller atomic radius of Ni than Pt [52]. Therefore, Ni-Pt alloys could be prepared in the Ni-Pt thin films for all elemental compositions.

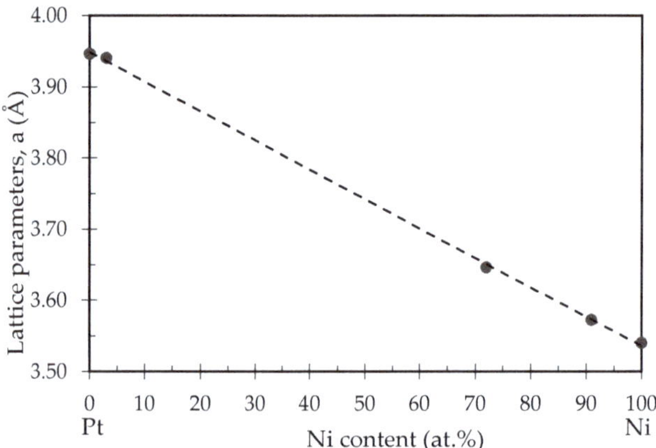

Figure 6. Lattice parameters as a function of Ni content in alloy films.

SEM was used to observe the surface morphology of the Ni-Pt alloy thin films. The top-surface SEM images of the Ni-Pt alloy thin films with different elemental composition deposited at 315 °C and 300 °C are shown in Figure 7 and Figure S2, respectively. The deposited films are smooth, continuous, and uniformly distributed on the substrate surface for all elemental compositions studied, including $Ni_{91}Pt_9$, $Ni_{72}Pt_{28}$, and Ni_3Pt_{97}. The Pt surface in the Ni region was populated with dense and bright structures as shown in Figure 7a–c and the Ni-Pt alloy films exhibited an increase in appreciable microstructure with increasing Pt content, as shown for the Ni_3Pt_{97} alloy film. In addition, a compositional map was created for the surface of the $Ni_{72}Pt_{28}$ alloy thin film by SEM-EDX (Figure 7d–f) which clearly showed a homogeneous distribution of Ni and Pt in the film.

The cross-sectional TEM image in Figure 8a illustrates the crystalline morphology of the $Ni_{72}Pt_{28}$ alloy thin film deposited by the SFCD technique at 315 °C. It is evident that the top surface of this Ni-Pt alloy thin films is polycrystalline and purely metallic. Furthermore, Figure 8b,c depicts the elemental distribution of Ni and Pt in the alloy film obtained by a high-angle annular dark-field scanning transmission electron microscopy system equipped with an energy-dispersive X-ray spectrometer (HAADF-STEM EDX). It can be clearly seen that a Ni-Pt alloy with the atomic ratio of 72:28 is uniformly distributed on the TiN/SiO_2 substrate. However, a layer a few nanometers thick covers the top surface of the thin films, which may be a surface oxide or contamination layer caused by handling of the final product after the deposition process.

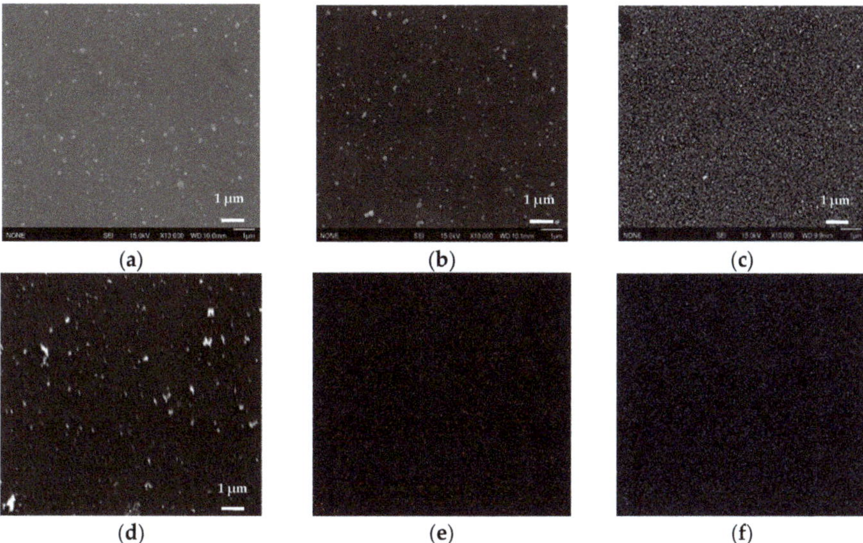

Figure 7. SEM images of deposited Ni-Pt alloy thin films: (**a**) $Ni_{91}Pt_9$; (**b**) $Ni_{72}Pt_{28}$; (**c**) Ni_3Pt_{97}; (**d**) EDX for $Ni_{72}Pt_{28}$ and corresponding elemental mapping images of (**e**) Ni; (**f**) Pt.

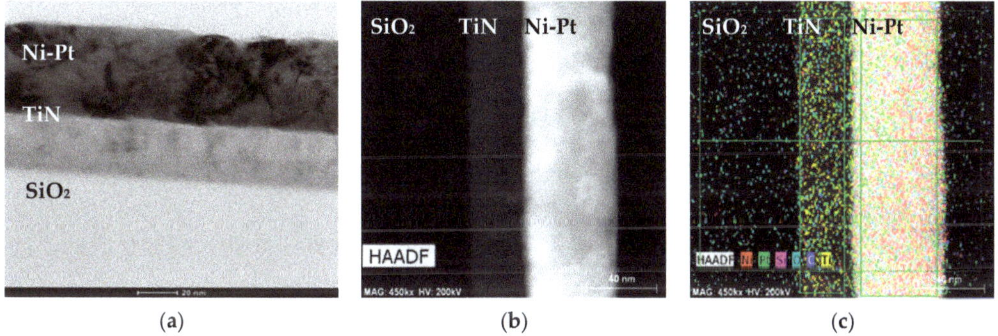

Figure 8. Cross-sectional (**a**) TEM; (**b**) -angle annular dark-field scanning transmission electron microscopy system (HAADF STEM); and (**c**) STEM EDX compositional map of deposited $Ni_{72}Pt_{28}$ alloy thin film.

To identify the chemical state of elements in the deposited Ni-Pt alloy thin films, XPS analysis was performed, as shown in Figure 9.

Figure 9a shows Ni 2p spectra of Ni-Pt alloy thin films with different elemental compositions deposited at 315 °C. In the Ni 2p spectra, the Ni_{100} and $Ni_{91}Pt_9$ films showed a Ni peak corresponding to metallic Ni at a binding energy of 852.9 eV [53]. However, the $Ni_{72}Pt_{28}$ and Ni_3Pt_{97} films showed no Ni peaks at the film surface because of the dominance of Pt in the elemental distribution. For the Pt 4f spectra, the deposited Ni-Pt alloy thin films with different elemental compositions are shown in Figure 9b. Regarding the Pt 4f spectra, the binding energies of metallic Pt 4f7/2 and Pt 4f5/2 peaks were obtained at 71.5 eV and 74.8 eV, respectively. It can be clearly seen that the intensity of the Pt 4f peaks for the films decreased as the Ni content increased. In addition, the Pt 4f peaks are shifted to higher binding energy than that of the original peak position of the pure Pt sample, indicating either a change in the oxidation state of the element or a change in the electron configuration caused by the alloying of Pt with Ni [32,54].

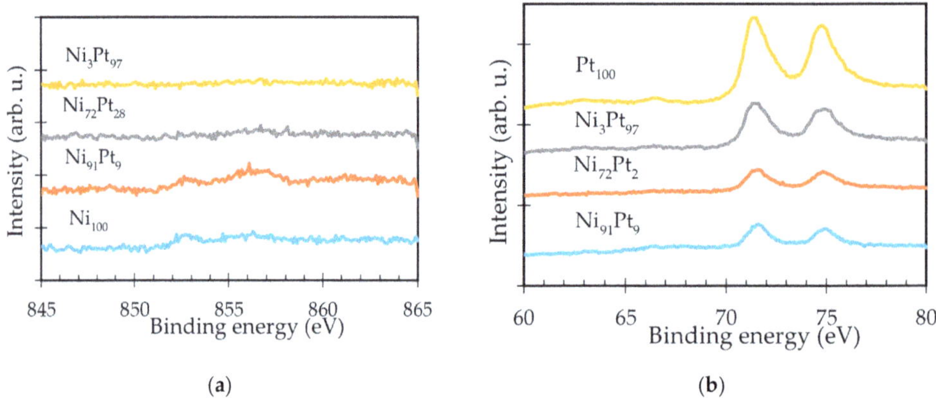

Figure 9. XPS spectra of deposited Ni-Pt alloy thin films with different elemental compositions: (**a**) Ni 2p; (**b**) Pt 4f.

The XPS depth profile of the Ni-Pt alloy thin film with atomic ratio of 72:28 deposited at 315 °C is shown in Figure 10. As can be seen in this figure, the top surface is dominated by Pt and Ni, and the intensity of the peaks corresponding to these elements decreased with increasing etching time. In addition, all the Ni was distributed near the TiN film substrate, which indicated that Ni deposition can be initiated at a lower temperature than Pt, in agreement with our previous result concerning the effect of temperature on the deposition rate (see Figure 2). Moreover, no atomic diffusion occurred between Pt and Ni within the etching time range of 0 to 3600 s. Such a Ni-Pt depth profile indicated alloy formation between Ni and Pt due to full intermixing of Ni with Pt [38]. However, oxygen and carbon species are present on the Ni-Pt alloy thin films, indicating that the surface was covered by organics [35]. The presence of O and C signals may be due to acetone or organic by-products remaining on the film surface after the deposition process. Meanwhile, no F signal was observed, indicating that there was no contamination by the fluoro-based ligand complex. As the etching time reached 2400 s, titanium and nitrogen signals were detected, associated with the TiN film on the substrate used in this work.

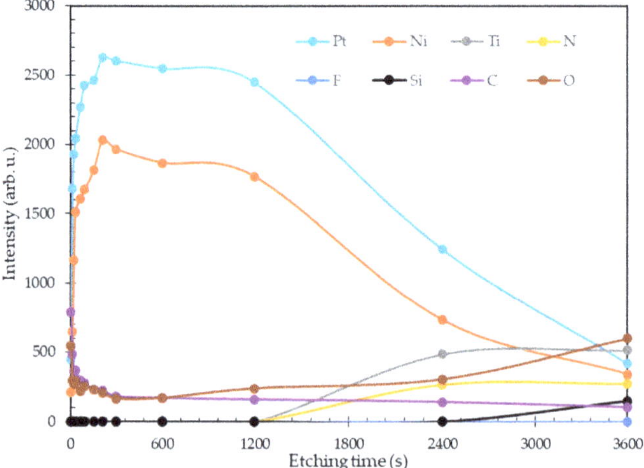

Figure 10. Depth profiles of $Ni_{72}Pt_{28}$ alloy thin films.

4. Conclusions

Ni-Pt alloy thin films prepared by the SFCD technique from $Ni(hfac)_2 \cdot 3H_2O$ and $Pt(hfac)_2$ precursors through hydrogen reduction in supercritical carbon dioxide solutions have been successfully synthesized and characterized. The deposition rate of the Ni-Pt alloy thin films decreased with increasing Ni content in the alloying systems, and gradually increased as the precursor concentration was increased to 15.3×10^{-3} (mole%). The film peaks shifted to lower diffraction angles with decreasing Ni content, as determined by XRD, indicating the presence of a single-phase Ni-Pt alloy formation. Obviously, with increasing deposition temperature from 300 °C to 330 °C (Figure S1c), a single-phase Ni-Pt alloy peaks became sharper, indicating that the crystallites were larger. The deposited films were smooth, continuous, and uniformly distributed on the substrate surface for all elemental compositions, as revealed by SEM-EDX and HAADF-STEM EDX analyses. In the XPS analysis, the intensity of the Pt 4f peaks of the films decreased as the Ni content was increased, and vice versa for the Ni 2p peak intensities. Furthermore, based on the depth profiles as determined by XPS, the top surface is dominated by Pt and Ni, and the intensity of peaks associated with those elements decreased with increasing etching time. In this profile, it was clear that no atomic diffusion occurred between Pt and Ni in the etching time range of 0 to 3600 s, indicating alloy formation in the film. Therefore, Ni-Pt alloy films deposited by the SFCD technique can be used as a suitable model for catalytic reactions such as ORR, HER, and methanol oxidation.

Supplementary Materials: The following are available online at https://www.mdpi.com/2079-4991/11/1/151/s1, Table S1: Thickness measurement for Ni-Pt alloy thin films, Figure S1: XRD patterns for deposited Ni-Pt alloy thin films at different elemental compositions: (a) T = 300 °C; (b) T = 330 °C; (c) Ni/Pt with precursor ratio of 50:50 (at.%) at different temperatures, Figure S2: SEM images of deposited Ni-Pt alloy thin films at temperature of 300 °C: (a) $Ni_{94}Pt_6$; (b) $Ni_{79}Pt_{21}$; (c) $Ni_{38}Pt_{62}$.

Author Contributions: Conceptualization, E.K. and Sudiyarmanto; methodology, Sudiyarmanto; validation, E.K.; formal analysis, E.K. and Sudiyarmanto; investigation, E.K. and Sudiyarmanto; resources, Sudiyarmanto; data curation, E.K.; writing—original draft preparation, Sudiyarmanto; writing—review and editing, E.K. and Sudiyarmanto; visualization, Sudiyarmanto; supervision, E.K.; project administration, E.K.; funding acquisition, E.K. All authors have read and agreed to the published version of the manuscript.

Funding: This research was funded by the Ministry of Education, Culture, Sports, Science, and Technology (MEXT) scholarship program in the fiscal year 2018 and the MEXT WISE Program for Power Energy Professionals (PEP) of Waseda University in the fiscal year 2019.

Institutional Review Board Statement: Not applicable.

Informed Consent Statement: Not applicable.

Data Availability Statement: Data sharing not applicable.

Acknowledgments: This work was supported financially by the Ministry of Education, Culture, Sports, Science, and Technology (MEXT) scholarship program of Japan during the fiscal year 2018 and the MEXT WISE Program for Power Energy Professionals (PEP) of Waseda University in the fiscal year 2019. The authors gratefully acknowledged J. Yamanaka, Y. Miyazawa, and C. Yamamoto for the useful discussions about TEM analysis.

Conflicts of Interest: The authors declare no conflict of interest.

References

1. Garcia-Garcia, F.J.; Gil-Rostra, J.; Yubero, F.; González-Elipe, A.R. Electrochromism in WOx and WxSiyOz thin films prepared by Magnetron sputtering at glancing angles. *Nanosci. Nanotechnol. Lett.* **2013**, *5*, 89–93. [CrossRef]
2. Garcia-Garcia, F.J.; Gil-Rostra, J.; Terriza, A.; González, J.C.; Cotrino, J.; Ferrer, F.J.; González-Elipe, A.R.; Yubero, F. Low refractive index SiOF thin films prepared by reactive magnetron sputtering. *Thin Solid Films* **2013**, *542*, 332–337. [CrossRef]
3. Nishikawa, H.; Yano, H.; Inukai, J.; Tryk, D.A.; Iiyama, A.; Uchida, H. Effects of sulfate on the oxygen reduction reaction activity on stabilized Pt Skin/PtCo alloy catalysts from 30 to 80 °C. *Langmuir* **2018**, *34*, 13558–13564. [CrossRef] [PubMed]
4. Wolff, S.; Crone, M.; Muller, T.; Enders, M.; Bräse, S.; Türk, M. Preparation of supported Pt nanoparticles by supercritical fluid reactive deposition: Influence of precursor, substrate and pressure on product properties. *J. Supercrit. Fluids* **2014**, *95*, 588–596. [CrossRef]

5. Wasiak, T.; Przypis, L.; Walczak, K.Z.; Janas, D. Nickel nanowires: Synthesis, characterization and application as effective catalyst for the reduction of nitroarenes. *Catalysts* **2018**, *8*, 566. [CrossRef]
6. Zhang, C.; Xu, L.; Yan, Y.; Chen, J. Controlled synthesis of pt nanowires with ordered large mesopores for methanol oxidation reaction. *Sci. Rep.* **2016**, *6*, 31440. [CrossRef]
7. Kariuki, N.N.; Khudhayer, W.J.; Karabacak, T.; Myers, D.J. GLAD Pt−Ni alloy nanorods for oxygen reduction reaction. *ACS Catal.* **2013**, *3*, 3123–3132. [CrossRef]
8. Yoon, J.; Park, J.; Sa, Y.J.; Yang, Y.; Baik, H.; Joo, S.H.; Lee, K. Synthesis of bare Pt_3Ni nanorod from PtNi@Ni core-shell nanorod by acid etching: One step surfactant removal and phase conversion for optimal electrochemical performance toward oxygen reduction reaction. *CrystEngComm* **2016**, *3*, 6002–6007. [CrossRef]
9. Pascu, O.; Marre, S.; Aymonier, C. Creation of interfaces in composite/hybrid nanostructured materials using supercritical fluids. *Nanotechnol. Rev.* **2015**, *4*, 487–515. [CrossRef]
10. Barim, S.B.; Uzunlar, E.; Bozbag, S.E.; Erkey, C. Review—Supercritical deposition: A powerful technique for synthesis of functional materials for electrochemical energy conversion and storage. *J. Electrochem. Soc.* **2020**, *167*, 054510. [CrossRef]
11. Türk, M.; Erkey, C. Synthesis of supported nanoparticles in supercritical fluids by supercritical fluid reactive deposition: Current state, further perspectives and needs. *J. Supercrit. Fluids* **2018**, *134*, 176–183. [CrossRef]
12. Kondoh, E. Kinetics of Ni thin film synthesis by supercritical fluid chemical deposition. *Jpn. J. Appl. Phys.* **2020**, *59*, SLLE02.
13. Ye, X.; Wai, C.M. Making nanomaterials in supercritical fluids: A review. *J. Chem. Educ.* **2003**, *80*, 198–204. [CrossRef]
14. Cabanas, A.; Blackburn, J.M.; Watkins, J.J. Deposition of Cu films from supercritical fluids using Cu(I)β-diketonate precursors. *Microelectron. Eng.* **2002**, *64*, 53–61. [CrossRef]
15. Zong, Y.; Watkins, J.J. Deposition of copper by the H_2-assisted reduction of $Cu(tmod)_2$ in supercritical carbon dioxide: Kinetics and reaction mechanism. *Chem. Mater.* **2005**, *17*, 560–565. [CrossRef]
16. Tenorio, M.J.; Cabanas, A.; Pando, C.; Renuncio, J.A.R. Solubility of $Pd(hfac)_2$ and $Ni(hfac)_2 \cdot 2H_2O$ in supercritical carbon dioxide pure and modified with ethanol. *J. Supercrit. Fluids* **2012**, *70*, 106–111. [CrossRef]
17. Matsubara, M.; Hirose, M.; Tamai, K.; Shimogaki, Y.; Kondoh, E. Kinetics of deposition of Cu thin films in supercritical carbon dioxide solutions from a F-free copper(II) β-diketone complex. *J. Electrochem. Soc.* **2009**, *156*, H443–H447. [CrossRef]
18. Watanabe, M.; Akimoto, T.; Kondoh, E. Synthesis of platinum-ruthenium alloy nanoparticles on carbon using supercritical fluid deposition. *J. Solid State Sci. Technol.* **2013**, *02*, M9–M12. [CrossRef]
19. Bozbag, S.E.; Erkey, C. Supercritical deposition: Current status and perspectives for the preparation of supported metal nanostructures. *J. Supercrit. Fluids* **2015**, *96*, 298–312. [CrossRef]
20. Majimel, M.; Marre, S.; Garrido, E.; Aymonier, C. Supercritical fluid chemical deposition as an alternative process to CVD for the surface modification of materials. *Chem. Vap. Depos.* **2011**, *17*, 342–352. [CrossRef]
21. Giroire, B.; Ahmad, M.A.; Aubert, G.; Teule-Gay, L.; Michau, D.; Watkins, J.J.; Aymonier, C.; Poulon-Quintin, A. A comparative study of copper thin films deposited using magnetron sputtering and supercritical fluid deposition techniques. *Thin Solid Films* **2017**, *643*, 53–59. [CrossRef]
22. Henry, L.; Roger, J.; Petitcorps, Y.L.; Aymonier, C.; Maille, L. Preparation of ceramic materials using supercritical fluid chemical deposition. *J. Supercrit. Fluids* **2018**, *141*, 113–119. [CrossRef]
23. Hunde, E.T.; Watkins, J.J. Reactive deposition of cobalt and nickel films from their metallocenes in supercritical carbon dioxide solution. *Chem. Mater.* **2004**, *16*, 498–503. [CrossRef]
24. Blackburn, J.M.; Long, D.P.; Cabanas, A.; Watkins, J.J. Deposition of conformal copper and nickel films from supercritical carbon dioxide. *Science* **2001**, *294*, 141–145. [CrossRef] [PubMed]
25. Kondoh, E. Deposition of Ru thin films from supercritical carbon dioxide fluids. *Jpn. J. Appl. Phys.* **2005**, *44*, 5799–5802. [CrossRef]
26. Kondoh, E.; Fukuda, J. Deposition kinetics and narrow-gap-filling in Cu thin film growth from supercritical carbon dioxide fluids. *J. Supercrit. Fluids* **2008**, *44*, 466–474. [CrossRef]
27. Karanikas, C.F.; Watkins, J.J. Kinetics of the ruthenium thin film deposition from supercritical carbon dioxide by the hydrogen reduction of $Ru(tmhd)_2cod$. *Microelectron. Eng.* **2010**, *87*, 566–572. [CrossRef]
28. Rasadujjaman, M.; Watanabe, M.; Sudoh, H.; Kondoh, E. Codeposition of Cu/Ni thin films from mixed precursors in supercritical carbon dioxide solutions. *Jpn. J. Appl. Phys.* **2014**, *53*, 05GA07. [CrossRef]
29. Zhao, B.; Momose, T.; Shimogaki, Y. Deposition of Cu–Ag alloy film by supercritical fluid deposition. *Jpn. J. Appl. Phys.* **2006**, *45*, L1296–L1299. [CrossRef]
30. Zhao, B.; Zhang, Y.; Yang, J. Deposition of Cu–Mn alloy film from supercritical carbon dioxide for advanced interconnects. *Mater. Sci. Mater. Electron.* **2013**, *24*, 4439–4444. [CrossRef]
31. Eiler, K.; Fornell, J.; Navarro-Sennent, C.; Pellicer, E.; Sort, J. Tailoring magnetic and mechanical properties of mesoporous single-phase Ni–Pt films by electrodeposition. *Nanoscale* **2020**, *12*, 7749–7758. [CrossRef] [PubMed]
32. Takahashi, H.; Hiromoto, T.; Taguchi, M. Preparation of Pt-Ni alloy thin films with various compositions by sputtering and their activity for oxygen reduction reaction. *Int. J. Soc. Mater. Eng. Resour.* **2017**, *22*, 1–7. [CrossRef]
33. Todoroki, N.; Kato, T.; Hayashi, T.; Takahashi, S.; Wadayama, T. Pt−Ni nanoparticle-stacking thin film: Highly active electrocatalysts for oxygen reduction reaction. *ACS Catal.* **2015**, *5*, 2209–2212. [CrossRef]
34. Liu, Y.; Hangarter, C.M.; Garcia, D.; Moffat, T.P. Self-terminating electrodeposition of ultrathin Pt films on Ni: An active, low-cost electrode for H_2 production. *Surf. Sci.* **2015**, *631*, 141–154. [CrossRef]

35. Eiler, K.; Surinach, S.; Sort, J.; Pellicer, E. Mesoporous Ni-rich Ni–Pt thin films: Electrodeposition, characterization and performance toward hydrogen evolution reaction in acidic media. *Appl. Catal. B Environ.* **2020**, *265*, 118597. [CrossRef]
36. Artal, R.; Serra, A.; Michler, J.; Philippe, L.; Gomez, E. Electrodeposition of mesoporous Ni-Rich Ni-Pt films for highly efficient methanol oxidation. *Nanomaterials* **2020**, *10*, 1435. [CrossRef] [PubMed]
37. Vinh, P.V. Preparation of nipt alloys by galvanic replacement reaction on Ni films for direct ethanol fuel cell. *Commun. Phys.* **2017**, *27*, 245–254. [CrossRef]
38. Park, K.W.; Choi, J.H.; Sung, Y.E. Structural, chemical, and electronic properties of Pt/Ni thin film electrodes for methanol electro oxidation. *J. Phys. Chem. B* **2003**, *107*, 5851–5856. [CrossRef]
39. De, S.; Zhang, J.; Luque, R.; Yan, N. Ni-based bimetallic heterogeneous catalysts for energy and environmental applications. *Energy Environ. Sci.* **2016**, *9*, 3314–3347. [CrossRef]
40. Nash, P.; Singleton, M.F. The Ni-Pt (Nickel-Platinum) system. *Bull. Alloy Phase Diagr.* **1989**, *10*, 258–262. [CrossRef]
41. Walker, R.A.; Darby, J.B., Jr. Thermodynamic properties of solid nickel-platinum alloys. *Acta Metall.* **1970**, *18*, 1261–1266. [CrossRef]
42. Cardarelli, F. *Materials Handbook*, 3rd ed.; Springer International Publishing AG: Cham, Switzerland, 2018; pp. 103, 566.
43. Premkumar, P.A.; Prakash, N.S.; Gaillard, F.; Bahlawane, N. CVD of Ru, Pt and Pt-based alloy thin films using ethanol as mild reducing agent. *Mater. Chem. Phys.* **2011**, *125*, 757–762. [CrossRef]
44. Ishikawa, M.; Muramoto, I.; Machida, H.; Imai, S.; Ogura, A.; Ohshita, Y. Chemical vapor deposition Pt-Ni alloy using $Pt(PF_3)_4$ and $Ni(PF_3)_4$. *ECS Trans.* **2008**, *13*, 433–439. [CrossRef]
45. Lagalante, A.F.; Hansen, B.N.; Bruno, T.J.; Sievers, R.E. Solubilities of copper(II) and chromium(III) β-diketonates in supercritical carbon-dioxide. *Inorg. Chem.* **1995**, *34*, 5781–5785. [CrossRef]
46. Saito, N.; Ikushima, Y.; Goto, T. Liquid/solid extraction of acetylacetonate chelates with supercritical carbon dioxide. *Bull. Chem. Soc. Jpn.* **1990**, *63*, 1532–1534. [CrossRef]
47. Lee, W.J.; Wan, Z.; Kim, C.M.; Oh, I.K.; Harada, R.; Suzuki, K.; Choi, E.A.; Kwon, S.H. Atomic layer deposition of Pt thin films using dimethyl (N,N-dimethyl-3-butene-1-amine-N) platinum and O_2 reactant. *Chem. Mater.* **2019**, *31*, 5056–5064. [CrossRef]
48. Battiston, G.A.; Gerbasi, R.; Rodriguez, A. A novel study of the growth and resistivity of nanocrystalline Pt films obtained from $Pt(acac)_2$ in the presence of oxygen or water vapor. *Chem. Vap. Depos.* **2005**, *11*, 130–135. [CrossRef]
49. Maruyama, T.; Tago, T. Nickel thin films prepared by chemical vapour deposition from nickel acetylacetonate. *J. Mater. Sci.* **1993**, *28*, 5345–5348. [CrossRef]
50. Teoh, W.H.; Mammucari, R.; Foster, N.R. Solubility of organometallic complexes in supercritical carbon dioxide: A review. *J. Organomet. Chem.* **2013**, *724*, 102–116. [CrossRef]
51. Plieth, W. Deposition of alloys. In *Electrochemistry for Materials Science*, 1st ed.; Chapter 8; Elsevier Science: Dresden, Germany, 2008; p. 231.
52. Zhang, S.; Johnson, D.D.; Shelton, W.A.; Xu, Y. Hydrogen adsorption on ordered and disordered Pt-Ni alloys. *Top. Catal.* **2020**, *63*, 714–727. [CrossRef]
53. Pauly, N.; Yubero, F.; García-García, F.J.; Tougaard, S. Quantitative analysis of Ni 2p photoemission in NiO and Ni diluted in a SiO2 matrix. *Surf. Sci.* **2016**, *644*, 46–52. [CrossRef]
54. Bozbag, S.E.; Unal, U.; Kurykin, M.A.; Ayala, C.J.; Aindow, M.; Erkey, C. Thermodynamic control of metal loading and composition of carbon aerogel supported Pt−Cu alloy nanoparticles by supercritical deposition. *J. Phys. Chem. C* **2013**, *117*, 6777–6787. [CrossRef]

Article

One-Pot Synthesis of W$_2$C/WS$_2$ Hybrid Nanostructures for Improved Hydrogen Evolution Reactions and Supercapacitors

Sajjad Hussain [1,2], **Iqra Rabani** [2], **Dhanasekaran Vikraman** [3], **Asad Feroze** [4], **Muhammad Ali** [5], **Young-Soo Seo** [2], **Hyun-Seok Kim** [3], **Seung-Hyun Chun** [4] and **Jongwan Jung** [1,2,*]

[1] Hybrid Materials Center (HMC), Sejong University, Seoul 05006, Korea; shussainawan@gmail.com
[2] Department of Nano and Advanced Materials Engineering, Sejong University, Seoul 05006, Korea; iqra.rabani@yahoo.com (I.R.); ysseo@sejong.ac.kr (Y.-S.S.)
[3] Division of Electronics and Electrical Engineering, Dongguk University-Seoul, Seoul 04620, Korea; v.j.dhanasekaran@gmail.com (D.V.); hyunseokk@dongguk.edu (H.-S.K.)
[4] Department of Physics, Sejong University, Seoul 05006, Korea; asadferoze55@gmail.com (A.F.); schun@sejong.ac.kr (S.-H.C.)
[5] Center of Research Excellence in Nanotechnology (CENT), King Fahd University of Petroleum and Minerals (KFUPM), Dhahran 31261, Saudi Arabia; muhammadaliaskari@gmail.com
* Correspondence: jwjung@sejong.ac.kr; Tel.: +82-2-3408-3688; Fax: +82-2-3408-4342

Received: 11 June 2020; Accepted: 12 August 2020; Published: 14 August 2020

Abstract: Tungsten sulfide (WS$_2$) and tungsten carbide (W$_2$C) are materialized as the auspicious candidates for various electrochemical applications, owing to their plentiful active edge sites and better conductivity. In this work, the integration of W$_2$C and WS$_2$ was performed by using a simple chemical reaction to form W$_2$C/WS$_2$ hybrid as a proficient electrode for hydrogen evolution and supercapacitors. For the first time, a W$_2$C/WS$_2$ hybrid was engaged as a supercapacitor electrode and explored an incredible specific capacitance of ~1018 F g^{-1} at 1 A g^{-1} with the outstanding robustness. Furthermore, the constructed symmetric supercapacitor using W$_2$C/WS$_2$ possessed an energy density of 45.5 Wh kg^{-1} at 0.5 kW kg^{-1} power density. For hydrogen evolution, the W$_2$C/WS$_2$ hybrid produced the low overpotentials of 133 and 105 mV at 10 mA cm^{-2} with the small Tafel slopes of 70 and 84 mV dec^{-1} in acidic and alkaline media, respectively, proving their outstanding interfaced electrocatalytic characteristics. The engineered W$_2$C/WS$_2$-based electrode offered the high-performance for electrochemical energy applications.

Keywords: hybrid; HER; WS$_2$; W$_2$C; symmetric; supercapacitors

---

1. Introduction

To overcome the ever-increasing energy necessities, researchers have devoted considerable attention to designing and developing new and eco-friendly materials for electrochemical energy production and storage uses [1,2]. Among the various electrochemical storage devices, supercapacitors (SCs) are highly favored, owing to their quick charge–discharge ability, great power density, robust cycling constancy, and simple configuration [3]. SCs are mainly divided into pseudocapacitive and electrical double-layer capacitive (EDLC) behaviors. In the EDLC, energy is filled by the accretion of ions at the junction of electrode/electrolyte. In pseudocapacitors, storage operation is ensued by rapid revocable Faradaic operation between the electro-active species of electrolyte and electrode [3–5]. The commonly used electrode materials for EDLCs have some limitations because of low conductivity or low specific capacitance. The improved capacitance and energy density are to be attained from pseudocapacitive property, owing to the rapid Faradic reaction. On the other hand, one of the water electrolysis processes of hydrogen evolution reaction (HER) is a prominently accomplished route for green energy. The efficient HER electrocatalysts are improve the rate of electrolysis and produce low overpotential, to reach a specific current density [6,7]. Precious platinum (Pt) is a highly proficient HER electrocatalyst, but costliness and low stability obstruct its commercial use [8,9]. Due to the economic issue, researchers are keen on developing inexpensive and earth-abundant electrode materials [3,10,11].

Transition metal dichalcogenides (TMD) materials, especially metal sulfides (MS_2, M = W, Mo, Co, etc.) are received greater consideration for energy storage and HER devices because of its many beneficial properties such as covalently bonded S−M−S with feeble van der Waals relations between the each layer, efficient mass transport, high specific area, robust edges, and chemical stability [1,12,13]. Moreover, the layered 2D TMD materials own excessive potential as the SC electrode materials, due to their excellent electronic structure, rapid ion intercalation, and preferred pseudocapacitive behavior. However, the low intrinsic conductivity and inactive basal planes are greatly limited to their widespread use [14,15]. In order to circumvent the electrical conductance limitations and to enhance the structural stability and efficiency, TMDs can be hybridized with other high-conducting materials. Carbonaceous materials (graphene or CNT) and transition metal carbides (TMCs) are suitable partner materials for hybridization. TMCs, such as Mo_2C and W_2C, which are newer than graphene and CNT, could be useful electrodes for supercapacitors and HER [16–18]; however, their performance was not widely appealing. The previous reported hybrid materials are $MoSe_2/Mo_2C$ [9], WS_2/reduced graphene oxide hybrids [14,19], $MoS_2/Ti_3C_2T_x$ hybrid [20], MoS_2/WS_2/graphene heterostructures [21], MoS_2/Mo_2C hybrid nanosheets [22], $WS_{(1-x)}Se_x$ decorated 3D graphene [12], graphene supported lamellar 1T'-$MoTe_2$ [23], and MoS_2/reclaimed carbon fiber [24]. Recently, the W_2C/WS_2 has been used as a potential candidate to perceive the enriched catalytic properties for high hydrogen evolution characteristics [25,26]. Furthermore, Wang et al. [27] fabricated $W_xC@WS_2$ heterostructure via carbonizing WS_2 nanotubes, which produced the overpotential of 146 mV at 10 mA cm^{-2} and Tafel slope of 61 mV dec^{-1}. Chen et al. [28] also claimed the Tafel slope of 59 mV dec^{-1} with overpotential of 75 mV at 10 mA cm^{-2} in alkaline medium for a novel eutectoid-structured WC/W_2C heterostructure. The detailed literatures are suggested to combine the W_2C and WS_2 and thereby to progress the electronic and electrochemical properties of resulted hybrid material.

Hence, in this work, W_2C was chosen as a partner material for the hybridization with WS_2, to form W_2C/WS_2 hybrid. This study focused on a simple one-pot strategy to synthesize W_2C/WS_2 as efficient and durable electrodes for electrochemical HER and SCs. So far, few reports are available for W_2C/WS_2 electrocatalysts for HER [26,27,29]. W_2C/WS_2 hybrid, for the first time, is employed as an SCs electrode in this work for improved storage behavior. W_2C/WS_2 possessed an excellent specific capacitance of ~1018 F g^{-1} at 1 A g^{-1} and high cycling stability with 94% retention. Furthermore, symmetric W_2C/WS_2 supercapacitor owned the high energy density of 45.5 Wh kg^{-1} at a low power density of 0.5 kW kg^{-1}. As HER electrocatalyst, low overpotential of 133 and 105 mV was exhibited in acidic and alkaline media, respectively. Synergetic chemical coupling effects between the conducting W_2C and semiconducting WS_2 are believed to contribute significantly improving the electrochemical properties.

2. Materials and Methods

2.1. Synthesis of WS_2 and W_2C Nanostructures

For WS_2 synthesis, 0.5 g tungsten chloride (WO_3) was dispersed in the 2:1 volume ratio of ethanol and DI water mixture, and then the solution was stirred for 30 min, using a magnetic stirrer. Then, 1 g of thiourea was dissolved in aqueous solution, blended, and placed on a hot plate, at 90 °C, with vigorous stirring for 2 h, and then endorsed, to realize the room temperature. Consequently, the dark solution was segregated out by centrifuge, and sediment was cleansed with deionized (DI) water and ethanol and parched at 100 °C, in oven, overnight. Lastly, the synthesized powder was sulfurized, using CVD tubular furnace using argon (Ar) carrier gas at 600 °C for 120 min.

The reported simple chemical reduction route was employed to produce the W_2C nanoparticles [18]. Briefly, 1 g of commercial W_2C powder was dispersed in the ethanol solution (50 mL), with stirring for

3 h, at room temperature. Subsequently, 25 mL of ammonia liquid was poured in the bath mixture and kept to vigorous stirring at 85 °C for 5 h. Then, resulted sediment, after being cleansed using DI water by centrifuge, was kept in an oven, overnight, at 60 °C. The product powder was kept under the gas mixture of H_2 (80 standard cubic centimeters per minute, sccm), CH_4 50 (sccm), and Ar environment in the tubular furnace at 850 °C for 3 h annealing process. Finally, the W_2C powder was collected after the tube attained room temperature.

2.2. Synthesis of W_2C/WS_2 Hybrids

About 1 g of commercial W_2C powder (Sigma Aldrich, Seoul, Korea; CAS number: 12070-12-1) was dispersed in the ethanol solution (50 mL), with stirring, at room temperature, for 3 h. Then 0.25 g WO_3 (Sigma Aldrich, Seoul, Korea; CAS number: 1314-35-8) was dispersed in ethanol and DI water mixture solution. Then 0.5 g of thiourea (Sigma Aldrich, Seoul, Korea; CAS number: 62-56-6) dissolved aqueous solution was blended with W_2C solution and stirred. After that, 2 mL of hydrazine solution and 30 mL of liquid ammonia were mixed with the one-pot solution and stirred for 5 h at 85 °C. The deposit was parted, cleansed with DI water, and dehydrated in a hot oven, overnight, at 60 °C. The powder was post-annealed in a furnace, at 850 °C, for 3 h, under the CH_4 (50 sccm), H_2 (80 sccm), and Ar atmosphere. Figure 1 shows the illustration for the synthesis of a hierarchically W_2C/WS_2 hybrid nanostructure.

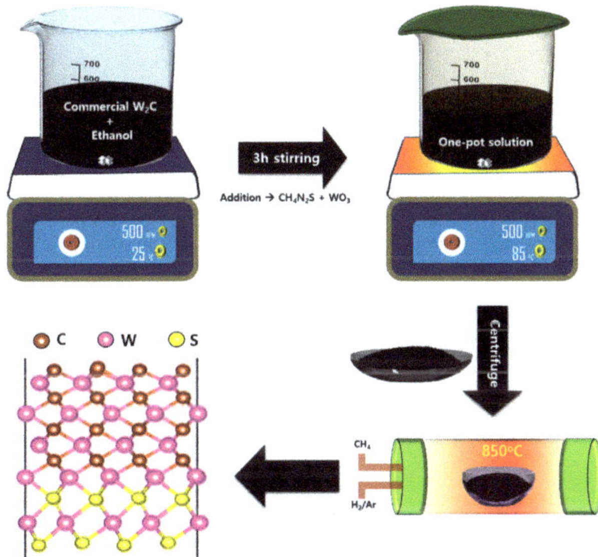

Figure 1. Graphic representation for the synthesis of W_2C/WS_2 hybrid with its derived structure.

2.3. HER Performance

For the working electrode preparation, polyvinylidene fluoride (PVDF), active material (W_2C, WS_2, and W_2C/WS_2), and carbon black at 10:80:10 mass ratio were mixed, and *N*-methyl-2-pyrrolidone (NMP) was added drop-wise. The paste was layered on Ni foam (NF) and overnight dehydrated at 100 °C. For the reference electrode, Ag/AgCl and Hg/HgO were used in acidic and alkaline media performance, respectively, with a graphite counter electrode. We recorded iR corrected linear sweep voltammetry (LSV) by using an electrochemical system (model: 660D; company: CH Instruments, Inc., Austin, TX, USA) in 1 M KOH and 0.5 M H_2SO_4 media, with a scan speed of 10 mV s^{-1}. Electrochemical impedance spectroscopy (EIS) results were noted at the frequencies of 0.01 Hz–100 kHz in acid and alkaline media. The HER potential values were converted for reversible hydrogen electrode (RHE) by the following formula: E (vs. RHE) = E (vs. Ag/AgCl) + E^0 (Ag/AgCl) + 0.0592 × pH for acidic medium and E (vs. RHE) = E (vs. Hg/HgO) + E^0 (Hg/HgO) + 0.0592 × pH for alkaline medium.

2.4. Supercapacitor Performance

A biologic science instrument (SP150, Seyssinet-Pariset, France) was used to analyze the supercapacitor properties. The fabrication process for working electrode was same as for HER. The electrochemical performance was employed in 2 M KOH aqueous for three-electrode (denoted as half-cell) and two-electrode (denoted as symmetric) measurements. The active materials (W_2C, WS_2, and W_2C/WS_2) loaded NFs were employed as the working electrodes, along with Ag/AgCl as a reference electrode and a Pt wire as a counter electrode for three electrode measurements. For symmetric, a cut of Whatman filter paper was socked for 2 h in 2 M KOH and then dried, to remove the excess water. As-prepared filter paper was positioned between the couple of similar W_2C/WS_2 electrodes and pressed to make a sandwich structure. Cyclic Voltammetry (CV) and galvanostatic charge–discharge (GCD) scans were noted from −0.8 to + 0.2 V (vs. Ag/AgCl), at different sweep rates and current densities. The capacitance (C, F g^{-1}), energy density (E, Wh kg^{-1}), and power density (P, Wkg^{-1}) were derived by using the Equations (1)–(3), respectively [3,30].

$$C = \frac{(I \times \Delta t)}{(m \times \Delta V)} \tag{1}$$

$$E = \frac{\left(C \times \Delta V^2\right)}{(2 \times 3.6)} \tag{2}$$

$$P = (E \times 3600)/\Delta t \tag{3}$$

where m is the mass, I is the current, Δt is the discharging time, and ΔV is the potential. EIS studies were accomplished at 10 mV AC amplitude, in an open circuit, in the frequency region of 0.01 Hz to 200 KHz.

2.5. Characterization Details

Field-emission scanning electron microscopy (FESEM) (HITACHI S-4700, Tokyo, Japan) was used to explore the morphological studies. The atomic structures were analyzed by a JEOL-2010F transmission electron microscopy (TEM) with an operation voltage of 200 keV. The Raman spectroscopy (Renishaw inVia RE04, Gloucestershire, UK) measurements were performed in ambient conditions, using the 512 nm Ar laser source with a laser spot size of 1 μm and a scan speed of 30 s. The structural properties were characterized by Rigaku X-ray diffractometer (XRD) (Tokyo, Japan) with Cu-K_α radiation (0.154 nm), at 40 kV and 40 mA, in the scanning range of 10–80° (2θ). For chemical composition and binding energy, the X-ray photoelectron spectroscopy (XPS) measurements were carried out by using an Ulvac PHI X-tool spectrometer (Kanagawa, Japan) with Al K_α X-ray radiation (1486.6 eV). A Brunauer–Emmet–Teller (BET) study to calibrate the surface area of nanostructures was performed, using a N_2 adsorption/desorption medium at 77 K (Micromeritics, Norcross, GA, USA). Pore size distribution was measured with the Barrett–Joyner–Halenda (BJH) analysis.

3. Results and Discussion

3.1. Materials Characteristics

The Raman spectroscopy were performed to measure the crystalline quality and phonon vibration mode properties of W_2C, WS_2, and W_2C/WS_2. Figure 2a displays Raman profiles for W_2C, WS_2, and W_2C/WS_2. For W_2C, Raman spectrum reveals the solid peaks at 693 and 808 cm^{-1}, which relates to the W–C mode of vibration [31,32]. The sp^2-hybridized graphitic G and defective carbon D related bands observed at 1582 and 1353 cm^{-1}, respectively [32]. For WS$_2$, the characteristic bands exhibit at 354 and 420 cm^{-1}, which relates to the E$_{2g}$ and A$_{1g}$ mode of vibration, respectively. The additional peaks in the lower frequency regions enabled at 137, 188, and 258 cm^{-1}, referring to J$_1$, J$_2$, and A$_g$ mode, support the formation of 1T' phase WS$_2$ [33,34]. Interestingly, the synchronized W$_2$C and WS$_2$ characteristic peaks appear for W$_2$C/WS$_2$ hybrid. The relatively higher intensity for A$_{1g}$ mode than the E$_{2g}$ mode credits to the edge exposed TMD structure, which facilitates the high electrocatalytic activity [35,36].

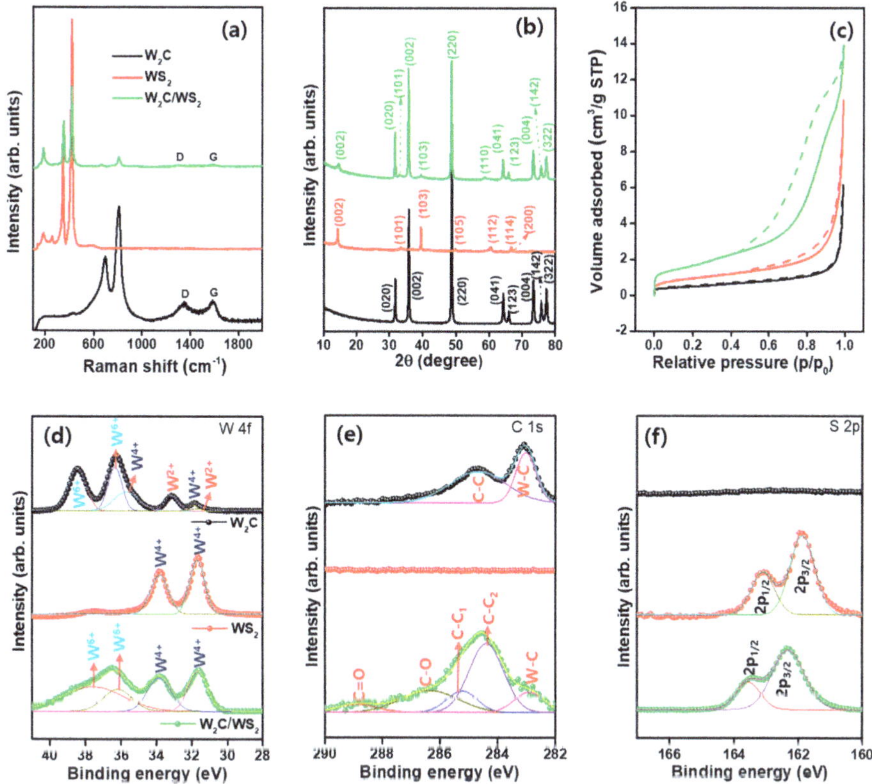

Figure 2. Structural studies of W$_2$C, WS$_2$, and W$_2$C/WS$_2$. (**a**) Raman profiles, (**b**) X-ray diffraction patterns, (**c**) N$_2$ sorption isotherms, (**d**–**f**) XPS curves, (**d**) W 4f, (**e**) C 1s, and (**f**) S 2p regions.

The materials structure was validated further by XRD. In Figure 2b, W$_2$C spectrum shows polycrystalline structure. The (020), (002), (220), (041), (123), (004), (142), and (322) lattice directions observe at 31.5°, 35.1°, 48.8°, 64.1°, 65.8°, 73.2°, 75.6°, and 77.1°, respectively (JCPDS: 89-2371). The diffraction signals at 2θ values of 14.2°, 33.2°, 43.0°, 49.6°, 60.4°, 66.7°, and 67.6° correspond to the (002), (101), (103), (105), (112), (114), and (200) lattices of hexagonal WS$_2$, respectively (JCPDS: 87-2417). Similar to Raman observation, XRD pattern of W$_2$C/WS$_2$ hybrid also produced the cumulative XRD peak positions from W$_2$C and WS$_2$ phases, which are indexed with black and red color, respectively. From the structural outcomes, the blended nature of W$_2$C and WS$_2$ phase is evidently proved in hybrid with preferential orientation of (220) and (002) lattice planes of W$_2$C, which might be originated by the dominating behavior of W$_2$C and their rich presence. Further, the full width at half maximum (FWHM) values were derived from the XRD peaks. The crystallite size was estimated by using FWHM by Scherrer relation [37,38]. The derived FWHM and crystallite values are provided in the Supplementary Materials Tables S1–S3 for W$_2$C, WS$_2$, and W$_2$C/WS$_2$, respectively. The estimated mean size (τ) are at 22.9, 18.7, and 18.1 nm for W$_2$C, WS$_2$, and W$_2$C/WS$_2$, respectively. The reduced-nanosize crystallites for hybrid can be originated by interconnection mechanism between W_2C and WS_2 and their modified crystallographic structure. The d-spacing values were estimated by the Bragg's law ($2d \sin\theta = n\lambda$), and their values are well correlated with the standard results [25]. The extracted values are provided in the Supplementary Materials Tables S1–S3 for W_2C, WS_2, and W_2C/WS_2, respectively. The observed d-spacing values are considerably strained for hybrid compared with pristine, which might be due to the interfacial bonding nature.

The BET surface area was examined to assess the area and pore size distributions by using an N_2 adsorption/desorption medium at 77 K. Figure 2c shows the N_2 isotherms. W_2C/WS_2 possessed a maximum surface area of 6.368 m^2g^{-1}, equated with WS_2 (3.405 m^2g^{-1}) and W_2C (1.75 m^2g^{-1}). The measured total pore volume was 0.009, 0.016, and 0.020 cm^3g^{-1} for W_2C, WS_2, and W_2C/WS_2, respectively (Supplementary Materials Figure S1). The high porosity of W_2C/WS_2, compared with pure, is believed to give the significant contribution of enhancing the electrocatalytic activity by promoting the electrolyte diffusion into the electrode.

XPS was employed to describe the composition and valence states of constructed material. The survey profiles are given in the Supplementary Materials Figure S2, to prove the coexistence of all the elements. The high-resolution XPS profiles for W 4f, C 1s, and S 2p states are provided in Figure 2d–f. From a W 4f XPS profile of W_2C (Figure 2d), the deconvoluted peaks reveal the W^{2+} (31.4 and 33.1 eV), W^{4+} (31.8 and 35.69 eV), and W^{6+} (37.0 and 38.3 eV) doublets. For WS_2, the W4f core level peaks are located at 31.7 and 33.9, due to $W4f_{7/2}$ and $W4f_{5/2}$, respectively. W 4f region of W_2C/WS_2 reveals the peaks at 31.7 eV ($W4f_{7/2}$) and 33.8 eV ($W4f_{5/2}$) with the W^{6+} couplets (37.6 and 36.2 eV) [39,40]. Figure 2e shows the C 1s profile and explores the graphitic sp^2 carbon peak at 284.7 eV and W-C peak at 283.0 eV for W_2C [6]. For the W_2C/WS_2 hybrid, C 1s spectrum produced the $C\text{-}C_1$ (284.4 eV), $C\text{-}C_2$ (285.2 eV), W-C (282.9 eV), C-O (286.3 eV), and C=O (288.9 eV) peaks [29,41,42]. Figure 2f deconvolution peaks reveal the S 2p couplets of $S2p_{1/2}$ and $S2p_{3/2}$ at 163.1 and 161.8 eV for WS_2, whereas, they are at 163.6 and 162.3 eV for the W_2C/WS_2 hybrid, respectively [43,44]. The atomic percentage of W_2C/WS_2 hybrid is determined to be 32.13%, 22.20%, and 45.67% for W, C, and S atoms, respectively, which is well correlated with EDX results (discussed later). The observed elemental confirmation proved the formation of W_2C/WS_2 hybrid.

Surface characteristics were further elaborated by FESEM and TEM examinations. Figure 3 shows the FESEM micrographs of W_2C, WS_2, and W_2C/WS_2. FESEM micrographs clearly picture the formation of different sizes of nanograins by chemical reduction process in the W_2C nanoparticles (Figure 3a). Nano-spherical-shaped agglomerated grains exhibit for WS_2 (Figure 3b). The sizes of the grains are considerably varied in the nanoscales, due to the bulk agglomeration process during the annealing. In the hybrid, the spherically shaped W_2C particles seem to cover the WS_2 particles (Figure 3c) due to the interconnected mechanism. Reduced sizes of the grains appear with cauliflower like agglomerated grain bunches and well-interconnected domain structure for W_2C/WS_2 hybrid. To prove the hybrid formation, EDX spectrum for W_2C/WS_2 clarifies the elemental composition, as shown in Figure 3d. Furthermore, the mapping images are provided to confirm the equal distribution of all the elements in the hybrid (Figure 3e–h).

TEM measurements were carried out for W_2C/WS_2 hybrid (Figure 4). The different magnification TEM images are provided in Figure 4a–c. Vertically aligned nano-stirpes-like structures are broadly exhibited for the hybrid. The interconnection between the layered fringes and finger-printed structures is clearly visualized. Due to the polycrystalline lattices for W_2C/WS_2 hybrid, the different widths of the lattice fringes are obviously demonstrated in the TEM images (Figure 4b,c). A higher magnification TEM image (Figure 4d) explores the layer structure with the cross-section of different lattice fringes in the W_2C/WS_2 hybrid (inset—fast Fourier transform (FFT), left panel). The phase profile spectrum, extracted by point mask mode and inverse FFT (iFFT, right panel) pattern of inset Figure 4d, shows 6.2 nm spacing which related to (002) WS_2 lattice orientation (Figure 4e). The fingerprint structured grains interface with layered WS_2 (Figure 4f–g) [44]. The phase profile spectrum, extracted from the iFFT pattern of inset Figure 4g, elevates 2.9 nm spacing, which is related to (020) W_2C lattice orientation (Figure 4h).

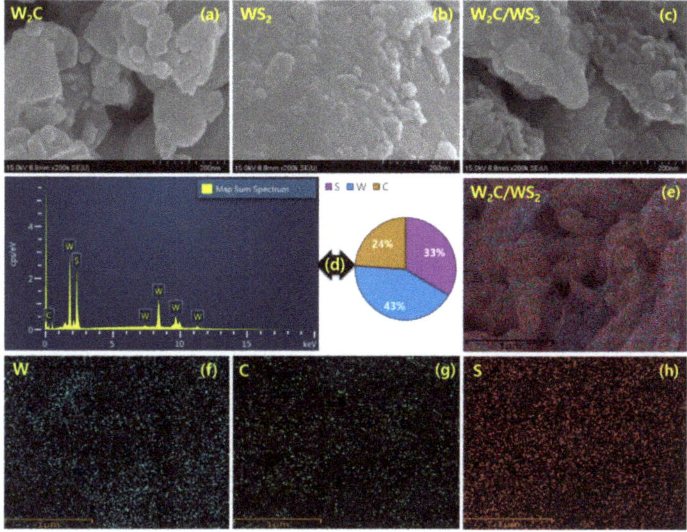

Figure 3. FESEM images of (**a**) W_2C, (**b**) WS_2, and (**c**) W_2C/WS_2 hybrid. (**d**) EDX spectrum of W_2C/WS_2 hybrid; (**e**) mapping image of W_2C/WS_2 hybrid and its elements, (**f**) W, (**g**) C, and (**h**) S.

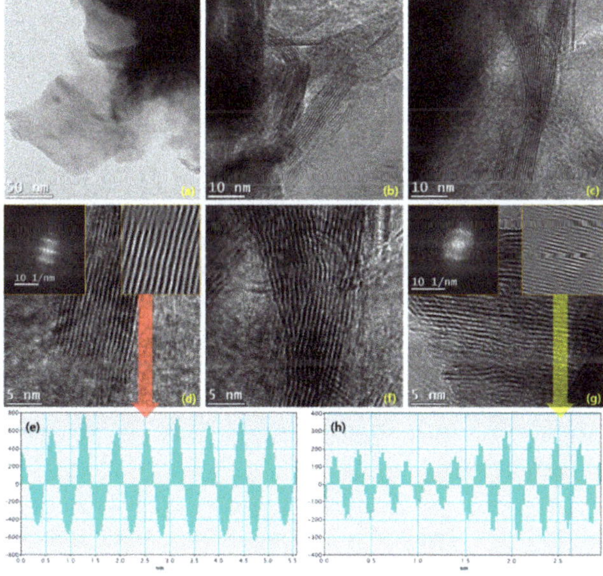

Figure 4. HRTEM images for W_2C/WS_2 hybrid. (**a**) Low- and (**b,c**) high-resolution TEM images. (**d**) Layered WS_2 structure with the inset of FFT and iFFT patterns. (**e**) Phase profile spectrum for (002) lattice orientation of WS_2 with 6.2 nm spacing in the W_2C/WS_2 hybrid. (**f,g**) High-resolution TEM images for W_2C related portion in the hybrid with inset of FFT and iFFT patterns. (**h**) Phase profile spectrum for (020) lattice orientation of W_2C with 2.9 nm spacing in the W_2C/WS_2 hybrid.

3.2. Hydrogen Evolution Studies

Active-materials-coated NFs were engaged as working electrodes to appraise the HER activities in 0.5 M H_2SO_4 and 1 M KOH electrolytes, at room temperature. Figure 5a explores iR-recompensed LSV polarization profiles in 0.5 M H_2SO_4, using 10 mV s^{-1} sweep speed. The W_2C and WS_2 produce 171 and 242 mV to attain 10 mA cm^{-2}, respectively. In contrast, the W_2C/WS_2 hybrid produces an overpotential of 133 mV at 10 mA cm^{-2} (51 mV @ 10 mA cm^{-2} for Pt/C). The exhibited low overpotential credits to interfacial active edges sharing and rapid electron conductivity in the W_2C/WS_2 which proves the importance of hybrid formation. In the 1 M KOH media (Figure 5b), the W_2C/WS_2 electrode also produces highly dynamic HER behavior with a small overpotential of 105 mV at 10 mA cm^{-2} than WS_2 (189 mV) and W_2C (123 mV). The HER performance of W_2C/WS_2 is superior to most of the hybrid-based electrodes (Figure 5c) [45–48]. Li et al. [29] have prepared the nanocomposite of N, S-decorated porous carbon matrix encapsulated WS_2/W_2C (WS_2/W_2C@NSPC), which delivered the small overpotential of 126 and 205 mV in 0.5 M H_2SO_4 and 1.0 M KOH, respectively. In addition, Nguyen et al. [26] have reported the 170 mV of onset potential with 55.4 mV dec^{-1} of Tafel slope in 0.5 M H_2SO_4 for W_2C@WS_2 nanoflowers synthesized by hydrothermal method.

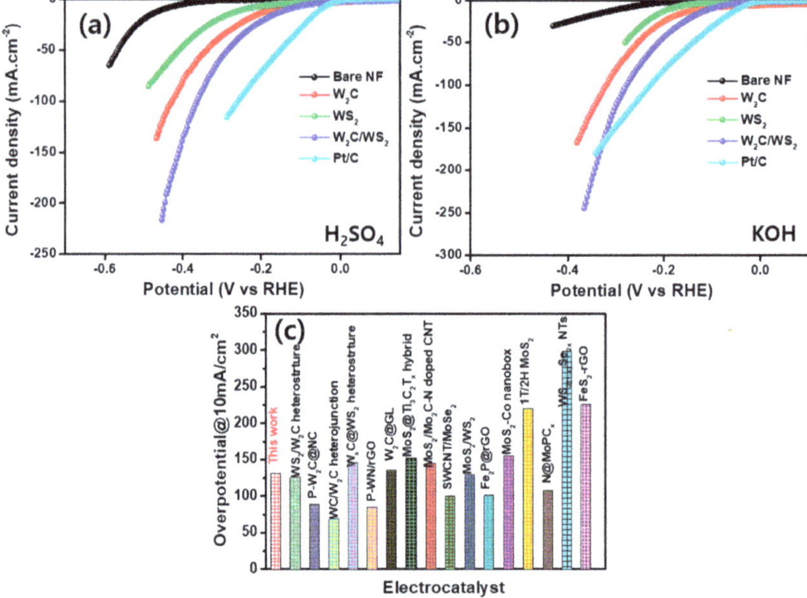

Figure 5. (**a**,**b**) Hydrogen evolution polarization profiles for Pt/C, bare NF, W_2C, WS_2, and W_2C/WS_2 at 10 mV s^{-1} sweep speed in (**a**) 0.5 M H_2SO_4 and (**b**) 1 M KOH media. (**c**) Overpotential comparison of different electrocatalysts.

Tafel slope is a factor to indicate the inherent electrocatalytic activity of the electrode. The Tafel slope values of Pt/C, bare NF, W_2C, WS_2, and W_2C/WS_2 electrocatalysts, are 36, 168, 86, 138, and 70 mV dec^{-1}, respectively, in H_2SO_4 medium (Figure 6a). The outputs prove the outstanding electrocatalytic activity of the W_2C/WS_2 hybrid. Exchange current densities (j_0) assessed by extrapolation Tafel lines to X-axis and their values observe at ~1.03, 1.02, 0.55, and 0.19 mA cm^{-2} for Pt/C, W_2C/WS_2, WS_2, and W_2C, respectively. W_2C, WS_2, and W_2C/WS_2 produce the Tafel slopes of 141, 127, and 84 mV dec^{-1}, respectively, in KOH medium (Figure 6b). The small Tafel slope of W_2C/WS_2 also supports high HER behavior of hybrid electrode in the KOH electrolyte. The extrapolated j_0 is 0.93, 0.72, and 0.38 mA cm^{-2} for W_2C, WS_2, and W_2C/WS_2, respectively, in KOH electrolyte. The exhibited Tafel slope range suggests that HER involves a Volmer–Heyrovsky mechanism for W_2C/WS_2 hybrid, with electrochemical desorption as the rate-regulatory direction [44,49–51]. Outstanding HER activity in the W_2C/WS_2 hybrid could be explained with electrode kinetics by accumulated electrocatalytic edge facets and a high ratio of charge transfer. The observed HER parameters are provided in Supplementary Materials Table S4 for all the measured electrodes. The j_0 and Tafel values are superior to most reported hybrid electrocatalysts (Figure 6c,d and Supplementary Materials Table S5) [46,48,52]. The low overpotential, large j_0 value, and small Tafel slope in the W_2C/WS_2 hybrid also confirm the importance of hybrid formation for efficient HER electrocatalytic activity in KOH and H_2SO_4 medium.

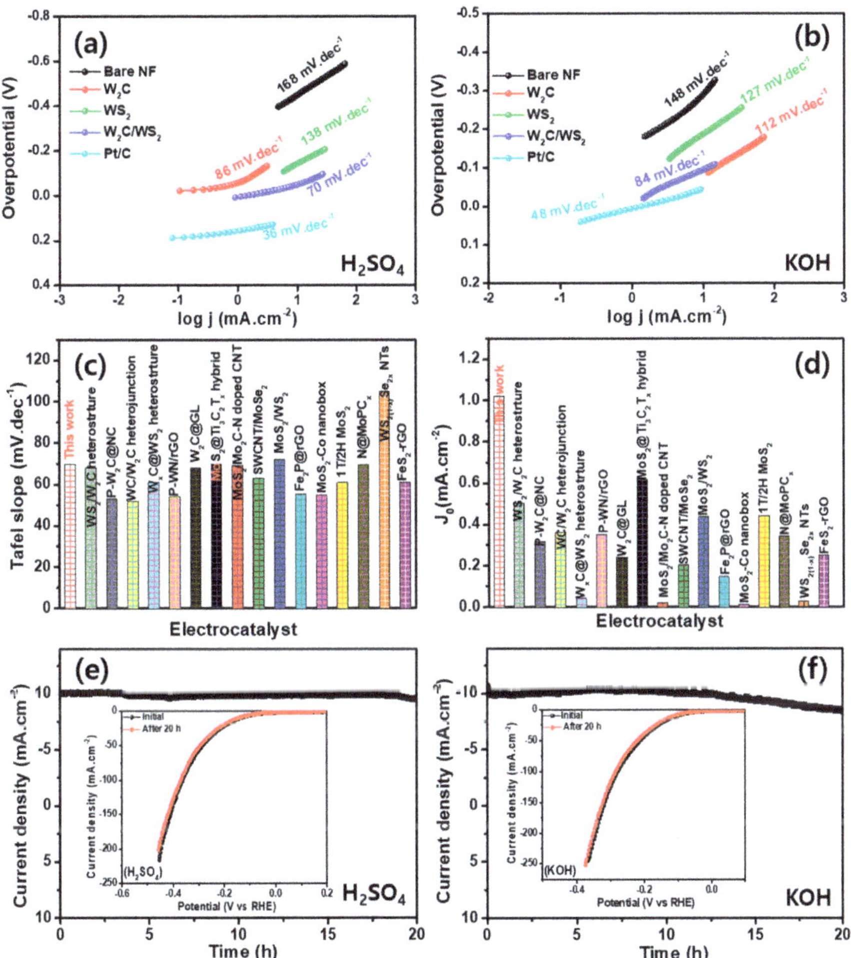

Figure 6. (**a,b**) Tafel plots for Pt/C, bare NF, W_2C, WS_2, and W_2C/WS_2 hybrid at 10 mV s^{-1} sweep speed in (**a**) 0.5 M H_2SO_4 and (**b**) 1 M KOH media; comparison of (**c**) Tafel slope and (**d**) exchange current density with different electrocatalysts; chronoamperometric profile of W_2C/WS_2 hybrid for 20 h continuous hydrogen evolution reaction (HER) operation in (**e**) 0.5 M H_2SO_4 and (**f**) 1 M KOH electrolyte (inset: LSV curves before and after 20 h operation).

CV profiles were acquired in the non-Faradaic region, to estimate the double-layer capacitance (C_{dl}) (Supplementary Materials Figure S3a–c). The C_{dl} by linear fitting (Supplementary Materials Figure S3c) are 3.81 mF cm^{-2} (in H_2SO_4) and 3.33 mF cm^{-2} (in KOH) for the W_2C/WS_2 hybrid.

Electrochemical surface areas are 108 and 83 cm^2 in H_2SO_4 and KOH, respectively. A stability and durability evaluation of W_2C/WS_2 electrode was carried out by using chronoamperometric response at a persistent 133 and 105 mV overpotential in H_2SO_4 and KOH (Figure 6e,f), respectively. No significant decline is observed over 20 h in H_2SO_4 the medium, whereas slight deterioration is exhibited for the KOH medium. Note the excellent robustness of W_2C/WS_2 electrode in the H_2SO_4, rather than KOH, medium for HER. LSV curves at initial and after 20 h of continuous HER operation are shown in the inset of Figure 6e,f, respectively.

To probe insights for electrocatalytic activity of materials, EIS was performed in the H_2SO_4 and KOH (Supplementary Materials Figure S4). The observed EIS plot revealed the low charge-transfer resistance (R_{ct}) and swift electron transfer via the electrolyte–electrode interface for W_2C/WS_2. The charge-transfer resistances, R_{ct}, of W_2C/WS_2 (~1.8–2.2 Ω) in the H_2SO_4 and KOH media are lower than those of W_2C and WS_2. Moreover, the small series of resistances (~1.5–2.5 Ω) of all the electrodes suggests that the active materials are well integrated with the porous NF. The high electrocatalytic properties and robust solidity of the W_2C/WS_2 hybrid support it as a potential material to substitute Pt in HER application.

3.3. Supercapacitor Performances

Electrochemical storage properties were elucidated by CV and GCD tests in 2 M KOH electrolyte, using three electrodes, as explained in the experimental part of the manuscript. The CVs were recorded with the potential interval of −0.8 to 0.2 V vs. Ag/AgCl at 10 mV s^{-1} sweep speed for the W_2C, WS_2, and W_2C/WS_2 electrodes (Figure 7a). All the electrodes produce the identical CV loops with Faradaic-adsorptions-blended EDLC operations [53]. The W_2C/WS_2 hybrid electrode shows a wider electrochemical area than the W_2C and WS_2. Moreover, the W_2C/WS_2 hybrid shows a couple of redox peaks (−0.61 and −0.25 V), indicating the reversible reaction from W^{4+} to W^{6+}, corresponding to the proton's absorption/desorption into the WS_2 interlayers [54]. Figure 7b shows CVs of various scan rates (10–50 mV s^{-1}), and their shapes are maintained, indicating good electrochemical capacitive characteristics and high-rate performance. The results for WS_2 and W_2C at various scan speeds are shown in Supplementary Figure S5a,b, respectively. Successive 100 CV cycles were executed in the potential region of −0.8 to 0.2 V, to assess the stability for the W_2C/WS_2 electrode, and its output is presented Figure 7c. Due to the EDLC-combined Faradaic storage mechanism, it possesses good stability with minimal degradation over the repeated cycles [55].

The electrochemical storage performance was further tested by GCD curves, as shown in Figure 8a–d. The slightly distorted triangle like the GCD curve is exhibited for the W_2C/WS_2 and WS_2 electrodes, due to the redox reaction. For the W_2C electrode, a square-structured GCD curve is exhibited, with a voltage drop, due to the easy oxidation characteristic. The W_2C/WS_2 electrode exposes an outstanding specific capacitance (estimated by using the Equation (1)) of 1018 F g^{-1}, as compare to the WS_2 (~158 F g^{-1}) and W_2C (~133 F g^{-1}), at the current density of 1 A g^{-1}. GCD analysis displays similar curves at the different current densities. Observe the specific capacitances of 133, 90, 66, and 50 F g^{-1} for W_2C (Figure 8b) and 158, 120, 87, and 80 F g^{-1} for WS_2 (Figure 8c) at 1, 2, 3, and 5 A g^{-1} current density, respectively. For W_2C/WS_2, a high specific capacitance of 1018, 866, 816, and 660 F g^{-1} exhibits at the 1, 2, 3, and 5 A g^{-1} current density, respectively (Figure 8d). The specific capacitance changes with current density, as shown in Supplementary Materials Figure S6. The significant enhancement of capacitances for W_2C/WS_2 electrode credits to the mutual interactions between W_2C and WS_2, large surface area, rich active edge facets of WS_2, and high electronic conductivity of W_2C to enable the rapid transference of electrons during a charge–discharge process. Supplementary Materials Table S6 provides the extended comparison of W_2C/WS_2 SCs performance with the previously reported hybrid electrodes.

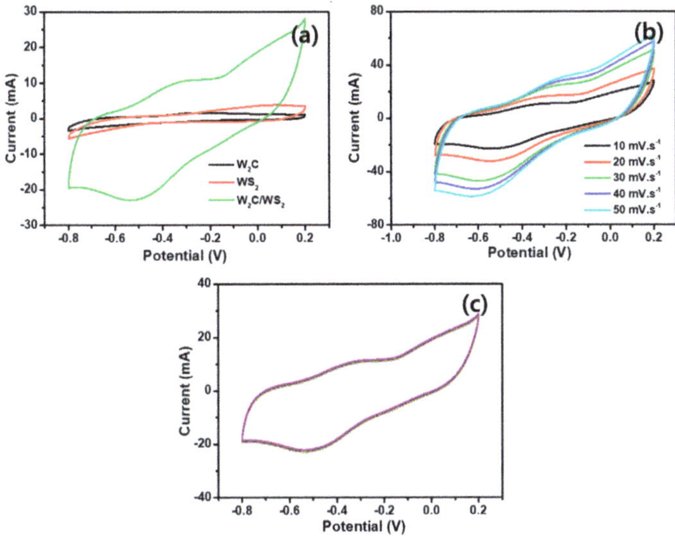

Figure 7. Supercapacitor Cyclic Voltammetry (CV) for three electrode measurement: (**a**) CV curves for W$_2$C, WS$_2$, and W$_2$C/WS$_2$ electrodes; (**b**) different scan rate CV curves for the W$_2$C/WS$_2$ hybrid; (**c**) multiple cycle CV curves for the W$_2$C/WS$_2$ hybrid.

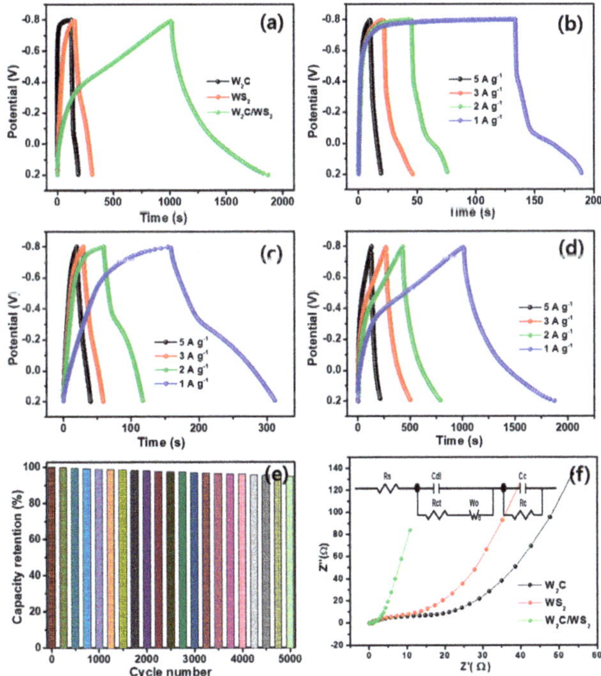

Figure 8. (**a–d**) Galvanostatic charge–discharge (GCD) profiles for three electrode measurement; (**a**) GCDs at 1 Ag^{-1} for W$_2$C, WS$_2$, and W$_2$C/WS$_2$; GCDs at different current densities for (**b**) W$_2$C, (**c**) WS$_2$, and (**d**) W$_2$C/WS$_2$; (**e**) stability performance of W$_2$C/WS$_2$ hybrid; (**f**) EIS curves for W$_2$C, WS$_2$, and W$_2$C/WS$_2$ (inset-fitted circuit).

Cyclic stability is an essential property for the supercapacitor electrodes. In the case of the W_2C/WS_2 electrode, 94% of primary capacitance was perceived after 5000 cycles (Figure 8e), suggesting long-term stability. EIS measurements were performed to prove the charge-transfer characteristics (Figure 8f). Fitted curve from the Nyquist profile is inserted in Figure 8f, where C_{dl} is the double-layer capacitance, R_{ct} is the charge-transfer resistance, R_s is the series resistance, W_o the Warburg impedance at open circuit voltage, and R_c and C_c are the capacitive resistance and capacitive capacitance, respectively. The output curve indicates that the charge-transfer resistance significantly reduces for W_2C/WS_2 hybrid electrode, as compared to their pristine. A lower R_{ct} ($\sim$1.0 Ω) was obtained by the hybrid electrode, as compared to the WS_2 ($\sim$13.9 Ω) and W_2C (10.1 Ω), respectively.

To assess the practical application, the symmetric supercapacitor (SSC) was assembled with two identical W_2C/WS_2 hybrid electrodes. The electrochemical CV measurement of SSC was measured by using the similar potential range of half-cell measurements. Figure 9a shows the CV curves for W_2C/WS_2 symmetric cell device. The modified rectangular shape of CV curve for the W_2C/WS_2 hybrid SSC reveals slightly shouldered redox reaction, confirming the key contribution of EDLC characteristics. The constructed SSC device gives an excellent current response and larger integral area, compared to previous TMDs-based materials [56,57]. The different scan rates, using performed CV curves, prove the high rate of capability of the prepared SSC devices. Figure 9b shows the different current densities, using prepared SSC GCD curves, indicating the excellent electrochemical rate capability. The charge/discharge time considerably decreases for the SSC device due to its direct intercalation and extraction of ions through the solid electrolyte, compared with its half-cell outcomes. The capacitance value of symmetric device was valued from the GCD curve, using Equation (1) [58]. The W_2C/WS_2 hybrid delivers the higher symmetric capacitance of 328, 306, 255, and 220 F g^{-1} at 1, 2, 3, and 5 A g^{-1} of current densities, respectively, as presented in Figure 9c. Interestingly, our symmetric device results show the enhancing capacitance, compared to other symmetric capacitor results [56,59,60]. The specific energy and specific power values are significant for the practical uses, which were weighed by relations 2 and 3, respectively. The symmetric device carries the energy densities of 45.5, 42.5, 35.4, and 30.5 Wh kg^{-1} at 0.5, 1.0, 1.5, and 2.0 kW kg^{-1} power density, respectively, as given in the Figure 9d.

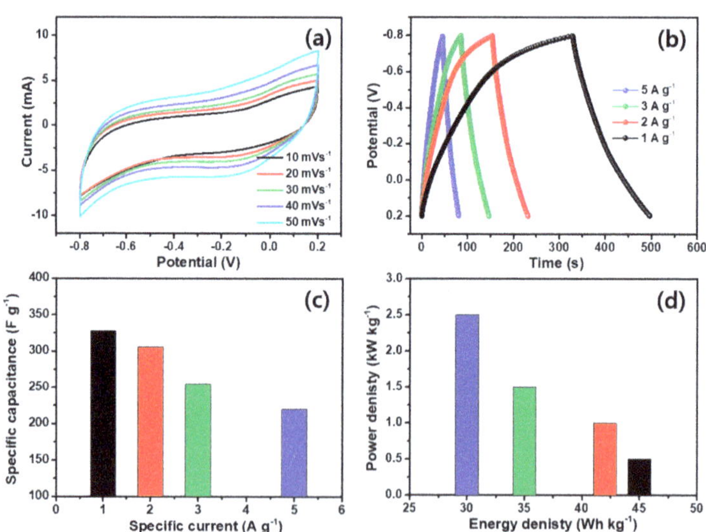

Figure 9. Symmetric supercapacitor performance of W_2C/WS_2. (**a**) Different scan rate CV curves for symmetric W_2C/WS_2 hybrid; (**b**) GCDs for symmetric W_2C/WS_2 at different current densities; (**c**) specific capacitance at various scan rates for W_2C/WS_2 by symmetric measurements; (**d**) Ragone plots of W_2C/WS_2 for symmetric device.

The exhibited energy density of symmetric W_2C/WS_2 capacitor is superior to the recently reported symmetric devices using TMDs and TMCs electrodes, MoS_2 sheets (18.43 Wh kg^{-1}) [60], $Ti_3C_2T_x$/MWCNT (3 Wh kg^{-1}) [61], s-MoS_2/CNS (7.4 Wh kg^{-1}) [62], 3D-graphene/MoS_2 (24.59 Wh kg^{-1}) [63], MoS_2/RCF (22.5 Wh kg^{-1}) [24], MoS_2/RGO/MoS_2@Mo (6.22 Wh kg^{-1}) [56], and MoS_2 sponge (6.15 Wh kg^{-1}) [59]. Overall, our findings demonstrate that the inclusion of carbide-based material with TMDs deliberately improves the conductance of the hybrid material, assists swift conveyance of electrons/ion, and improves the stability of electrode material.

4. Conclusions

We have successfully engineered the W_2C/WS_2 hybrid electrode by a simple cost-effective one-pot chemical reaction. Highly conductive W_2C-supported WS_2 hybrids were designed to promote high electrocatalytic activity for HER and SCs by accumulating the number of active edges and facilitating the swift electron transport. In the case of HER, the interfaced W_2C/WS_2 hybrid produced the small overpotentials of 133 and 105 mV, to achieve the 10 mA cm^{-2} current density with the Tafel slope of 70 and 84 mV dec^{-1} in H_2SO_4 and KOH media, respectively, which proved the outstanding electrocatalytic HER characteristics. Half-cell measurements unveiled the remarkable specific capacitance of ~1018 F g^{-1} at 1 A g^{-1} with the rate competency nature and robust responses for W_2C/WS_2 hybrid electrode. W_2C/WS_2-based symmetric supercapacitor exposed the specific energy of 45.5 Wh kg^{-1} at 0.5 kW kg^{-1} specific power with a capacitance of 328 F g^{-1} at 1 A g^{-1} current density. The suggested low-cost methodology of one-pot reaction is highly feasible to fabricate the efficacious nanostructured hybrids and has larger-scale production capability because of its controlled synthesis process. Hence, the developed hybrid material and methodology have a broad scope for the future electrochemical applications.

Supplementary Materials: The following are available online at http://www.mdpi.com/2079-4991/10/8/1597/s1. Figure S1: Pore diameter versus pore volume variations for the nanostructures. Figure S2: Survey XPS spectrum of (a) W_2C, (b) WS_2, and (c) W_2C/WS_2. Figure S3: CV spectra in the non-Faradaic region with different scan rates for W_2C/WS_2 hybrid HER electrodes in (a) 0.5 M H_2SO_4 and (b) 1 M KOH electrolyte and (c) their current differences. Figure S4: EIS spectra of Pt, W_2C, WS_2, and W_2C/WS_2 hybrids electrodes in (a) 0.5 M H_2SO_4 and (b) 1M KOH electrolyte. Figure S5: CV curves for the (a) WS_2 and (c) W_2C electrodes at various scan rates, using half-cell (10–50 mVs^{-1}). Figure S6: The specific capacitance variations at different current densities for W_2C, WS_2, and W_2C/WS_2 electrodes by half-cell measurements. Table S1: Microstructural parameters of W_2C. Table S2: Microstructural parameters of WS_2. Table S3: Microstructural parameters of W_2C/WS_2. Table S4: Comparison of electrochemical parameters for different electrocatalysts. Table S5: HER catalytic performances TMDs and TMCs-based electrocatalysts. Table S6: Performances of TMDs- and TMCs-based electrodes for supercapacitors.

Author Contributions: Conceptualization, S.H. and J.J.; methodology, S.H.; formal analysis, I.R. and A.F.; software, M.A.; validation and investigation, D.V. and S.H.; resources, Y.-S.S., H-S.K., and S.-H.C.; data curation, M.A., I.R., and A.F.; writing—original draft preparation, S.H.; writing—review and editing, D.V., H.-S.K., and J.J.; supervision and funding acquisition, J.J. All authors have read and agreed to the published version of the manuscript.

Funding: This research was supported by the Nano Material Technology Development Program and Basic Science Research Program through the National Research Foundation of Korea (NRF) funded by the Ministry of Education, and the Science and ICT (2017R1C1B5076952, 2016M3A7B4909942, and 2020R1A6A1A03043435). This research was also sponsored by the Korea Institute of Energy Technology Evaluation and Planning and the Ministry of Trade, Industry and Energy of the Republic of Korea (20172010106080).

Conflicts of Interest: The authors declare no conflict of interest. The funders had no role in the design of the study; in the collection, analyses, or interpretation of data; in the writing of the manuscript; or in the decision to publish the results.

References

1. Vikraman, D.; Akbar, K.; Hussain, S.; Yoo, G.; Jang, J.-Y.; Chun, S.-H.; Jung, J.; Park, H.J. Direct synthesis of thickness-tunable MoS_2 quantum dot thin layers: Optical, structural and electrical properties and their application to hydrogen evolution. *Nano Energy* **2017**, *35*, 101–114. [CrossRef]

2.	Sarma, P.V.; Kayal, A.; Sharma, C.H.; Thalakulam, M.; Mitra, J.; Shaijumon, M. Electrocatalysis on edge-rich spiral WS_2 for hydrogen evolution. *ACS Nano* **2019**, *13*, 10448–10455. [CrossRef] [PubMed]

3.	Vikraman, D.; Karuppasamy, K.; Hussain, S.; Kathalingam, A.; Sanmugam, A.; Jung, J.; Kim, H.-S. One-pot facile methodology to synthesize MoS_2-graphene hybrid nanocomposites for supercapacitors with improved electrochemical capacitance. *Compos. Part. B Eng.* **2019**, *161*, 555–563. [CrossRef]

4.	Liu, Y.; Peng, X. Recent advances of supercapacitors based on two-dimensional materials. *Appl. Mater. Today* **2017**, *8*, 104–115. [CrossRef]

5.	Wazir, M.B.; Daud, M.; Ullah, N.; Hai, A.; Muhammad, A.; Younas, M.; Rezakazemi, M. Synergistic properties of molybdenum disulfide (MoS_2) with electro-active materials for high-performance supercapacitors. *Int. J. Hydrog Energy* **2019**, *44*, 17470–17492. [CrossRef]

6.	Zhang, H.; Pan, Q.; Sun, Z.; Cheng, C. Three-dimensional macroporous W_2C inverse opal arrays for the efficient hydrogen evolution reaction. *Nanoscale* **2019**, *11*, 11505–11512. [CrossRef]

7.	Attanayake, N.H.; Thenuwara, A.C.; Patra, A.; Aulin, Y.V.; Tran, T.M.; Chakraborty, H.; Borguet, E.; Klein, M.L.; Perdew, J.P.; Strongin, D.R. Effect of intercalated metals on the electrocatalytic activity of 1T-MoS_2 for the hydrogen evolution reaction. *ACS Energy Lett.* **2017**, *3*, 7–13. [CrossRef]

8.	Hussain, S.; Vikraman, D.; Akbar, K.; Naqvi, B.A.; Abbas, S.M.; Kim, H.-S.; Chun, S.-H.; Jung, J. Fabrication of $MoSe_2$ decorated three-dimensional graphene composites structure as a highly stable electrocatalyst for improved hydrogen evolution reaction. *Renew. Energy* **2019**, *143*, 1659–1669. [CrossRef]

9.	Vikraman, D.; Hussain, S.; Karuppasamy, K.; Feroze, A.; Kathalingam, A.; Sanmugam, A.; Chun, S.-H.; Jung, J.; Kim, H.-S. Engineering the novel $MoSe_2$-Mo_2C hybrid nanoarray electrodes for energy storage and water splitting applications. *Appl. Catal. B Environ.* **2020**, *264*, 118531. [CrossRef]

10.	Huang, M.; Li, F.; Dong, F.; Zhang, Y.X.; Zhang, L.L. Mno_2-based nanostructures for high-performance supercapacitors. *J. Mater. Chem. A* **2015**, *3*, 21380–21423. [CrossRef]

11.	Ramesh, S.; Vikraman, D.; Karuppasamy, K.; Yadav, H.M.; Sivasamy, A.; Kim, H.-S.; Kim, J.-H.; Kim, H.-S. Controlled synthesis of SnO_2@$NiCo_2O_4$/nitrogen doped multiwalled carbon nanotube hybrids as an active electrode material for supercapacitors. *J. Alloys Compd.* **2019**, *794*, 186–194. [CrossRef]

12.	Hussain, S.; Akbar, K.; Vikraman, D.; Afzal, R.A.; Song, W.; An, K.S.; Farooq, A.; Park, J.Y.; Chun, S.H.; Jung, J. $Ws_{(1-x)}se_x$ nanoparticles decorated three-dimensional graphene on nickel foam: A robust and highly efficient electrocatalyst for the hydrogen evolution reaction. *Nanomaterials* **2018**, *8*, 929. [CrossRef] [PubMed]

13.	Hussain, S.; Chae, J.; Akbar, K.; Vikraman, D.; Truong, L.; Naqvi, A.B.; Abbas, Y.; Kim, H.-S.; Chun, S.-H.; Kim, G.; et al. Fabrication of robust hydrogen evolution reaction electrocatalyst using Ag_2Se by vacuum evaporation. *Nanomaterials* **2019**, *9*, 1460. [CrossRef] [PubMed]

14.	Tu, C.-C.; Lin, L.-Y.; Xiao, B.-C.; Chen, Y.-S. Highly efficient supercapacitor electrode with two-dimensional tungsten disulfide and reduced graphene oxide hybrid nanosheets. *J. Power Sources* **2016**, *320*, 78–85. [CrossRef]

15.	Vikraman, D.; Hussain, S.; Akbar, K.; Karuppasamy, K.; Chun, S.H.; Jung, J.; Kim, H.S. Design of basal plane edges in metal-doped nanostripes-structured $MoSe_2$ atomic layers to enhance hydrogen evolution reaction activity. *ACS Sustain. Chem. Eng.* **2019**, *7*, 458–469. [CrossRef]

16.	Esposito, D.V.; Hunt, S.T.; Kimmel, Y.C.; Chen, J.G. A new class of electrocatalysts for hydrogen production from water electrolysis: Metal monolayers supported on low-cost transition metal carbides. *J. Am. Chem. Soc.* **2012**, *134*, 3025–3033. [CrossRef]

17.	Hunt, S.T.; Milina, M.; Alba-Rubio, A.C.; Hendon, C.H.; Dumesic, J.A.; Román-Leshkov, Y. Self-assembly of noble metal monolayers on transition metal carbide nanoparticle catalysts. *Science* **2016**, *352*, 974–978. [CrossRef]

18.	Hussain, S.; Vikraman, D.; Feroze, A.; Song, W.; An, K.-S.; Kim, H.-S.; Chun, S.-H.; Jung, J. Synthesis of Mo_2C and W_2C nanoparticle electrocatalysts for the efficient hydrogen evolution reaction in alkali and acid electrolytes. *Front. Chem.* **2019**, *7*, 716. [CrossRef]

19.	Ratha, S.; Rout, C.S. Supercapacitor electrodes based on layered tungsten disulfide-reduced graphene oxide hybrids synthesized by a facile hydrothermal method. *ACS Appl. Mater. Interfaces* **2013**, *5*, 11427–11433. [CrossRef]

20. Liu, J.; Liu, Y.; Xu, D.; Zhu, Y.; Peng, W.; Li, Y.; Zhang, F.; Fan, X. Hierarchical "nanoroll" like MoS$_2$/Ti$_3$C$_2$T$_x$ hybrid with high electrocatalytic hydrogen evolution activity. *Appl. Catal. B Environ.* **2019**, *241*, 89–94. [CrossRef]

21. Lin, T.W.; Sadhasivam, T.; Wang, A.Y.; Chen, T.Y.; Lin, J.Y.; Shao, L.D. Ternary composite nanosheets with MoS$_2$/WS$_2$/graphene heterostructures as high-performance cathode materials for supercapacitors. *Chem. Electro. Chem.* **2018**, *5*, 1024–1031. [CrossRef]

22. Zhao, Z.; Qin, F.; Kasiraju, S.; Xie, L.; Alam, M.K.; Chen, S.; Wang, D.; Ren, Z.; Wang, Z.; Grabow, L.C. Vertically aligned MoS$_2$/Mo$_2$C hybrid nanosheets grown on carbon paper for efficient electrocatalytic hydrogen evolution. *ACS Catal.* **2017**, *7*, 7312–7318. [CrossRef]

23. Liu, M.; Wang, X.; Huang, Z.; Guo, P.; Wang, Z. In-situ solution synthesis of graphene supported lamellar 1T′-MoTe$_2$ for enhanced pseudocapacitors. *Mater. Lett.* **2017**, *206*, 229–232. [CrossRef]

24. Zhao, C.; Zhou, Y.; Ge, Z.; Zhao, C.; Qian, X. Facile construction of MoS$_2$/RCF electrode for high-performance supercapacitor. *Carbon* **2018**, *127*, 699–706. [CrossRef]

25. Nguyen, T.P.; Choi, K.S.; Kim, S.Y.; Lee, T.H.; Jang, H.W.; Van Le, Q.; Kim, I.T. Strategy for controlling the morphology and work function of W2C/WS$_2$ nanoflowers. *J. Alloys Compd.* **2020**, *829*, 154582. [CrossRef]

26. Nguyen, T.P.; Kim, S.Y.; Lee, T.H.; Jang, H.W.; Le, Q.V.; Kim, I.T. Facile synthesis of W$_2$C@WS$_2$ alloy nanoflowers and their hydrogen generation performance. *Appl. Surf. Sci.* **2020**, *504*, 144389. [CrossRef]

27. Wang, F.; He, P.; Li, Y.; Shifa, T.A.; Deng, Y.; Liu, K.; Wang, Q.; Wang, F.; Wen, Y.; Wang, Z. Interface engineered W$_x$C@WS$_2$ nanostructure for enhanced hydrogen evolution catalysis. *Adv. Funct. Mater.* **2017**, *27*, 1605802. [CrossRef]

28. Chen, Z.; Gong, W.; Cong, S.; Wang, Z.; Song, G.; Pan, T.; Tang, X.; Chen, J.; Lu, W.; Zhao, Z. Eutectoid-structured WC/W$_2$C heterostructures: A new platform for long-term alkaline hydrogen evolution reaction at low overpotentials. *Nano Energy* **2020**, *68*, 104335. [CrossRef]

29. Li, Y.; Wu, X.; Zhang, H.; Zhang, J. Interface designing over WS$_2$/W$_2$C for enhanced hydrogen evolution catalysis. *ACS Appl. Energy Mater.* **2018**, *1*, 3377–3384. [CrossRef]

30. Javed, M.S.; Dai, S.; Wang, M.; Guo, D.; Chen, L.; Wang, X.; Hu, C.; Xi, Y. High performance solid state flexible supercapacitor based on molybdenum sulfide hierarchical nanospheres. *J. Power Sources* **2015**, *285*, 63–69. [CrossRef]

31. Dash, T.; Nayak, B. Preparation of WC–W$_2$C composites by arc plasma melting and their characterisations. *Ceram. Int.* **2013**, *39*, 3279–3292. [CrossRef]

32. Yan, G.; Wu, C.; Tan, H.; Feng, X.; Yan, L.; Zang, H.; Li, Y. N-carbon coated pw 2 c composite as efficient electrocatalyst for hydrogen evolution reactions over the whole ph range. *J. Mater. Chem. A* **2017**, *5*, 765–772. [CrossRef]

33. Voiry, D.; Yamaguchi, H.; Li, J.; Silva, R.; Alves, D.C.; Fujita, T.; Chen, M.; Asefa, T.; Shenoy, V.B.; Eda, G. Enhanced catalytic activity in strained chemically exfoliated WS$_2$ nanosheets for hydrogen evolution. *Nat. Mater.* **2013**, *12*, 850–855. [CrossRef] [PubMed]

34. Liu, Z.; Li, N.; Su, C.; Zhao, H.; Xu, L.; Yin, Z.; Li, J.; Du, Y. Colloidal synthesis of 1T′phase dominated WS$_2$ towards endurable electrocatalysis. *Nano Energy* **2018**, *50*, 176–181. [CrossRef]

35. Gao, M.-R.; Chan, M.K.; Sun, Y. Edge-terminated molybdenum disulfide with a 9.4-Å interlayer spacing for electrochemical hydrogen production. *Nat. Commun.* **2015**, *6*, 1–8. [CrossRef]

36. Wang, H.; Lu, Z.; Kong, D.; Sun, J.; Hymel, T.M.; Cui, Y. Electrochemical tuning of MoS$_2$ nanoparticles on three-dimensional substrate for efficient hydrogen evolution. *ACS Nano* **2014**, *8*, 4940–4947. [CrossRef]

37. Vikraman, D.; Park, H.J.; Kim, S.I.; Thaiyan, M. Magnetic, structural and optical behavior of cupric oxide layers for solar cells. *J. Alloys Compd.* **2016**, *686*, 616–627. [CrossRef]

38. Sebastian, S.; Kulandaisamy, I.; Valanarasu, S.; Yahia, I.S.; Kim, H.-S.; Vikraman, D. Microstructural and electrical properties evaluation of lead doped tin sulfide thin films. *J. Sol. Gel Sci. Technol.* **2020**, *93*, 52–61. [CrossRef]

39. Gong, Q.; Wang, Y.; Hu, Q.; Zhou, J.; Feng, R.; Duchesne, P.N.; Zhang, P.; Chen, F.; Han, N.; Li, Y. Ultrasmall and phase-pure W$_2$C nanoparticles for efficient electrocatalytic and photoelectrochemical hydrogen evolution. *Nat. Commun.* **2016**, *7*, 1–8. [CrossRef]

40. Zhang, L.-N.; Ma, Y.-Y.; Lang, Z.-L.; Wang, Y.-H.; Khan, S.U.; Yan, G.; Tan, H.-Q.; Zang, H.-Y.; Li, Y.-G. Ultrafine cable-like WC/W$_2$C heterojunction nanowires covered by graphitic carbon towards highly efficient electrocatalytic hydrogen evolution. *J. Mater. Chem. A* **2018**, *6*, 15395–15403. [CrossRef]

41. Lewin, E.; Persson, P.Å.; Lattemann, M.; Stüber, M.; Gorgoi, M.; Sandell, A.; Ziebert, C.; Schäfers, F.; Braun, W.; Halbritter, J. On the origin of a third spectral component of C1s XPS-spectra for nc-TiC/a-C nanocomposite thin films. *Surf. Coat. Technol.* **2008**, *202*, 3563–3570. [CrossRef]

42. Li, J.; Zhou, C.; Mu, J.; Yang, E.-C.; Zhao, X.-J. In situ synthesis of molybdenum carbide/n-doped carbon hybrids as an efficient hydrogen-evolution electrocatalyst. *RSC Adv.* **2018**, *8*, 17202–17208. [CrossRef]

43. Chen, R.; Zhao, T.; Wu, W.; Wu, F.; Li, L.; Qian, J.; Xu, R.; Wu, H.; Albishri, H.M.; Al-Bogami, A. Free-standing hierarchically sandwich-type tungsten disulfide nanotubes/graphene anode for lithium-ion batteries. *Nano Lett.* **2014**, *14*, 5899–5904. [CrossRef] [PubMed]

44. Vikraman, D.; Hussain, S.; Truong, L.; Karuppasamy, K.; Kim, H.-J.; Maiyalagan, T.; Chun, S.-H.; Jung, J.; Kim, H.-S. Fabrication of MoS_2/WSe_2 heterostructures as electrocatalyst for enhanced hydrogen evolution reaction. *Appl. Surf. Sci.* **2019**, *480*, 611–620. [CrossRef]

45. Fan, X.; Zhou, H.; Guo, X. Wc nanocrystals grown on vertically aligned carbon nanotubes: An efficient and stable electrocatalyst for hydrogen evolution reaction. *ACS Nano* **2015**, *9*, 5125–5134. [CrossRef]

46. Xu, K.; Wang, F.; Wang, Z.; Zhan, X.; Wang, Q.; Cheng, Z.; Safdar, M.; He, J. Component-controllable $WS_{2(1-x)}Se_{2x}$ nanotubes for efficient hydrogen evolution reaction. *ACS Nano* **2014**, *8*, 8468–8476. [CrossRef]

47. Ma, L.; Ting, L.R.L.; Molinari, V.; Giordano, C.; Yeo, B.S. Efficient hydrogen evolution reaction catalyzed by molybdenum carbide and molybdenum nitride nanocatalysts synthesized via the urea glass route. *J. Mater. Chem. A* **2015**, *3*, 8361–8368. [CrossRef]

48. Zhang, K.; Zhao, Y.; Fu, D.; Chen, Y. Molybdenum carbide nanocrystal embedded n-doped carbon nanotubes as electrocatalysts for hydrogen generation. *J. Mater. Chem. A* **2015**, *3*, 5783–5788. [CrossRef]

49. Vikraman, D.; Hussain, S.; Akbar, K.; Truong, L.; Kathalingam, A.; Chun, S.-H.; Jung, J.; Park, H.J.; Kim, H.-S. Improved hydrogen evolution reaction performance using MoS_2–WS_2 heterostructures by physicochemical process. *ACS Sustain. Chem. Eng.* **2018**, *6*, 8400–8409. [CrossRef]

50. Li, Y.; Wang, H.; Xie, L.; Liang, Y.; Hong, G.; Dai, H. MoS_2 nanoparticles grown on graphene: An advanced catalyst for the hydrogen evolution reaction. *J. Am. Chem. Soc.* **2011**, *133*, 7296–7299. [CrossRef]

51. Zhang, K.; Li, C.; Zhao, Y.; Yu, X.; Chen, Y. Porous one-dimensional Mo_2C–amorphous carbon composites: High-efficient and durable electrocatalysts for hydrogen generation. *Phys. Chem. Chem. Phys.* **2015**, *17*, 16609–16614. [CrossRef] [PubMed]

52. Šljukić, B.; Vujković, M.; Amaral, L.; Santos, D.; Rocha, R.; Sequeira, C.; Figueiredo, J.L. Carbon-supported Mo_2C electrocatalysts for hydrogen evolution reaction. *J. Mater. Chem. A* **2015**, *3*, 15505–15512. [CrossRef]

53. Soon, J.M.; Loh, K.P. Electrochemical double-layer capacitance of MoS_2 nanowall films. *Electrochem. Solid State Lett.* **2007**, *10*, A250–A254. [CrossRef]

54. Chen, W.; Yu, X.; Zhao, Z.; Ji, S.; Feng, L. Hierarchical architecture of coupling graphene and 2D WS_2 for high-performance supercapacitor. *Electrochim. Acta* **2019**, *298*, 313–320. [CrossRef]

55. Zhai, Y.; Dou, Y.; Zhao, D.; Fulvio, P.F.; Mayes, R.T.; Dai, S. Carbon materials for chemical capacitive energy storage. *Adv. Mater.* **2011**, *23*, 4828–4850. [CrossRef] [PubMed]

56. Zhang, Y.; Ju, P.; Zhao, C.; Qian, X. In-situ grown of $MoS_2/rGO/$ MoS_2@Mo nanocomposite and its supercapacitor performance. *Electrochim. Acta* **2016**, *219*, 693–700. [CrossRef]

57. Huang, K.-J.; Wang, L.; Zhang, J.-Z.; Wang, L.-L.; Mo, Y.-P. One-step preparation of layered molybdenum disulfide/multi-walled carbon nanotube composites for enhanced performance supercapacitor. *Energy* **2014**, *67*, 234–240. [CrossRef]

58. Cao, Z.; Liu, C.; Huang, Y.; Gao, Y.; Wang, Y.; Li, Z.; Yan, Y.; Zhang, M. Oxygen-vacancy-rich $NiCo_2O_4$ nanoneedles electrode with poor crystallinity for high energy density all-solid-state symmetric supercapacitors. *J. Power Sources* **2020**, *449*, 227571. [CrossRef]

59. Balasingam, S.K.; Lee, M.; Kim, B.H.; Lee, J.S.; Jun, Y. Freeze-dried MoS_2 sponge electrodes for enhanced electrochemical energy storage. *Dalton Trans.* **2017**, *46*, 2122–2128. [CrossRef]

60. Pazhamalai, P.; Krishnamoorthy, K.; Manoharan, S.; Kim, S.-J. High energy symmetric supercapacitor based on mechanically delaminated few-layered MoS_2 sheets in organic electrolyte. *J. Alloys Compd.* **2019**, *771*, 803–809. [CrossRef]

61. Navarro-Suárez, A.M.; Van Aken, K.L.; Mathis, T.; Makaryan, T.; Yan, J.; Carretero-González, J.; Rojo, T.; Gogotsi, Y. Development of asymmetric supercapacitors with titanium carbide-reduced graphene oxide couples as electrodes. *Electrochim. Acta* **2018**, *259*, 752–761. [CrossRef]

62. Khawula, T.N.; Raju, K.; Franklyn, P.J.; Sigalas, I.; Ozoemena, K.I. Symmetric pseudocapacitors based on molybdenum disulfide (MoS_2)-modified carbon nanospheres: Correlating physicochemistry and synergistic interaction on energy storage. *J. Mater. Chem. A* **2016**, *4*, 6411–6425. [CrossRef]
63. Singh, K.; Kumar, S.; Agarwal, K.; Soni, K.; Gedela, V.R.; Ghosh, K. Three-dimensional graphene with MoS_2 nanohybrid as potential energy storage/transfer device. *Sci. Rep.* **2017**, *7*, 1–12.

 *nanomaterials*

MDPI

Article

Synthesis and Characterization of Li-C Nanocomposite for Easy and Safe Handling

Subash Sharma [1,*], Tetsuya Osugi [1], Sahar Elnobi [1,2], Shinsuke Ozeki [1], Balaram Paudel Jaisi [1], Golap Kalita [1], Claudio Capiglia [1] and Masaki Tanemura [1,*]

[1] Department of Physical Science and Engineering, Nagoya Institute of Technology Gokiso-cho, Showa-ku, Nagoya 466-8555, Japan; t.osugi.676@stn.nitech.ac.jp (T.O.); sahar.elnobi@sci.svu.edu.eg (S.E.); s.ozeki.976@stn.nitech.ac.jp (S.O.); poudel_balaram@yahoo.com (B.P.J.); kalita.golap@nitech.ac.jp (G.K.); claudio@mvc.biglobe.ne.jp (C.C.)

[2] Department of Physics, Faculty of Science, South Valley University, Qena 83523, Egypt

*   Correspondence: sharmasubash2006@gmail.com (S.S.); tanemura.masaki@nitech.ac.jp (M.T.)

Received: 16 July 2020; Accepted: 27 July 2020; Published: 29 July 2020

Abstract: Metallic lithium (Li) anode batteries have attracted considerable attention due to their high energy density value. However, metallic Li is highly reactive and flammable, which makes Li anode batteries difficult to develop. In this work, for the first time, we report the synthesis of metallic Li-embedded carbon nanocomposites for easy and safe handling by a scalable ion beam-based method. We found that vertically standing conical Li-C nanocomposite (Li-C NC), sometimes with a nanofiber on top, can be grown on a graphite foil commonly used for the anodes of lithium-ion batteries. Metallic Li embedded inside the carbon matrix was found to be highly stable under ambient conditions, making transmission electron microscopy (TEM) characterization possible without any sophisticated inert gas-based sample fabrication apparatus. The developed ion beam-based fabrication technique was also extendable to the synthesis of stable Li-C NC films under ambient conditions. In fact, no significant loss of crystallinity or change in morphology of the Li-C film was observed when subjected to heating at 300 °C for 10 min. Thus, these ion-induced Li-C nanocomposites are concluded to be interesting as electrode materials for future Li-air batteries.

Keywords: lithium-ion battery (LIB); anode; lithium-carbon nanocomposites; ion beam

---

1. Introduction

With a global focus on clean and renewable energy, storage and transfer of electric energy at the grid scale is a challenge. The lithium-ion battery (LIB) has been a topic of great research interest in recent years [1]. LIBs are superior to conventional batteries due to their high energy density, portability and safety [2–5]. Sony was the first company to commercialize the lithium-ion battery in the 1990s. LIBs consist of Li oxide cathode and carbonaceous materials as the anode. Li salts are used as electrolytes with a separator separating the electrode. Graphite is the most common anode material used in almost all LIBs. Graphite is useful due to its intercalation property, storing a large amount of Li atoms. Graphite is also stable due to less volumetric change (10%) during charge and discharge cycles. However, due to its limited energy density value (372 mA·h·g^{-1}) [6] and high weight, alternative lightweight and high-energy-density anode material is a necessity.

Li has been realized as the ideal anode material for LIBs due to its lightweight nature and very high energy density (3860 mA·h·g^{-1}) [7–9]. Previously, various attempts have been made to utilize metallic Li as the anode material. Due to its high reactivity to the electrolyte, there is continuous loss of active Li from the anode in Li anode batteries. As observed in cases of Ni and other metals, the formation of nanoscale structures, such as wire and dendrites, is common in electrochemical reactions. Li also suffers from the problem of dendrite formation during the electroplating process, leading to manyfold increases in the effective area of the anode. Degradation of the solid electrolyte interphase (SEI) during discharge followed by formation of a new SEI in the next charging leads to the loss of Li from the anode, also called dead Li. An increase in dead Li, as well as a decrease in the quantity of the electrolyte, effectively reduces the energy density and life cycle of the LIB with poor coulombic efficiency. Dendrite formation also leads to short-circuiting of the electrode, leading to thermal runaway and fire in the battery [10].

Safety is the main concern while using Li as the anode; hence, the battery market has settled with LIBs with reduced energy density but with higher safety standards. Various works have demonstrated the use of metallic Li as an anode material. These works generally focus on finding a stable host material for Li storage [11–13], minimization of Li dendrite formation by electrolyte engineering [14], use of solid electrolytes [15,16] and interface engineering [17,18]. Stabilization of highly reactive Li metal by surface coating seems promising for stable SEI formation. Coating the Li anode with conductive oxides and sulfides has shown promising results [19,20]. However, use of chemical reagents for surface coating requires perfect control to minimize the non-uniformity of the surface. Use of three-dimensional structures such as wires, fibers and tubes as current collectors has demonstrated a drastic reduction of Li dendrite formation [21,22]. Yang et al. showed that uniform electroplating of Li could be obtained on Cu nanowires, leading to a dendrite-less anode [21]. A carbon matrix has been traditionally used to encapsulate metal/metal oxides, such as NiO, $NiCo_2O_4$, Fe_2O_3, Fe_3O_4, MnO_2, CuO, ZnO, Ge and Sn/SnO_2, forming carbon–metal composites used as anode materials for LIBs. A carbon-based scaffold prevents the large volumetric change during charge/discharge cycles, improving the performance of the battery [23–32]. Here, we demonstrate the synthesis of highly stable conical and fibrous amorphous carbon with encapsulated Li nano-domains (Li-C nanocomposite; Li-C NC) on graphite foil as a candidate material for LIBs for easy and safe handling.

2. Materials and Methods

Li-C NC was fabricated on the edge of a graphite foil. Figure 1a shows the schematics of the sample fabrication method. Commercially available graphite foils, PERMA-FOIL®, TOYO TANSO Co. Ltd. (Nishiyodogawa-ku, Osaka, Japan), with dimensions of 10 mm × 20 mm × 0.1 mm, were used as substrates. The graphite foil was placed vertically on a thick graphite sheet. Three to four shots of Li were placed in front of the standing graphite foil. Ar^+ ions with energy of 700 or 1000 eV were continuously bombarded on the graphite foil and Li shots at 45 °C for at least 30 min to form Li-embedded NC. During Ar^+ ion bombardment, C and Li atoms were ejected and redeposited in the form of conical structures, with Li nano-crystallites embedded in the C matrix to form conical Li-C NC (Figure 1b,c). As seen in Figure 1c, carbon-based nanofibers (CNFs) sometimes grew on the respective conical tips. A Kaufmann-type ion gun (Iontech. Inc. Ltd., model 3-1500-100FC, (Veeco Instruments Inc., New York, NY, USA) was used for the Ar^+ irradiation. Further information on the detailed fabrication process can be found elsewhere [33–37]. Characterization of synthesized Li-C NC was done using TEM JEM ARM 200F operated at 200 kV. A double-tilt TEM holder (JEOL; EM-Z02154T, Chiyoda, Tokyo, Japan) was used without a liquid N_2-based cooling system for transmission electron microscopy (TEM) observations. It is to be noted that no inert gas or vacuum system was used for the preparation and transfer of TEM samples. As synthesized, the sample was directly mounted onto a TEM holder for the TEM characterization.

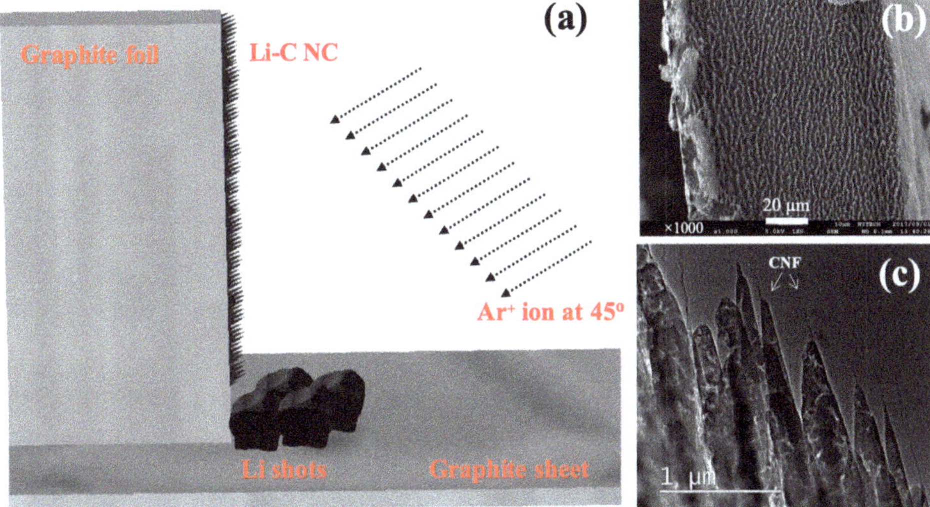

Figure 1. (**a**) Schematics showing sample setup for Li-C nanocomposite (NC) preparation. (**b**) Scanning electron microscope (SEM) image showing the top view of Li-C NC grown on the edge of graphite foil. (**c**) Low Mag transmission electron microscopy (TEM) image showing Li-C NC fabricated by Ar$^+$ ion irradiation at 700 eV. Carbon-based nanofibers (CNFs) grew on top of several conical structures, as exemplified by arrows.

3. Results

Figure 1b,c shows the scanning electron microscope (SEM) and transmission electron microscope (TEM) images of the as-synthesized Li-C NC, respectively. The conical structures are almost unidirectional and freestanding. Their length and density can be controlled by changing the Ar$^+$ irradiation angle and the time of irradiation. As synthesized, Li-C NC was further characterized at higher magnification to confirm the presence of Li atoms. Figure 2a shows the typical Li-C NC with a short CNF on top. CNFs of longer and slenderer size are unstable under an electron beam, making atomic-scale characterization difficult. Therefore, we selected the Li-C NC with a short CNF. Figure 2b, shows the high magnification TEM image taken around the red squared area of Figure 2a, showing the presence of Li atoms. The fast Fourier transform (FFT) corresponding to the electron diffraction pattern taken around the white square clearly shows sharp diffraction points, indicating the presence of crystalline Li. A high magnification TEM image was taken on the crystalline lattice of Figure 2b, as shown in Figure 2c. Individual Li atoms were clearly visible, with a lattice distance of 0.24 nm corresponding to the (110) plane of Li [38]. It is to be noted that this the first report on the atomic-level observation of Li atoms under 200 keV TEM operating voltage without the use of a cooling holder and special sample fabrication methods, as reported in previous works [39]. During TEM observation, no damage to the Li lattice was observed, implying stability of Li nano-domains embedded in an amorphous carbon matrix. Elemental analysis of the Li-C NC was carried out using electron energy loss spectroscopy (EELS). EELS spectra of Figure 2d clearly show the Li peak at 57 eV, which is close to the reported value [28,29]. It is clearly observed that Li-C NC can act as a host material to store Li, even in ambient conditions.

After the optimization of the experimental conditions for the synthesis of Li-C NC, the effect of Ar$^+$ ion energy on the morphology and Li storage of Li-C NC was analyzed in the second set of experiments. To study the effect of the higher Ar$^+$ energy, Li-C NC was synthesized at 1 keV of beam energy. Figure 3a shows the low magnification TEM image of the as-fabricated sample with Ar$^+$ ions sputtered at 1 keV. Interestingly, no particular array of Li-C NC was observed, contrary to the samples synthesized using 700 eV. The presence of a conical structure is observed at only a few places (inside the rectangular area), with most of the samples showing minor protrusions or the deposition of amorphous carbon. However, a large amount of Li was confirmed, as shown in Figure 3b. During TEM observation, the expansion of a Li ball-like structure was observed, as indicated by arrows in Figure 3b. Figure 3c shows an example of rarely present conical NC, showing a large amount of Li attached to the surface of the NC (of the rectangular area of Figure 3a). The inset shows the presence of Li on the tip of the conical NC. Interestingly, no Li lattice was observed inside the C matrix, but a Li-related lattice was visible on the surface of the conical NC, as shown in Figure 3d. As seen in the inset, the line profile taken for the air-exposed atomic lattice shows a higher separation value of 0.257 nm, corresponding to Li_2O_2 (JCPDS card No. 01-074-0115). From this observation, it can be concluded that under the higher ion beam energy of 1 keV, an excess of Li, which is reactive in air, is sputtered and supplied onto the Li-C NC, suggesting that the Li-C NC formed at higher ion energy densities would be less suitable for battery applications. Previous works have also shown that for the fabrication of metal-embedded CNF, the quantity of metal should be optimum. It can be clearly observed that 700 eV is the preferable energy for the fabrication of Li-embedded NC.

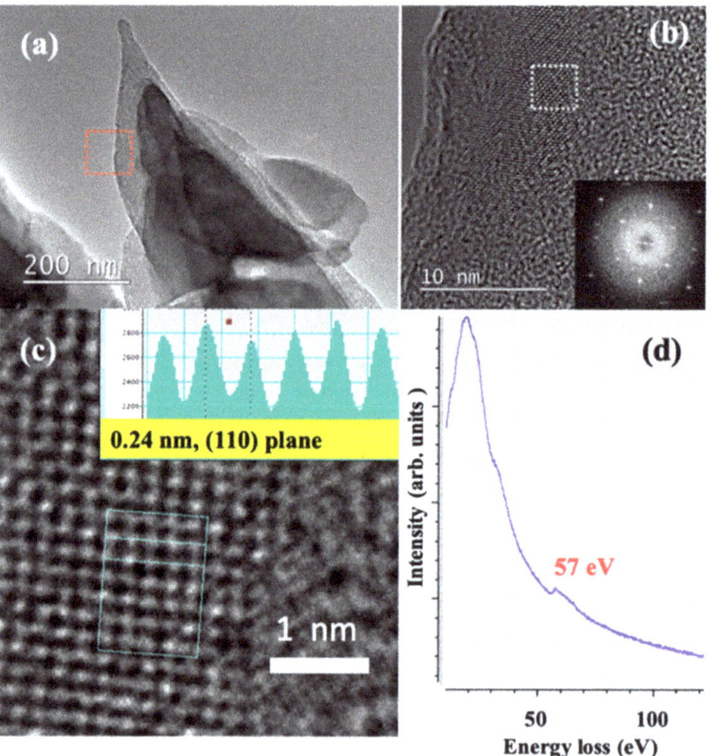

Figure 2. (**a**) TEM image of a typical Li-C NC. (**b**) High mag TEM image of the squared area of Figure 2a (the inset shows the fast Fourier transform (FFT) pattern of the area indicated by the white square). (**c**) High mag TEM of the squared area of Figure 2b showing Li lattice embedded in the C matrix. (**d**) Energy loss spectroscopy (EELS) spectra showing the presence of Li at 57 eV.

Next, we deposited Li-C NC film on a microgrid TEM mesh so that the deposited film could be characterized without any transfer steps. Figure 4a shows the low magnification TEM image of Li-C NC film on a Cu microgrid TEM mesh. Figure 4b shows the patches of light and dark contrasted micro-areas, indicating different Li and C distributions. C is higher in atomic weight compared to Li, yielding darker contrast, whereas the Li-rich area appears as bright micro-areas. A high magnification TEM image clearly shows the presence of the Li lattice (Figure 4c).

Figure 3. (**a**) Low mag TEM image showing Li-C NC fabricated at 1keV of Ar^+. (**b**) Red arrows showing Li balls observed during TEM observation (**c**) Conical Li-C NC synthesized at 1 keV (of the area indicated by the rectangular selection of Figure 3a). (**d**) Edge of the conical Li-C NC showing the presence of the atomic lattice. The inset shows the presence of the Li_2O_2 lattice in the area indicated by the red rectangle.

In order to study the role of the C matrix in protecting embedded Li crystallites, we heated a Li-C NC film deposited on a microgrid TEM mesh over a hot plate at 300 °C for 10 min in ambient conditions. Figure 4d shows a typical TEM image for the Li-C NC film after heating for 10 min. It is observed that Li lattices are protected in ambient conditions without apparent oxidation or change in morphology. The inset of Figure 4d shows a high magnification TEM image, revealing the pristine Li lattice with a lattice distance of 0.24 nm, corresponding to the (110) plane, and the preservation of Li was further confirmed by an EELS spectrum peaking at about 58 eV (Figure 4e). This shows that amorphous C acts as a buffer, protecting Li from oxidation or exposure to moisture.

This simple ion beam-based Li-C NC fabrication can be adopted for the industrial-scale development of pre-lithiated anode material for battery applications. Li-C NC should be accessible for the electrochemical processes, as carbon in Li-C NC is mostly amorphous carbon, which is not much different from carbonaceous materials that are commonly used in commercial batteries. In order to confirm this, we are now planning to measure its charge/discharge property by assembling a prototype of a battery, and to also carry out the in-situ TEM observation of this charge/discharge process using Li-C NC and Li-C composite films based on our in-situ TEM technique using CNFs [33,34,40]. The results will be reported in forthcoming papers.

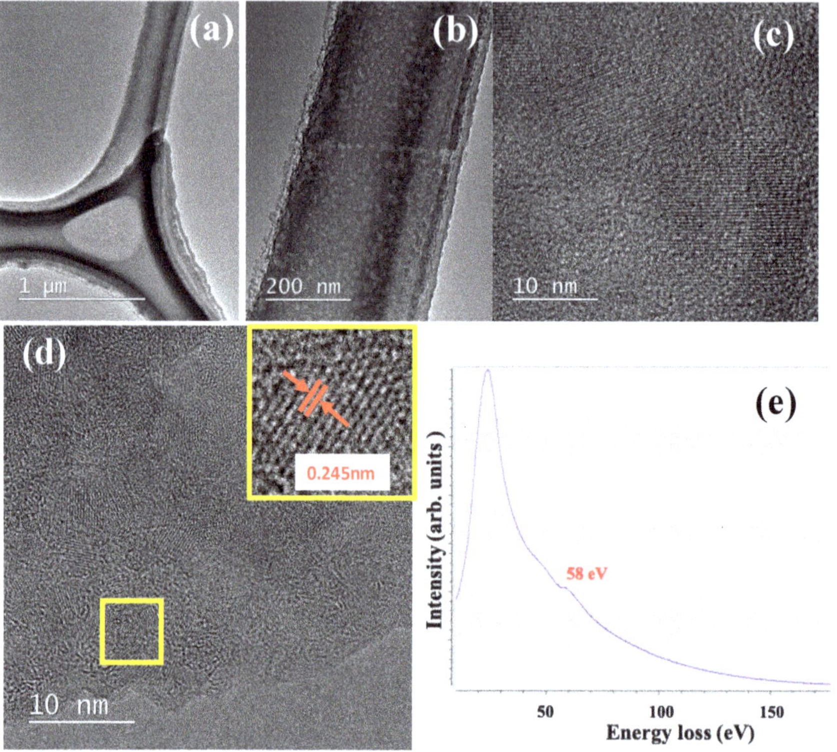

Figure 4. (**a**) Low mag TEM showing Li-C NC film deposited on TEM mesh. (**b**) Typical Li-C NC film showing dark and bright spots indicating different distributions of Li and C. (**c**) High mag TEM image showing Li lattice in C matrix before annealing. (**d**) Li-C NC film after annealing for 10 min at 300 °C. The inset presents the squared area showing the intact Li lattice (**e**) EELS spectra taken on the annealed sample showing the presence of pristine Li.

4. Conclusions

In summary, we developed a novel method for the fabrication of Li-embedded NC using an ion beam setup. We found that Li inside NC is well preserved from oxidation during handling in ambient conditions. Further, no significant electron beam-related damage was observed on the Li lattice during TEM measurements. Since Li is pre-stored inside the NC, Li-C NC can be attractive as the anode material for the LIB battery. Further, the method was extended to fabricate Li-C NC films on microgrid TEM meshes, which can be easily extended to deposit Li-C NC films on any arbitrary substrates for the fabrication of battery electrodes.

Author Contributions: Conceptualization, S.S. and M.T.; investigation, S.S., T.O., S.E., S.O. and B.P.J.; writing—original draft preparation, S.S.; writing—review and editing, G.K., C.C., M.T.; supervision, M.T. All authors have read and agreed to the published version of the manuscript.

Funding: This work was partially supported by JSPS Grant-in-Aid for Scientific Research (B) Grant No. 20H02618.

Conflicts of Interest: The authors declare no conflict of interest.

References

1. Nitta, N.; Wu, F.; Lee, J.T.; Yushin, G. Li-ion battery materials: Present and future. *Mater. Today* **2015**, *18*, 252–264. [CrossRef]
2. Yang, Z.; Zhang, J.; Kintner-Meyer, M.C.; Lu, X.; Choi, D.; Lemmon, J.P.; Liu, J. Electrochemical energy storage for green grid. *Chem. Rev.* **2011**, *111*, 3577–3613. [CrossRef]
3. Dunn, B.; Kamath, H.; Tarascon, J.-M. Electrical energy storage for the grid: A battery of choices. *Science* **2011**, *334*, 928–935. [CrossRef]
4. Goodenough, J.B.; Kim, Y. Challenges for rechargeable Li batteries. *Chem. Mater.* **2009**, *22*, 587–603. [CrossRef]
5. Bruce, P.G.; Freunberger, S.A.; Hardwick, L.J.; Tarascon, J.-M. Li–O_2 and Li–S batteries with high energy storage. *Nat. Mater.* **2012**, *11*, 19. [CrossRef]
6. Buqa, H.; Goers, D.; Holzapfel, M.; Spahr, M.E.; Novák, P. High rate capability of graphite negative electrodes for lithium-ion batteries. *J. Electrochem. Soc.* **2005**, *152*, A474–A481. [CrossRef]
7. Qian, J.; Adams, B.D.; Zheng, J.; Xu, W.; Henderson, W.A.; Wang, J.; Bowden, M.E.; Xu, S.; Hu, J.; Zhang, J.-G. Anode-free rechargeable lithium metal batteries. *Adv. Funct. Mater.* **2016**, *26*, 7094–7102. [CrossRef]
8. Xu, W.; Wang, J.; Ding, F.; Chen, X.; Nasybulin, E.; Zhang, Y.; Zhang, J.-G. Lithium metal anodes for rechargeable batteries. *Energy Environ. Sci.* **2014**, *7*, 513–537. [CrossRef]
9. Neudecker, B.J.; Dudney, N.J.; Bates, J.B. "Lithium-free" Thin-film battery with in situ plated Li anode. *J. Electrochem. Soc.* **2000**, *147*, 517–523. [CrossRef]
10. Roy, P.; Srivastava, S.K. Nanostructured anode materials for lithium ion batteries. *J. Mater. Chem. A* **2015**, *3*, 2454–2484. [CrossRef]
11. Lin, D.; Liu, Y.; Liang, Z.; Lee, H.-W.; Sun, J.; Wang, H.; Yan, K.; Xie, J.; Cui, Y. Layered reduced graphene oxide with nanoscale interlayer gaps as a stable host for lithium metal anodes. *Nat. Nanotechnol.* **2016**, *11*, 626. [CrossRef] [PubMed]
12. Liu, Y.; Lin, D.; Liang, Z.; Zhao, J.; Yan, K.; Cui, Y. Lithium-coated polymeric matrix as a minimum volume-change and dendrite-free lithium metal anode. *Nat. Commun.* **2016**, *7*, 10992. [CrossRef] [PubMed]
13. Liang, Z.; Lin, D.; Zhao, J.; Lu, Z.; Liu, Y.; Liu, C.; Lu, Y.; Wang, H.; Yan, K.; Tao, X. Composite lithium metal anode by melt infusion of lithium into a 3D conducting scaffold with lithiophilic coating. *Proc. Natl. Acad. Sci. USA* **2016**, *113*, 2862–2867. [CrossRef] [PubMed]
14. Lu, Y.; Tu, Z.; Archer, L.A. Stable lithium electrodeposition in liquid and nanoporous solid electrolytes. *Nat. Mater.* **2014**, *13*, 961. [CrossRef] [PubMed]
15. Kamaya, N.; Homma, K.; Yamakawa, Y.; Hirayama, M.; Kanno, R.; Yonemura, M.; Kamiyama, T.; Kato, Y.; Hama, S.; Kawamoto, K. A lithium superionic conductor. *Nat. Mater.* **2011**, *10*, 682. [CrossRef]
16. Han, X.; Gong, Y.; Fu, K.K.; He, X.; Hitz, G.T.; Dai, J.; Pearse, A.; Liu, B.; Wang, H.; Rubloff, G. Negating interfacial impedance in garnet-based solid-state Li metal batteries. *Nat. Mater.* **2017**, *16*, 572. [CrossRef]
17. Zheng, G.; Lee, S.W.; Liang, Z.; Lee, H.-W.; Yan, K.; Yao, H.; Wang, H.; Li, W.; Chu, S.; Cui, Y. Interconnected hollow carbon nanospheres for stable lithium metal anodes. *Nat. Nanotechnol.* **2014**, *9*, 618. [CrossRef]
18. Li, N.-W.; Yin, Y.-X.; Yang, C.-P.; Guo, Y.-G. An artificial solid electrolyte interphase layer for stable lithium metal anodes. *Adv. Mater.* **2016**, *28*, 1853–1858. [CrossRef]
19. Kozen, A.C.; Lin, C.-F.; Pearse, A.J.; Schroeder, M.A.; Han, X.; Hu, L.; Lee, S.-B.; Rubloff, G.W.; Noked, M. Next-generation lithium metal anode engineering via atomic layer deposition. *Acs Nano* **2015**, *9*, 5884–5892. [CrossRef]
20. Cao, Y.; Meng, X.; Elam, J.W. Atomic layer Deposition of Li_xAl_yS solid-state electrolytes for stabilizing lithium-metal anodes. *Chem. Electro. Chem.* **2016**, *3*, 858–863.
21. Yang, C.-P.; Yin, Y.-X.; Zhang, S.-F.; Li, N.-W.; Guo, Y.-G. Accommodating lithium into 3D current collectors with a submicron skeleton towards long-life lithium metal anodes. *Nat. Commun.* **2015**, *6*, 1–9. [CrossRef] [PubMed]
22. Zhang, A.; Fang, X.; Shen, C.; Liu, Y.; Zhou, C. A Carbon nanofiber network for stable lithium metal anodes with high coulombic efficiency and long cycle life. *Nano Res.* **2016**, *9*, 3428–3436. [CrossRef]
23. Cheng, M.Y.; Hwang, B.J. Mesoporous carbon-encapsulated NiO nanocomposite negative electrode materials for high-rate Li-ion battery. *J. Power Sources* **2010**, *195*, 4977–4983. [CrossRef]
24. Fan, Z.; Wang, B.; Xi, Y.; Xu, X.; Li, M.; Li, J.; Yang, G. A $NiCo_2O_4$ nanosheet-mesoporous carbon composite electrode for enhanced reversible lithium storage. *Carbon* **2016**, *99*, 633–641. [CrossRef]

25. Nagao, M.; Otani, M.; Tomita, H.; Kanzaki, S.; Yamada, A.; Kanno, R. New three-dimensional electrode structure for the lithium battery: Nano-sized γ-Fe$_2$O$_3$ in a mesoporous carbon matrix. *J. Power Sources* **2011**, *196*, 4741–4746. [CrossRef]

26. Zeng, L.; Zheng, C.; Deng, C.; Ding, X.; Wei, M. MoO$_2$-ordered mesoporous carbon nanocomposite as an anode material for lithium-ion batteries. *ACS Appl. Mater. Interfaces* **2013**, *5*, 2182–2187. [CrossRef]

27. Wu, F.; Huang, R.; Mu, D.; Shen, X.; Wu, B. A novel composite with highly dispersed Fe$_3$O$_4$ nanocrystals on ordered mesoporous carbon as an anode for lithium ion batteries. *J. Alloys Compd.* **2014**, *585*, 783–789. [CrossRef]

28. Yang, M.; Gao, Q. Copper oxide and ordered mesoporous carbon composite with high performance using as anode material for lithium-ion battery. *Microporous Mesoporous Mater.* **2011**, *143*, 230–235. [CrossRef]

29. Guo, R.; Zhao, L.; Yue, W. Assembly of core–shell structured porous carbon–graphene composites as anode materials for lithium-ion batteries. *Electrochim. Acta* **2014**, *152*, 338–344. [CrossRef]

30. Shen, L.; Uchaker, E.; Yuan, C.; Nie, P.; Zhang, M.; Zhang, X.; Cao, G. Three-dimensional coherent titania–mesoporous carbon nanocomposite and its lithium-ion storage properties. *ACS Appl. Mater. Interfaces* **2012**, *4*, 2985–2992. [CrossRef]

31. Ma, J.; Xiang, D.; Li, Z.; Li, Q.; Wang, X.; Yin, L. TiO$_2$ nanocrystal embedded ordered mesoporous carbons as anode materials for lithium-ion batteries with highly reversible capacity and rate performance. *CrystEngComm* **2013**, *15*, 6800–6807. [CrossRef]

32. Xu, X.; Fan, Z.; Yu, X.; Ding, S.; Yu, D.; Lou, X.W. A Nanosheets-on-channel architecture constructed from MoS$_2$ and CMK-3 for high-capacity and long-cycle-life lithium storage. *Adv. Energy Mater.* **2014**, *4*, 1400902. [CrossRef]

33. Sharma, S.; Rosmi, M.S.; Yaakob, Y.; Yusop, M.Z.M.; Kalita, G.; Kitazawa, M.; Tanemura, M. In situ TEM synthesis of Y-junction carbon nanotube by electromigration induced soldering. *Carbon* **2018**, *132*, 165–171. [CrossRef]

34. Yusop, M.Z.M.; Ghosh, P.; Yaakob, Y.; Kalita, G.; Sasase, M.; Hayashi, Y.; Tanemura, M. In situ TEM observation of Fe-included carbon nanofiber: Evolution of structural and electrical properties in field emission process. *ACS Nano* **2012**, *6*, 9567–9573. [CrossRef] [PubMed]

35. Yaakob, Y.; Kuwataka, Y.; Yusop, M.Z.M.; Tanaka, S.; Rosmi, M.S.; Kalita, G.; Tanemura, M. Room-temperature growth of ion-induced Si-and Ge-incorporated carbon nanofibers. *Phys. Status Solidi (b)* **2015**, *252*, 1345–1349. [CrossRef]

36. Tanemura, M.; Okita, T.; Yamauchi, H.; Tanemura, S.; Morishima, R. Room-temperature growth of a carbon nanofiber on the tip of conical carbon protrusions. *Appl. Phys. Lett.* **2004**, *84*, 3831–3833. [CrossRef]

37. Tanemura, M.; Hatano, H.; Kitazawa, M.; Tanaka, J.; Okita, T.; Lau, S.P.; Yang, H.Y.; Yu, S.F.; Huang, L.; Miao, L. Room-temperature growth of carbon nanofibers on plastic substrates. *Surf. Sci.* **2006**, *600*, 3663–3667. [CrossRef]

38. Li, Y.; Li, Y.; Pei, A.; Yan, K.; Sun, Y.; Wu, C.-L.; Joubert, L.-M.; Chin, R.; Koh, A.L.; Yu, Y.; et al. Atomic structure of sensitive battery materials and interfaces revealed by cryo–electron microscopy. *Science* **2017**, *358*, 506–510. [CrossRef]

39. Liu, X.H.; Huang, J.Y. In situ TEM electrochemistry of anode materials in lithium ion batteries. *Energy Environ. Sci.* **2011**, *4*, 3844–3860. [CrossRef]

40. Rosmi, M.S.; Yusop, M.Z.; Kalita, G.; Yaakob, Y.; Takahashi, C.; Tanemura, M. Visualizing copper assisted graphene growth in nanoscale. *Sci. Rep.* **2015**, *4*, 7563. [CrossRef]

 *nanomaterials*

MDPI

Article

Intelligent Identification of MoS$_2$ Nanostructures with Hyperspectral Imaging by 3D-CNN

Kai-Chun Li [1], Ming-Yen Lu [2], Hong Thai Nguyen [1], Shih-Wei Feng [3], Sofya B. Artemkina [4,5], Vladimir E. Fedorov [4,5] and Hsiang-Chen Wang [1,*

[1] Department of Mechanical Engineering and Advanced Institute of Manufacturing with High Tech Innovations, National Chung Cheng University, 168, University Rd., Min Hsiung, Chia Yi 62102, Taiwan; zuoo549674@gmail.com (K.-C.L.); nguyenhongthai194@gmail.com (H.T.N.)

[2] Department of Materials Science and Engineering, National Tsing Hua University, 101, Sec. 2, Kuang-Fu Road, Hsinchu 30013, Taiwan; mingyenlu@gmail.com

[3] Department of Applied Physics, National University of Kaohsiung, 700 Kaohsiung University Rd., Nanzih District, Kaohsiung 81148, Taiwan; swfeng@nuk.edu.tw

[4] Nikolaev Institute of Inorganic Chemistry, Siberian Branch of Russian Academy of Sciences, 630090 Novosibirsk, Russia; artem@niic.nsc.ru (S.B.A.); fed@niic.nsc.ru (V.E.F.)

[5] Department of Natural Sciences, Novosibirsk State University, 1, Pirogova str., 630090 Novosibirsk, Russia

[*] Correspondence: hcwang@ccu.edu.tw

Received: 23 April 2020; Accepted: 11 June 2020; Published: 13 June 2020

Abstract: Increasing attention has been paid to two-dimensional (2D) materials because of their superior performance and wafer-level synthesis methods. However, the large-area characterization, precision, intelligent automation, and high-efficiency detection of nanostructures for 2D materials have not yet reached an industrial level. Therefore, we use big data analysis and deep learning methods to develop a set of visible-light hyperspectral imaging technologies successfully for the automatic identification of few-layers MoS$_2$. For the classification algorithm, we propose deep neural network, one-dimensional (1D) convolutional neural network, and three-dimensional (3D) convolutional neural network (3D-CNN) models to explore the correlation between the accuracy of model recognition and the optical characteristics of few-layers MoS$_2$. The experimental results show that the 3D-CNN has better generalization capability than other classification models, and this model is applicable to the feature input of the spatial and spectral domains. Such a difference consists in previous versions of the present study without specific substrate, and images of different dynamic ranges on a section of the sample may be administered via the automatic shutter aperture. Therefore, adjusting the imaging quality under the same color contrast conditions is unnecessary, and the process of the conventional image is not used to achieve the maximum field of view recognition range of ~1.92 mm^2. The image resolution can reach ~100 nm and the detection time is 3 min per one image.

Keywords: hyperspectral imagery; deep learning; 3D-CNN; MoS$_2$; automated optical inspection

---

1. Introduction

The most frequently observed material among all of the two-dimensional (2D) transition metal chalcogenides for the next generation of electronic and optoelectronic components is molybdenum disulphide [1–6]. Materials related to nanometer-scale electronic and optoelectronic components, such as field effect transistor [7–10], prospective memory component [11], light-emitting diode [12,13], and sensors [14–19], have been produced due to the excellent spin-valley coupling and flexural and optoelectronic properties of MoS$_2$. However, the development of high-performance and large-area characterization techniques has been a major obstacle to the basic and commercial applications of 2D nanostructures.

In the prior art optical film measurement, the atomic force microscope (AFM) has various disadvantages, such as relatively limited scan range and time consuming; thus, it is unsuitable for large-area quick measurements [20,21]. Raman spectroscopy (Raman) is usually only capable of local characterization within the spot, which results in a limited measurement rate; hence, it is unsuitable for large-area analysis. Transmission electron microscopy (TEM) and scanning tunneling microscopy can be characterized at a high spatial resolution of up to atomic scale [22,23]. However, both techniques have the disadvantages of low throughput and complex sample preparation. The use of machine learning, as compared with the abovementioned techniques, in image or visual recognition is a mature application field. The integration of machine learning (SVM, KNN, BGMM-DP, and K-means) with optical microscopes has only begun in recent years. Thus, artificial intelligence has great potential in the recognition of microscopic images, especially nanostructures [24,25]. In 2019, Hong et al. demonstrated the machine learning algorithm to identify local atomic structures by reactive molecular dynamics [26]. In 2020, Masubuchi et al. showed the real-time detection of 2D materials by deep-learning-based image segmentation algorithm [27]. In the same year, Yang et al. presented the identification of 2D material flakes of different layers from the optical microscope images via machine learning-based model [28]. However, the following three shortcomings remain: (1) optical microscope image quality often depends on the user's experience and will pass through image processing. (2) Only the color space with few feature dimensions will have the possibility of underfitting. (3) Angular illumination asymmetry (ANILAS) of the field of view (FOV) is an important factor that is largely ignored, which results in a certain loss of pixel accuracy [29–31].

Here, we used big data analysis and deep learning methods, combined with the hyperspectral imagery common in the remote sensing field, to solve the difficulties that are encountered in the previous literature (e.g., uneven light intensity distribution, image dynamic range correction, and image noise filtering process). The first attempt was made by analyzing the eigenvalues in other dimensions that were not previously used (e.g., morphological features) to improve the prediction accuracy [29–31]. An intelligent detection can be achieved to identify the layer number of 2D materials.

2. Materials and Methods

2.1. Growth Mechanism and Surface Morphology of MoS_2

The majority of 2D material layer identification studies focus on film synthesis using mechanical stripping [32–34]. Although an improved quality of molybdenum disulfide can be obtained, this method cannot synthesize a large-area molybdenum disulfide film with a few layers. The chemical vapor deposition method [35–38] can produce high-quality and large-area molybdenum disulfide on a suitable substrate surface under a stable gas flux and temperature environment, and it is suitable for current device fabrication [38–40]. In 2014, Shanshan et al. explored the sensitivity of the MoS_2 region growth in a relatively uniform temperature range [41]. In 2017, Lei et al. found that temperature is one of the main factors controlling MoS_2 morphology [42]. Wei et al. maintained the state for a long time to observe the change of MoS_2 type when the precursor was heated to a constant temperature interval [43]. In 2018, Dong et al. discussed the nucleation and growth mechanism of MoS_2 [44]. On the basis of their results, two types of film growth dynamic paths were established: one is a central nanoparticle with a multi-layered MoS_2 structure and the other is a single-layered triangular dominated or double-layered structure. The conclusion of reference [45] explained the effect of adjusting the growth temperature and carrier gas flux. Understanding the growth pattern mechanism can help us to obtain the initial requirements and judgments of database collection and data tagging.

2.2. CVD Sample Preparation

The sample was grown on a sapphire substrate by using CVD to form a MoS_2 film. The precursor used had Sulfur (99.98%, Echo Chemical Co., Miaoli county, Taiwan) and MoO_3 (99.95%, Echo Chemical Co., Miaoli county, Taiwan), each placed in the appropriate position of the inner part of the quartz tube.

The substrate was placed over the MoO_3 crucible and in the center of the furnace tube, in a windward position. During growth, different parameters, such as ventilation, heating rate, temperature holding time, and maximum temperature, were set. The MoS_2 sample was obtained at the end of the growth process. Figure S1 shows the schematic of the experimental structure and position of the precursor. Some new pending data indicate the growth of the periodic structure of MoS_2, which is crucial for the large-scale controllable molybdenum disulfide synthesis and it will greatly benefit the future production of electronic components. In comparison with that of other studies [45–49], this sample was fabricated via laser processing to make periodic holes (hole diameter and depth of approximately 10 µm and 300 nm, respectively), followed by CVD to grow MoS_2.

2.3. Optical Microscope Image Acquisition

MoS_2 on the sapphire substrate was observed through an optical microscope (MM40, Nikon, Lin Trading Co., Taipei, Taiwan) at 10×, 40×, and 100× magnification rates. For the experimental sample, we recorded the shooting area code (Figure S1c to store the images captured via optical microscopy (OM) systematically in the image database. The image had a size of 1600 pixels × 1200 pixels with a depth of 32 bits. The portable network graphics (PNG) file, including the gain image at different dynamic range intervals, was acquired, but color calibration and denoising action were not performed. We aimed to replace the cumbersome image processing with deep learning. On the basis of the amount of data that were collected in this experiment, approximately 90 pieces of 2×2 cm^2 growth samples were obtained, and ~2000 images were used as sources of data exploration.

2.4. System Equipment and Procedures

This study aims to construct a system for the automated analysis of different layers of MoS_2 film. The system architecture is mainly divided into four parts, as shown in Figure S2. The program flow is as follows: (1) database: the prepared molybdenum disulfide sample is measured using a Raman microscope to determine the position of each layer distribution, and an image is taken using an optical microscope and CCD to capture the same position (Image Capture System). (2) Offline Training: the obtained CCD image is combined with hyperspectral imaging technology (VIS-HSI) to convert the spectral characteristics of each layer of molybdenum disulfide, and data preprocessing is performed. (3) Model Design: the data are further trained in deep learning, thereby completing the establishment of our classification algorithm. (4) Online Service: when a new sample of molybdenum disulfide is to be analyzed, it is placed under an optical microscope to capture the surface image by using CCD. Subsequently, the spectral characteristic value of each pixel is obtained through the hyperspectral imaging technique, and the model is predicted by training. The number of layers of the molybdenum disulfide film at the pixel position is determined, and the molybdenum disulfide film with different layers in the image is visualized while using different colors.

2.5. Tag Analysis and Feature Extraction Workflow

Figure 1 shows the processing step before the data enters the model. Layer number labeling of the manual area circle mask (Mask) is performed and the data are divided when the captured image is converted into a spectrum image by using a hyperspectral image technology; we will measure the result on the basis of the Raman spectrum (Figure S3) [50–54]. The categories are substrate, monolayer, bilayer, trilayer, bulk, and residues, which are our ground truths. Two types of data are available for model training, namely "feature" and "label". Feature has two types: one is hyperspectral vector as input for deep neural network (DNN) and one-dimensional (1D) convolutional neural network (1D-CNN), and the other is spatial domain hyperspectral cube as input for three-dimensional (3D) convolutional neural network (3D-CNN). After data preprocessing, we divide the dataset into three parts. We randomly select 80% of the labeled samples as training data, 20% as the verification set, and the remaining unmarked parts for the test set. As the light intensity distribution is incompletely covered in the dataset in the training and validation sets, we need a test set to help us measure whether the model has this capability.

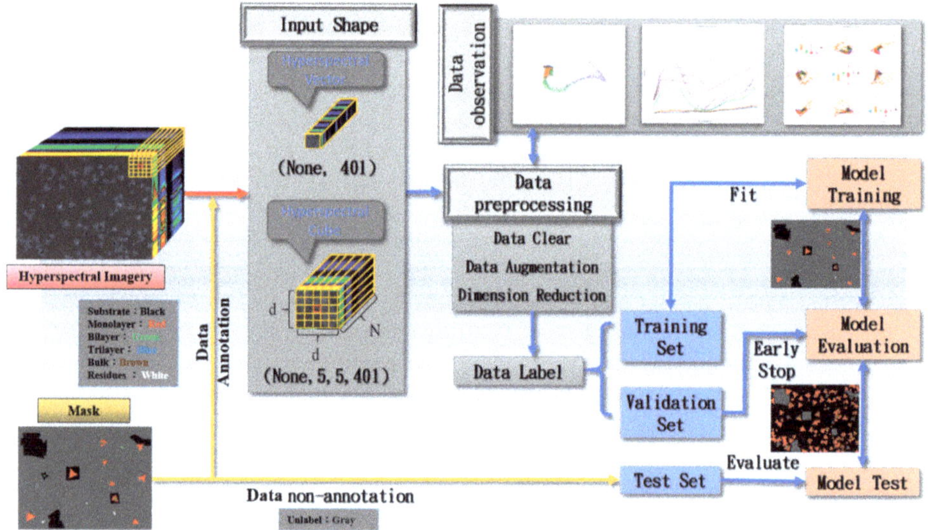

Figure 1. Data feature and label processing.

2.6. Visible Hyperspectral Imaging Algorithm

The VIS-HSI used in this study is a combination of CCD (Sentech, STC-620PWT) and visible hyperspectral algorithm (VIS-HSA). The calculated wavelength range is 380–780 nm and the spectral resolution is 1 nm. The core concept of this technology is to input the image captured by the CCD in the OM to the spectrometer, such that each pixel of the captured image has spectrum information [55]. Figure S4 shows the flow of the technology, while using MATLAB. A custom algorithm is created.

2.7. Data Preprocessing and Partitioning

All of the MoS_2 samples were obtained at different substrate locations and uneven illumination distribution to form a dataset. Two types of features were available: one is the 1×401 hyperspectral vector central pixel, which extracts the marker data as DNN and 1D-CNN inputs, and the other is the $1 \times 5 \times 5 \times 401$ hyperspectral cube, which regards the spatial region as the center of the cubes. The category labels of the pixels extracted the tag data as the 3D-CNN input. With single hyperspectral image data as an example (Figure S5), the marked area was divided into 80% training set and 20% validation set, whereas the unmarked area was the test set. Finally, the new pending data were generated. The generalization capability of the model was evaluated. The training set part had oversampling and data enhancement for the hyperspectral cube, because it contained spatial features and it was processed via horizontal mirroring and left and right flipping.

2.8. Software and Hardware

The model usage environment was implemented in the Microsoft Windows 10 operating system while using TensorFlow Framework version 1.10.0 and Python version 3.6. Open-source data analysis platforms, namely, Jupyter notebook, SciPy, NumPy, Matplotlib, Pandas, Scikit-learn, Keras, and Spectral, were used to analyze the feature values. The training hardware used a consumer-grade desktop computer with a GeForce GTX1070 graphics card (NVDIA, Hong Kong, China) and Core i5-3470 CPU @ 3.2GHz (Intel, Taipei, Taiwan).

3. Results

The model must have a certain recognition capability, because ensuring that the quality of each captured image is the same is impossible. Therefore, we add several image quality features to the hyperspectral image data of different imaging qualities to make the model close to the practical classification performance. We will explore the models with different magnification rates in order to discuss the spatial resolution and mixed pixel issues [56,57].

3.1. Model Framework for Deep Learning

The classification prediction model uses three models, namely, DNN, 1D-CNN, and 3D-CNN, as shown in Figure 2. Among the model parameters, the learning rate is adjusted to 1×10^{-6} to 5×10^{-6} and the batch size (batch) is based on the difference between the model and data. The size and dropout are 24 and 0.25, respectively, and the selected optimizer is RMSprop. Figure 2a illustrates a schematic of the basic DNN model architecture. The input is a hyperspectral vector feature belonging to single-pixel spectral information. The model only contains three layers of fully connected layer. Additional outer neuron nodes are expected to extract relatively shallow features [58]. The six categories that we classify are the outputs. Figure 2b displays a schematic of the 1D-CNN model architecture. The input is consistent with the DNN model. The model consists of four convolutional layers, which include two pooling layers and two fully connected layers. CNN convolution kernel has rights-sharing characteristics. We believe that convolving of spectral features is equivalent to chopping different frequencies. Given the 1 nm resolution of the band, the correlation with adjacent feature points is high. The pooling layer can help to eliminate features that are too similar in the neighborhood and reduce the redundant dimensions of features [59]. Figure 2c shows a schematic of the 3D-CNN model architecture. The input belongs to a hyperspectral cube type of the space-spectral domain. It extracts a feature cube consisting of pixels as $d \times d \times N$ in a small spatial neighborhood (not the entire image) along the entire spectral band as input data and convolves with the 3D kernel to learn the spectral spatial features. The reason for using neighboring pixels is based on the observation that pixels in the small spatial neighborhood often reflect similar characteristics [60]. This observation is proven in Ref [55,61], which indicated that the small 3×3 core is the best option for spatial features; thus, only two convolution operations are performed and the sample space size is set to 5×5, which can be reduced to only two convolutional layers (1×1). The spatial domain of each layer is extracted. The first 3D convolutional layers, namely, C1 and C2, each contain a 3D kernel. The kernel size is $K_1^1 \times K_2^1 \times K_3^1$, resulting in two sizes of 3D feature cubes as $(d - K_1^1 + 1) \times (d - K_2^1 + 1) \times (N - K_3^1 + 1)$. Two 3D feature cubes, namely, C1 and C2, as $(d - K_1^1 + 1) \times (d - K_2^1 + 1) \times (N - K_3^1 + 1)$, are used as inputs. The second 3D convolutional layer, namely, C3, involves four 3D cores (with a size of $K_1^2 \times K_2^2 \times K_3^2$) and produces eight 3D data cubes, each as $(d - K_1^1 - K_1^2 + 2) \times (d - K_2^1 - K_2^2 + 2) \times (N - K_3^1 - K_3^2 + 1)$ [62].

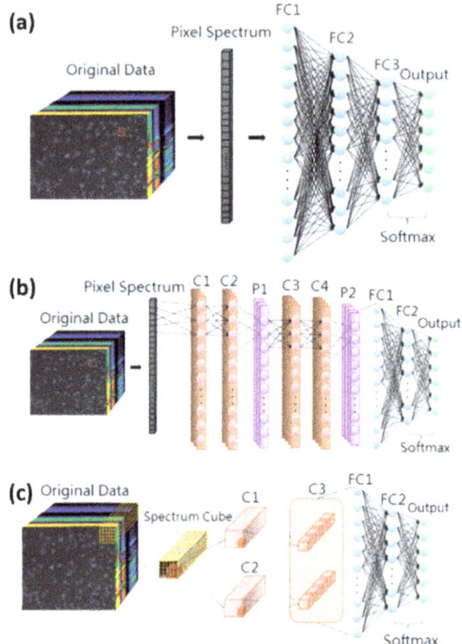

Figure 2. Schematic of the model construction. (**a**) Deep neural network (DNN), (**b**) one-dimensional (1D) convolutional neural network (1D-CNN), and (**c**) three-dimensional (3D) convolutional neural network (3D-CNN), where the inputs in (**a**,**b**) are hyperspectral vectors, and the input in (**c**) is hyperspectral cube. For the outputs of the three models, Softmax is used as a classifier for six categories: substrate, monolayer, bilayer, trilayer, bulk, and residues.

3.2. Training Results under Three Deep Learning Models with 10× Magnification

Figure 3 shows the calculation results of the sapphire substrate sample via three models, namely, DNN, 1D-CNN, and 3D-CNN, under 10× magnification. Figure 3a–c exhibit the convergence curve and training time of the loss and accuracy in the three algorithms, respectively. 3D-CNN has a longer training time and epoch than the first two models, and its input feature variable has more space domain parts; thus, it takes more time to start the convergence process. Figure 3d–f display the results of the confusion matrix of the verification set in the three algorithms, respectively. Table 1 presents the evaluation results of each category. Classifier precision determines how many of the positive categories of all the samples are true positive. The recall rate indicates how many of the true-positive category samples are judged as positive category samples by the classifier. F1-score is the harmonic mean of the accuracy and recall rate. Macro-average refers to the arithmetic mean of each statistical indicator value of all categories. The micro-average is used to establish a global confusion matrix for each model example in the dataset without category and then calculate the corresponding indicators. 3D-CNN accuracy is better from the indicators. Figures S6 and S7 exhibit the remaining training procedures at 40× and 100× magnification rates. Table 1 and Figures S6 and S7 show that the evaluation results under a small magnification are relatively poor. Therefore, the mixed pixels may cause the pixels to contain additional mixing or noise factors.

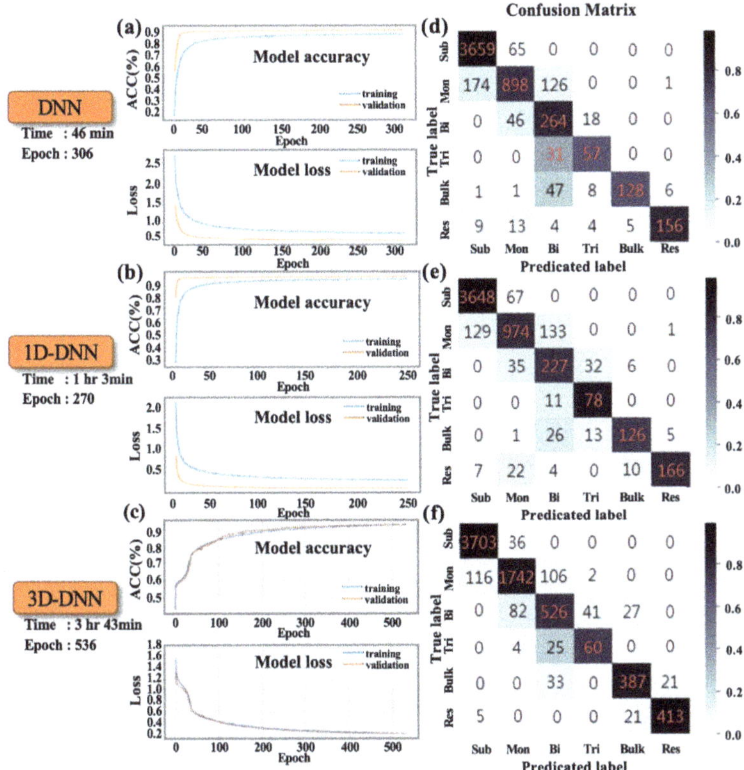

Figure 3. At 10× magnification condition: accuracy (ACC) and loss in the convergence process in (**a**) DNN, (**b**) 1D-CNN, and (**c**) 3D-CNN; confusion matrix results of the validation set in (**d**) DNN, (**e**) 1D-CNN, and (**f**) 3D-CNN.

Table 1. Model evaluation indicators for the three models at 10× magnification.

| | Final Accuracy (Validation Data) | | | | | | | | |
| | DNN | | | 1D-CNN | | | 3D-CNN | | |
	Precision	Recall	F1-Score	Precision	Recall	F1-Score	Precision	Recall	F1-Score
Substrate	0.9521	0.9825	0.9671	0.9641	0.9820	0.9729	0.9684	0.9904	0.9792
Monolayer	0.8778	0.7490	0.8083	0.8863	0.7874	0.8339	0.9345	0.8861	0.9097
Bilayer	0.5593	0.8049	0.6600	0.5661	0.7567	0.6476	0.7623	0.7781	0.7701
Tri-layer	0.6552	0.6477	0.6514	0.6341	0.8764	0.7358	0.6504	0.7339	0.6897
Bulk	0.9624	0.6702	0.7901	0.8873	0.7368	0.8051	0.8897	0.8776	0.8836
Residues	0.9571	0.8168	0.8814	0.9651	0.7943	0.8714	0.9516	0.9408	0.9462
micro average	0.9023	0.9023	0.9023	0.9123	0.9123	0.9123	0.9296	0.9296	0.9296
macro average	0.8273	0.7785	0.7930	0.8172	0.8223	0.8111	0.8595	0.8678	0.8631
weighted average	0.9100	0.9023	0.9026	0.9190	0.9123	0.9134	0.9300	0.9296	0.9295

3.3. Prediction Results at 10× Magnification

Figure 4 presents a randomly selected sample at 10× magnification, which is predicted by three models. Figure 4a displays the OM image under the corresponding range of the prediction data. Figure 4e exhibits the OM image of new pending data. Figure 4b–d show the training data and Figure 4f–h present the prediction results for the new data (new pending data) under three models (color classification image), respectively. The results from the region of interest (ROI) in the training data indicate that the DNN and 1D-CNN models cannot accurately predict the damaged region when the MoS_2 film encounters external force damage and the crystal structure is missing. 3D-CNN can be clearly judged under the destruction of the region, and Supplementary Figure S8 exhibits the remaining new pending data predictions. In other predictions for 40× and 100× magnification rates (Figures S9–S12), the color classification image (false-color composite) can be easily found for each model, but it will be reduced in the opposite FOV detection range.

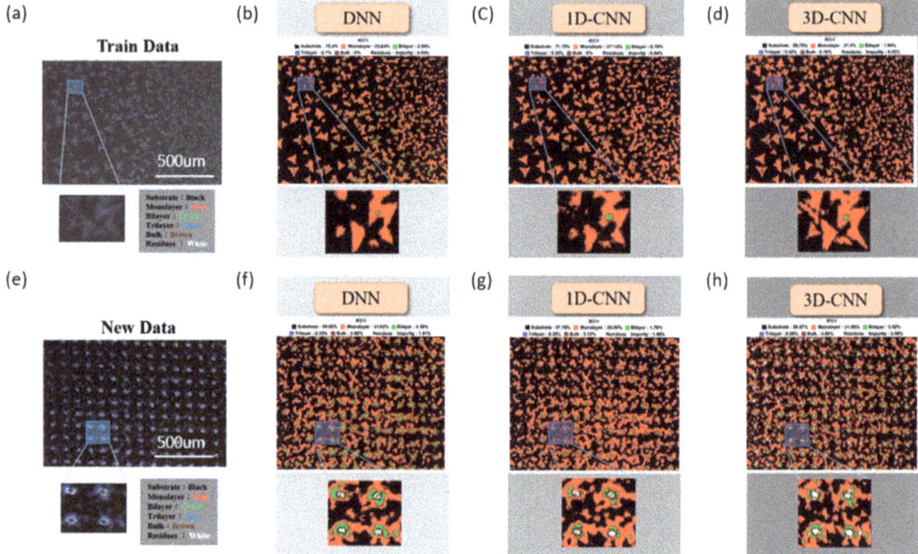

Figure 4. At 10× magnification: (**a,e**) optical microscopy (OM) images of the training (train data) and new test (new pending data) data, respectively; predicted results of the color classification image (false-color composite) under three models for the training data (**b–d**), namely, (**b**) DNN, (**c**) 1D-CNN, and (**d**) 3D-CNN, and for the new pending data (**f–h**), namely, (**f**) DNN, (**g**) 1D-CNN, and (**h**) 3D-CNN.

3.4. Differences in Models at Three Magnification Rates

In this experiment, the classification algorithm is based on the pixel unit data in the image; thus, determining how to obtain the quantitative value of the enhanced precision for the OM is crucial. The majority of existing microscopes achieve uniform spatial irradiance through Köhler illumination [63]. However, some shortcomings remain in the need for quantitative measurement and analysis, for example, FOV nonuniform illumination asymmetry (ANILAS) [64,65] is a largely important factor that is ignored (Figure S13). This phenomenon leads to a certain loss in pixel accuracy. Thus, we attempt to understand the various feature data types through deep learning in the case of uneven illumination distribution.

The FOV sizes of the 10×, 40×, and 100× magnification rates are 1.6 mm × 1.2 mm, 0.4 mm × 0.3 mm, and 0.17 mm × 0.27 mm, respectively, which are the actual detection sizes at the time of prediction. Figure 5 presents the optimal loss values of the (a) training and (b) verification sets for three models at three magnification rates. The optimal train loss is higher than the validation loss, partly because of the use of data enhancements in the training set, which makes the model difficult to learn when the data are increased in diversity. Figure 5 shows that the three models have low loss values under 100× magnification. When considering the difference in spatial resolution, the pixel resolutions at 10×, 40×, and 100× magnification rates are 0.5, 0.25, and 0.1 μm, respectively. Therefore, the problem of mixed pixels is further serious at a small magnification. 3D-CNN is the best among the three models at different magnification rates.

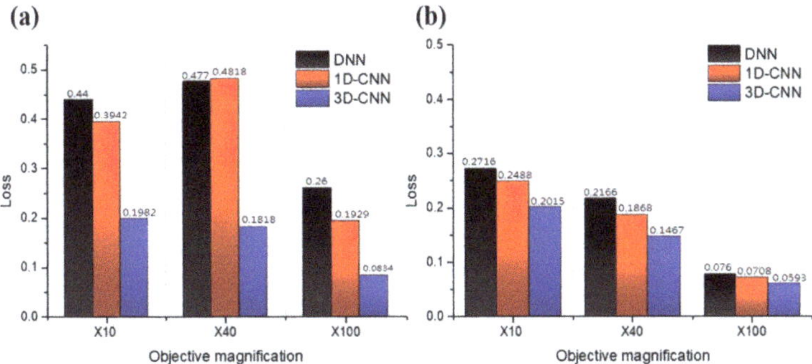

Figure 5. Optimal loss of the three models in the (**a**) training and (**b**) validation sets at 10×, 40×, and 100× magnification rates.

As 3D-CNN demonstrates the best generalization capability at different magnification rates, we will only discuss the results of this algorithm. Figure 6a shows an OM image of a large-area periodically grown single-layer MoS_2 on a sapphire substrate (also defined as ROI-3). The actual corresponding size is 1×1 mm, and the microscope magnification is 10×. We can observe that a single layer of MoS_2 is distributed in a star shape around the hole. Figure 6a–f display the analysis of the color classification image (false-color composite). Figure 6b–d present the OM images of the predicted results under three magnification rates, namely, 100× (ROI-1), 40× (ROI-2), and 10× (ROI-3), respectively.

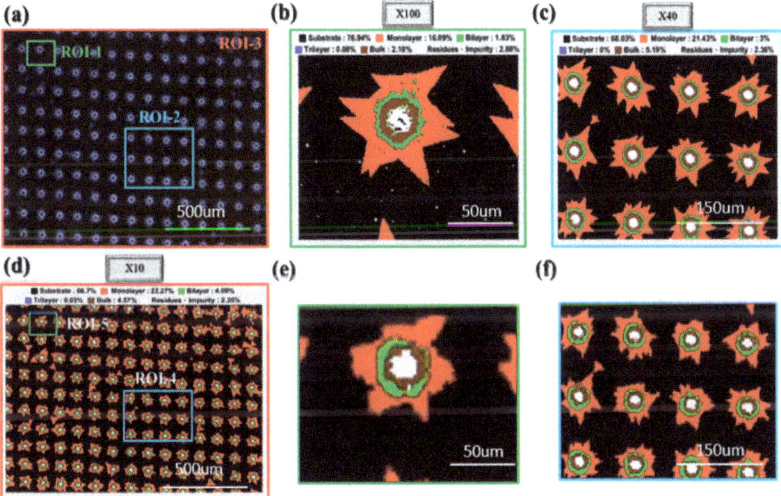

Figure 6. (**a**) OM images of the large-area periodic growth of single-layer MoS_2 on sapphire substrates at (**b**) 100× (ROI-1), (**c**) 40× (ROI-2), and (**d**) 10× (ROI-3) magnification rates. Color classification map predicted at 10× magnification: (**e,f**) are the amplification results of ROI-4 and ROI-5 ranges, respectively, corresponding to ROI-1 and ROI-2 circle selection.

We obtain the magnified images of the ROI-4 and ROI-5 regions from the color classification image under 10× magnification (Figure 6d) as Figure 6e,f, respectively. Corresponding to other magnification rates (Figure 6b,c) in the same region of the color classification image, we observe that a small magnification will be limited by the spatial resolution. Consequently, the fine type of features will be blurred or impurity points are further difficult to identify. Supplementary Figure S14 discusses the probability of class prediction confidence in images with various magnification rates.

This finding is different from previous research arguments [66]. Previous studies have considered that the poor image quality is due to several noise points and the surrounding blur at a large magnification or that fine impurities are caused by deposition, resulting in re-traditional classification algorithms. The effect is not good, but the small impurities will be ignored when the spatial resolution is low. However, in the present study, the cognition of the model in deep learning solves the bottleneck of the previous problem.

3.5. Instrument Measurement Verification

In the new pending data section, we observe the accuracy of the samples from Raman spectroscopy mapping, as shown in Figure S15. We show two oscillation modes of in-plane (E_{2g}^1) and out-of-plane (A_{1g}) in Raman mapping analysis. The classification of the number of layers is determined by two peak differences. Figure S16 presents the SEM measurement results. The instrument can be judged by the gray level. However, the instrument measurement cannot determine the number of layers and it can only be observed from relative contrast. The PL spectroscopy results (Figure S17) indicate that the mapping diagram is either 625 or 667 nm, and the periodic growth of the single-layer to multilayer distribution of MoS_2 has good uniformity. In Figure S18, the material of the sample profile is divided via HRTEM [67,68].

4. Conclusions

This study is aimed at the layer number discrimination of molybdenum disulfide film on sapphire. In comparison with the current measurement instruments, such as Raman, SEM, AFM, and TEM, the proposed equipment has a large detection area, less time, and low cost. The equipment can detect the low number of layers of molybdenum disulfide film. Unlike in the past, we use deep learning, but not image processing, for analysis, and experimentally confirm that 3D-CNN has the best precision and generalization capability. The reason is that the 3D-CNN model initially adds the spatial domain of the morphological features of MoS_2 to learn to avoid the misjudgment caused by the difference in imaging quality due to noise or to make further accurate judgments on the fuzzy regions between the category regions. For the problem that the low magnification is limited by the spatial resolution, which results in fine contaminants and edge blur morphology, the GAN model can be used to achieve the super-resolution method in low magnification [69,70]. In future research, we hope to integrate all types of 2D materials and various substrates, such as in the case of heterogeneous stack, in order to easily distinguish different 2D materials and their different layers easily. The future ideas are to determine the inference of the growth pattern of MoS_2 by detecting the image in real time and to avoid machine termination for reducing time and related costs under impurity intrusion

Supplementary Materials: The following are available online at http://www.mdpi.com/2079-4991/10/6/1161/s1, Figure S1: Experimental device. (a) Three-zone furnace tube used in this experiment (Lindberg/Blue HTF55347C), (b) schematic of the experimental structure and position of the precursor. (c) Area-coded position and image of the sample under OM shooting; Figure S2: Deep learning is applied to construct a flowchart for the number detection system of the optical MoS_2 layer. The gray, blue, orange, and green block colors correspond to (1) database, (2) offline training, (3) model design, and (4) online service, respectively; Figure S3: Data labeling is assisted by the Raman spectroscopy instrument (MRI-1532A). This instrument is mainly used to inject a 532 nm laser light into the sample. The photons in the laser will collide with the molecules in the sample material, namely, E_{2g}^1 and A_{1g}. The peak difference of the vibration mode is the signal of the main judgment layer of MoS2, and the two vibration modes have a high dependence on the thickness of MoS2. We select two 30 μm×30 μm Raman mapping results in the database and wait for ~45 min, especially in the ground truth mark, which is a considerable time; Figure S4: Flowchart of visible hyperspectral image algorithm; Figure S5: In the blue area, the data in the offline training section of Figure 2 are used. The ground truth and our label data are set as the training and validation sets, respectively. The rest of the VIS-HIS feature data are used as the test set. The green area is the predicted result of new pending data in the (4) online service architecture in Figure S2; Figure S6: OM image and prediction results of the new pending data at 10× magnification; Figure S7: At 40× magnification, (a) and (e) are the OM images of train data and new pending data, respectively. The predicted results of the three models are the false-color composite predicted by (b) DNN, (c) 1D-CNN, and (d) 3D-CNN in the training data, and DNN and 1D-CNN can be observed. In the CNN, impurities in the single layer are undetected. New pending data predict the color classification image (false-color composite) by (f) DNN, (g) 1D-CNN, and (h) 3D-CNN and can observe errors in a single layer in DNN and 1D-CNN. The result of misclassification of impurity classification; Figure S8: OM image and prediction results of the new pending data at 40× magnification; Figure S9: At 100× magnification, (a) and (e) are the OM images of the training (train data) and new test (new pending data) sets. (b) DNN, (c) 1D-CNN, and (d) 3D-CNN indicate the prediction results of the color classification image (false-color composite) in the training data. (f) DNN, (g) 1D-CNN, and (h) 3D-CNN reflect the prediction results of the color classification image (false-color composite) in the new pending data, from DNN and 1D-CNN in the single layer with the wrong double-layer classification misjudgment results; Figure S10: OM image and prediction results of the new pending data at 100× magnification; Figure S11: At 40× magnification, the accuracy (Accuracy) and loss (Loss) of the convergence process in (a) DNN, (b) 1D-CNN, and (c) 3D-CNN; validation set in (d) DNN, (e) 1D-CNN, and (f) 3D-CNN confusion matrix results; Figure S12: At 100× magnification, the accuracy (Accuracy) and loss (Loss) of the convergence process in (a) DNN, (b) 1D-CNN, and (c) 3D-CNN; validation set in (d) DNN, (e) 1D-CNN, and (f) 3D-CNN confusion matrix results; Figure S13: Optical microscope light intensity distribution analysis. (a,d,g) OM images of sapphire substrate taken at 10×, 40×, and 100× magnification rates (50, 15, and 10μm in the lower right corner, respectively). (b,e,h) V channels in the HSV color space of the microscope image (lightness) light intensity distribution. (c,f,i) Scatter plots of each pixel point RGB channel in the OM image; Figure S14: Prediction rate of each classification type at the shooting magnification rates of (a) 10×, (b) 40×, and (c) 100×; Figure S15: Raman measurement of the hyperspectral image of the new test data (new data). (a) OM image at 100× magnification, (b) prediction result compared with OM image at 100× magnification, (c) Raman spectrum at three specific points in (d), and (d) Raman measurement result corresponding to (b). The number of layers in the ROI circle is consistent, the triangle is a single-layer structure, and the intermediate core point is a double layer; Figure S16: SEM measurement of the hyperspectral image of the new test data (new data). (a,b) are OM images at 100× and 40× magnification rates, (c,d) are prediction results corresponding to the (a,b) OM image ranges, and (e,f) correspond to (c,d) ROI measurement results; Figure S17: (a) Photoluminescence spectrometry. (b) OM image of selected PL mapping range. (c) PL mapping with a wavelength of 625 nm (d) PL mapping with a wavelength of 667 nm. (e) The blue measurement in (d) indicates the PL measurement result of the test piece; Figure S18: (a) is the OM image after growing MoS_2, (b) is the cross-sectional TEM image of the selected area of (a), (c) is the magnified TEM image of the red arrow of (b). (d) is the HRTEM image at the red box in (c). (e) is the HRTEM image at the orange box in (c), and (f) is the HRTEM image at the yellow box in (c). (g) is the SAED diagram of (d), (h) is the SAED diagram of (e), and (i) is the SAED diagram of (f).

Author Contributions: Conceptualization, M.-Y.L. and H.-C.W.; methodology, H.-C.W.; software, K.-C.L.; validation, S.-W.F. and S.B.A.; resources, V.E.F. and H.-C.W.; data curation, K.-C.L. and M.-Y.L.; writing-original draft preparation, H.T.N.; writing-review and editing, V.E.F. and H.-C.W.; supervision, V.E.F. and H.-C.W.; project administration, H.-C.W.; funding acquisition, H.-C.W. All authors have read and agreed to the published version of the manuscript.

Funding: This research was supported by the Ministry of Science and Technology, The Republic of China under the Grants MOST 105-2923-E-194-003-MY3, 106-2112-M-194-002, and 108-2823-8-194-002.

Conflicts of Interest: The authors declare no conflict of interest.

References

1. Geim, A.K.; Novoselov, K.S. The rise of graphene. In *Nanoscience and Technology: A Collection of Reviews from Nature Journals*; World Scientific: Singapore, 2010; pp. 11–19.

2. Sarma, S.D.; Adam, S.; Hwang, E.; Rossi, E. Electronic transport in two-dimensional graphene. *Rev. Mod. Phys.* **2011**, *83*, 407. [CrossRef]

3. Lin, X.; Su, L.; Si, Z.; Zhang, Y.; Bournel, A.; Zhang, Y.; Klein, J.-O.; Fert, A.; Zhao, W. Gate-driven pure spin current in graphene. *Phys. Rev. Appl.* **2017**, *8*, 034006. [CrossRef]

4. Wu, C.; Zhang, J.; Tong, X.; Yu, P.; Xu, J.Y.; Wu, J.; Wang, Z.M.; Lou, J.; Chueh, Y.L. A Critical Review on Enhancement of Photocatalytic Hydrogen Production by Molybdenum Disulfide: From Growth to Interfacial Activities. *Small* **2019**, *15*, 1900578. [CrossRef] [PubMed]

5. Yang, D.; Wang, H.; Luo, S.; Wang, C.; Zhang, S.; Guo, S. Cut Flexible Multifunctional Electronics Using MoS2 Nanosheet. *Nanomaterials* **2019**, *9*, 922. [CrossRef]

6. Zhang, Y.; Wan, Q.; Yang, N. Recent Advances of Porous Graphene: Synthesis, Functionalization, and Electrochemical Applications. *Small* **2019**, *15*, 1903780. [CrossRef]

7. Radisavljevic, B.; Radenovic, A.; Brivio, J.; Giacometti, V.; Kis, A. Single-layer MoS_2 transistors. *Nat. Nanotechnol.* **2011**, *6*, 147. [CrossRef]

8. Han, T.; Liu, H.; Wang, S.; Chen, S.; Xie, H.; Yang, K. Probing the field-effect transistor with monolayer MoS_2 prepared by APCVD. *Nanomaterials* **2019**, *9*, 1209. [CrossRef]

9. Roh, J.; Ryu, J.H.; Baek, G.W.; Jung, H.; Seo, S.G.; An, K.; Jeong, B.G.; Lee, D.C.; Hong, B.H.; Bae, W.K. Threshold voltage control of multilayered MoS_2 field-effect transistors via octadecyltrichlorosilane and their applications to active matrixed quantum dot displays driven by enhancement-mode logic gates. *Small* **2019**, *15*, 1803852. [CrossRef]

10. Yang, K.; Liu, H.; Wang, S.; Li, W.; Han, T. A horizontal-gate monolayer MoS_2 transistor based on image force barrier reduction. *Nanomaterials* **2019**, *9*, 1245. [CrossRef]

11. Choi, M.S.; Lee, G.-H.; Yu, Y.-J.; Lee, D.-Y.; Lee, S.H.; Kim, P.; Hone, J.; Yoo, W.J. Controlled charge trapping by molybdenum disulphide and graphene in ultrathin heterostructured memory devices. *Nat. Commun.* **2013**, *4*, 1–7.

12. Lu, G.Z.; Wu, M.J.; Lin, T.N.; Chang, C.Y.; Lin, W.L.; Chen, Y.T.; Hou, C.F.; Cheng, H.J.; Lin, T.Y.; Shen, J.L. Electrically pumped white-light-emitting diodes based on histidine-doped MoS_2 quantum dots. *Small* **2019**, *15*, 1901908. [CrossRef] [PubMed]

13. Lee, C.-H.; Lee, G.-H.; Van Der Zande, A.M.; Chen, W.; Li, Y.; Han, M.; Cui, X.; Arefe, G.; Nuckolls, C.; Heinz, T.F. Atomically thin p–n junctions with van der Waals heterointerfaces. *Nat. Nanotechnol.* **2014**, *9*, 676. [CrossRef] [PubMed]

14. Sarkar, D.; Liu, W.; Xie, X.; Anselmo, A.C.; Mitragotri, S.; Banerjee, K. MoS_2 field-effect transistor for next-generation label-free biosensors. *ACS Nano* **2014**, *8*, 3992–4003. [CrossRef] [PubMed]

15. Zhao, J.; Li, N.; Yu, H.; Wei, Z.; Liao, M.; Chen, P.; Wang, S.; Shi, D.; Sun, Q.; Zhang, G. Highly sensitive MoS2 humidity sensors array for noncontact sensation. *Adv. Mater.* **2017**, *29*, 1702076. [CrossRef]

16. Park, Y.J.; Sharma, B.K.; Shinde, S.M.; Kim, M.-S.; Jang, B.; Kim, J.-H.; Ahn, J.-H. All MoS2-based large area, skin-attachable active-matrix tactile sensor. *ACS Nano* **2019**, *13*, 3023–3030. [CrossRef]

17. Shin, M.; Yoon, J.; Yi, C.; Lee, T.; Choi, J.-W. Flexible HIV-1 biosensor based on the Au/MoS_2 nanoparticles/Au nanolayer on the PET substrate. *Nanomaterials* **2019**, *9*, 1076. [CrossRef]

18. Yadav, V.; Roy, S.; Singh, P.; Khan, Z.; Jaiswal, A. 2D MoS_2-based nanomaterials for therapeutic, bioimaging, and biosensing applications. *Small* **2019**, *15*, 1803706. [CrossRef]

19. Zhang, P.; Yang, S.; Pineda-Gómez, R.; Ibarlucea, B.; Ma, J.; Lohe, M.R.; Akbar, T.F.; Baraban, L.; Cuniberti, G.; Feng, X. Electrochemically exfoliated high-quality 2H-MoS_2 for multiflake thin film flexible biosensors. *Small* **2019**, *15*, 1901265. [CrossRef]

20. Tu, Q.; Lange, B.R.; Parlak, Z.; Lopes, J.M.J.; Blum, V.; Zauscher, S. Quantitative subsurface atomic structure fingerprint for 2D materials and heterostructures by first-principles-calibrated contact-resonance atomic force microscopy. *ACS Nano* **2016**, *10*, 6491–6500. [CrossRef]

21. Wastl, D.S.; Weymouth, A.J.; Giessibl, F.J. Atomically resolved graphitic surfaces in air by atomic force microscopy. *ACS Nano* **2014**, *8*, 5233–5239. [CrossRef]

22. Zhao, W.; Xia, B.; Lin, L.; Xiao, X.; Liu, P.; Lin, X.; Peng, H.; Zhu, Y.; Yu, R.; Lei, P. Low-energy transmission electron diffraction and imaging of large-area graphene. *Sci. Adv.* **2017**, *3*, e1603231. [CrossRef] [PubMed]

23. Meyer, J.C.; Geim, A.K.; Katsnelson, M.I.; Novoselov, K.S.; Booth, T.J.; Roth, S. The structure of suspended graphene sheets. *Nature* **2007**, *446*, 60–63. [CrossRef] [PubMed]

24. Nolen, C.M.; Denina, G.; Teweldebrhan, D.; Bhanu, B.; Balandin, A.A. High-throughput large-area automated identification and quality control of graphene and few-layer graphene films. *ACS Nano* **2011**, *5*, 914–922. [CrossRef] [PubMed]

25. Konstantopoulos, G.; Koumoulos, E.P.; Charitidis, C.A. Testing novel portland cement formulations with carbon nanotubes and intrinsic properties revelation: Nanoindentation analysis with machine learning on microstructure identification. *Nanomaterials* **2020**, *10*, 645. [CrossRef] [PubMed]

26. Hong, S.; Nomura, K.-I.; Krishnamoorthy, A.; Rajak, P.; Sheng, C.; Kalia, R.K.; Nakano, A.; Vashishta, P. Defect healing in layered materials: A machine learning-assisted characterization of MoS_2 crystal phases. *J. Phys. Chem. Lett.* **2019**, *10*, 2739–2744. [CrossRef]

27. Masubuchi, S.; Watanabe, E.; Seo, Y.; Okazaki, S.; Sasagawa, T.; Watanabe, K.; Taniguchi, T.; Machida, T. Deep-learning-based image segmentation integrated with optical microscopy for automatically searching for two-dimensional materials. *NPJ 2D Mater. Appl.* **2020**, *4*, 1–9. [CrossRef]

28. Yang, J.; Yao, H. Automated identification and characterization of two-dimensional materials via machine learning-based processing of optical microscope images. *Extrem. Mech. Lett.* **2020**, *39*, 100771. [CrossRef]

29. Li, Y.; Kong, Y.; Peng, J.; Yu, C.; Li, Z.; Li, P.; Liu, Y.; Gao, C.-F.; Wu, R. Rapid identification of two-dimensional materials via machine learning assisted optic microscopy. *J. Mater.* **2019**, *5*, 413–421. [CrossRef]

30. Masubuchi, S.; Machida, T. Classifying optical microscope images of exfoliated graphene flakes by data-driven machine learning. *NPJ 2D Mater. Appl.* **2019**, *3*, 1–7. [CrossRef]

31. Lin, X.; Si, Z.; Fu, W.; Yang, J.; Guo, S.; Cao, Y.; Zhang, J.; Wang, X.; Liu, P.; Jiang, K. Intelligent identification of two-dimensional nanostructures by machine-learning optical microscopy. *Nano Res.* **2018**, *11*, 6316–6324. [CrossRef]

32. Zhao, Y.; Luo, X.; Li, H.; Zhang, J.; Araujo, P.T.; Gan, C.K.; Wu, J.; Zhang, H.; Quek, S.Y.; Dresselhaus, M.S. Interlayer breathing and shear modes in few-trilayer MoS_2 and WSe_2. *Nano Lett.* **2013**, *13*, 1007–1015. [CrossRef] [PubMed]

33. Li, H.; Wu, J.; Yin, Z.; Zhang, H. Preparation and applications of mechanically exfoliated single-layer and multilayer MoS_2 and WSe_2 nanosheets. *Acc. Chem. Res.* **2014**, *47*, 1067–1075. [CrossRef] [PubMed]

34. Li, H.; Lu, G.; Yin, Z.; He, Q.; Li, H.; Zhang, Q.; Zhang, H. Optical identification of single-and few-layer MoS_2 sheets. *Small* **2012**, *8*, 682–686. [CrossRef] [PubMed]

35. Najmaei, S.; Liu, Z.; Zhou, W.; Zou, X.; Shi, G.; Lei, S.; Yakobson, B.I.; Idrobo, J.-C.; Ajayan, P.M.; Lou, J. Vapour phase growth and grain boundary structure of molybdenum disulphide atomic layers. *Nat. Mater.* **2013**, *12*, 754–759. [CrossRef]

36. Van Der Zande, A.M.; Huang, P.Y.; Chenet, D.A.; Berkelbach, T.C.; You, Y.; Lee, G.-H.; Heinz, T.F.; Reichman, D.R.; Muller, D.A.; Hone, J.C. Grains and grain boundaries in highly crystalline monolayer molybdenum disulphide. *Nat. Mater.* **2013**, *12*, 554–561. [CrossRef]

37. Lee, Y.H.; Zhang, X.Q.; Zhang, W.; Chang, M.T.; Lin, C.T.; Chang, K.D.; Yu, Y.C.; Wang, J.T.W.; Chang, C.S.; Li, L.J. Synthesis of large-area MoS_2 atomic layers with chemical vapor deposition. *Adv. Mater.* **2012**, *24*, 2320–2325. [CrossRef]

38. Jeon, J.; Jang, S.K.; Jeon, S.M.; Yoo, G.; Jang, Y.H.; Park, J.-H.; Lee, S. Layer-controlled CVD growth of large-area two-dimensional MoS_2 films. *Nanoscale* **2015**, *7*, 1688–1695. [CrossRef]

39. Dumcenco, D.; Ovchinnikov, D.; Marinov, K.; Lazic, P.; Gibertini, M.; Marzari, N.; Sanchez, O.L.; Kung, Y.-C.; Krasnozhon, D.; Chen, M.-W. Large-area epitaxial monolayer MoS_2. *ACS Nano* **2015**, *9*, 4611–4620. [CrossRef]

40. Yu, Y.; Li, C.; Liu, Y.; Su, L.; Zhang, Y.; Cao, L. Controlled scalable synthesis of uniform, high-quality monolayer and few-layer MoS_2 films. *Sci. Rep.* **2013**, *3*, 1866. [CrossRef]

41. Wang, S.; Rong, Y.; Fan, Y.; Pacios, M.; Bhaskaran, H.; He, K.; Warner, J.H. Shape evolution of monolayer MoS_2 crystals grown by chemical vapor deposition. *Chem. Mater.* **2014**, *26*, 6371–6379. [CrossRef]

42. Wang, L.; Chen, F.; Ji, X. Shape consistency of MoS_2 flakes grown using chemical vapor deposition. *Appl. Phys. Express* **2017**, *10*, 065201. [CrossRef]

43. Zheng, W.; Qiu, Y.; Feng, W.; Chen, J.; Yang, H.; Wu, S.; Jia, D.; Zhou, Y.; Hu, P. Controlled growth of six-point stars MoS_2 by chemical vapor deposition and its shape evolution mechanism. *Nanotechnology* **2017**, *28*, 395601. [CrossRef] [PubMed]

44. Zhou, D.; Shu, H.; Hu, C.; Jiang, L.; Liang, P.; Chen, X. Unveiling the growth mechanism of MoS_2 with chemical vapor deposition: From two-dimensional planar nucleation to self-seeding nucleation. *Cryst. Growth Des.* **2018**, *18*, 1012–1019. [CrossRef]

45. Zhu, D.; Shu, H.; Jiang, F.; Lv, D.; Asokan, V.; Omar, O.; Yuan, J.; Zhang, Z.; Jin, C. Capture the growth kinetics of CVD growth of two-dimensional MoS_2. *NPJ 2D Mater. Appl.* **2017**, *1*, 1–8. [CrossRef]

46. Sun, D.; Nguyen, A.E.; Barroso, D.; Zhang, X.; Preciado, E.; Bobek, S.; Klee, V.; Mann, J.; Bartels, L. Chemical vapor deposition growth of a periodic array of single-layer MoS_2 islands via lithographic patterning of an SiO_2/Si substrate. *2D Mater.* **2015**, *2*, 045014. [CrossRef]

47. Li, Y.; Hao, S.; DiStefano, J.G.; Murthy, A.A.; Hanson, E.D.; Xu, Y.; Wolverton, C.; Chen, X.; Dravid, V.P. Site-specific positioning and patterning of MoS_2 monolayers: The role of Au seeding. *ACS Nano* **2018**, *12*, 8970–8976. [CrossRef]

48. Han, G.H.; Kybert, N.J.; Naylor, C.H.; Lee, B.S.; Ping, J.; Park, J.H.; Kang, J.; Lee, S.Y.; Lee, Y.H.; Agarwal, R. Seeded growth of highly crystalline molybdenum disulphide monolayers at controlled locations. *Nat. Commun.* **2015**, *6*, 6128. [CrossRef]

49. Wang, X.; Kang, K.; Chen, S.; Du, R.; Yang, E.-H. Location-specific growth and transfer of arrayed MoS2 monolayers with controllable size. *2D Materials* **2017**, *4*, 025093. [CrossRef]

50. Lee, C.; Yan, H.; Brus, L.E.; Heinz, T.F.; Hone, J.; Ryu, S. Anomalous lattice vibrations of single-and few-layer MoS_2. *ACS Nano* **2010**, *4*, 2695–2700. [CrossRef]

51. Li, H.; Zhang, Q.; Yap, C.C.R.; Tay, B.K.; Edwin, T.H.T.; Olivier, A.; Baillargeat, D. From bulk to monolayer MoS_2: Evolution of Raman scattering. *Adv. Funct. Mater.* **2012**, *22*, 1385–1390. [CrossRef]

52. Lei, S.; Ge, L.; Najmaei, S.; George, A.; Kappera, R.; Lou, J.; Chhowalla, M.; Yamaguchi, H.; Gupta, G.; Vajtai, R. Evolution of the electronic band structure and efficient photo-detection in atomic layers of InSe. *ACS Nano* **2014**, *8*, 1263–1272. [CrossRef] [PubMed]

53. Ye, G.; Gong, Y.; Lin, J.; Li, B.; He, Y.; Pantelides, S.T.; Zhou, W.; Vajtai, R.; Ajayan, P.M. Defects engineered monolayer MoS_2 for improved hydrogen evolution reaction. *NanoLett.* **2016**, *16*, 1097–1103. [CrossRef] [PubMed]

54. Han, T.; Liu, H.; Wang, S.; Chen, S.; Li, W.; Yang, X.; Cai, M.; Yang, K. Probing the optical properties of MoS_2 on SiO_2/Si and sapphire substrates. *Nanomaterials* **2019**, *9*, 740. [CrossRef] [PubMed]

55. Tran, D.; Bourdev, L.; Fergus, R.; Torresani, L.; Paluri, M. Learning spatiotemporal features with 3d convolutional networks. In Proceedings of the IEEE International Conference on Computer Vision, Araucano Park, Las Condes, Chille, 11–18 December 2015; pp. 4489–4497.

56. Wang, Z.; Hu, G.; Yao, S. Decomposition mixed pixel of remote sensing image based on tray neural network model. In Proceedings of the International Symposium on Intelligence Computation and Applications, Wuhan, China, 21–23 September 2007; pp. 305–309.

57. Foschi, P.G. A geometric approach to a mixed pixel problem: Detecting subpixel woody vegetation. *Remote Sens. Environ.* **1994**, *50*, 317–327. [CrossRef]

58. Lin, Z.; Chen, Y.; Zhao, X.; Wang, G. Spectral-spatial classification of hyperspectral image using autoencoders. In Proceedings of the 9th International Conference on Information, Communications & Signal Processing, Tainan, Taiwan, 10–13 December 2013; pp. 1–5.

59. Chen, Y.; Jiang, H.; Li, C.; Jia, X.; Ghamisi, P. Deep feature extraction and classification of hyperspectral images based on convolutional neural networks. *IEEE Trans. Geosci. Remote Sens.* **2016**, *54*, 6232–6251. [CrossRef]

60. Shen, L.; Jia, S. Three-dimensional Gabor wavelets for pixel-based hyperspectral imagery classification. *IEEE Trans. Geosci. Remote Sens.* **2011**, *49*, 5039–5046. [CrossRef]

61. Simonyan, K.; Zisserman, A. Very deep convolutional networks for large-scale image recognition. *arXiv* **2014**, arXiv:1409.1556.

62. Li, Y.; Zhang, H.; Shen, Q. Spectral–spatial classification of hyperspectral imagery with 3D convolutional neural network. *Remote Sens.* **2017**, *9*, 67. [CrossRef]

63. Evennett, P. Kohler illumination: A simple interpretation. *Proce. R. Microsc. Soc.* **1983**, *28*, 10–13.

64. Attota, R.K. Step beyond Kohler illumination analysis for far-field quantitative imaging: Angular illumination asymmetry (ANILAS) maps. *Opt. Express* **2016**, *24*, 22616–22627. [CrossRef]

65. Attota, R.; Silver, R. Optical microscope angular illumination analysis. *Opt. Express* **2012**, *20*, 6693–6702. [CrossRef] [PubMed]

66. KaiáLee, M. Large-area few-layered graphene film determination by multispectral imaging microscopy. *Nanoscale* **2015**, *7*, 9033–9039.

67. Chen, J.; Wei, Q. Phase transformation of molybdenum trioxide to molybdenum dioxide: An in-situ transmission electron microscopy investigation. *Int. J. Appl. Ceram. Technol.* **2017**, *14*, 1020–1025. [CrossRef]

68. Wang, Q.H.; Kalantar-Zadeh, K.; Kis, A.; Coleman, J.N.; Strano, M.S. Electronics and optoelectronics of two-dimensional transition metal dichalcogenides. *Nat. Nanotechnol.* **2012**, *7*, 699. [CrossRef] [PubMed]

69. Wang, H.; Rivenson, Y.; Jin, Y.; Wei, Z.; Gao, R.; Günaydın, H.; Bentolila, L.A.; Kural, C.; Ozcan, A. Deep learning enables cross-modality super-resolution in fluorescence microscopy. *Nat. Methods* **2019**, *16*, 103–110. [CrossRef]

70. Rivenson, Y.; Ceylan Koydemir, H.; Wang, H.; Wei, Z.; Ren, Z.; Günaydın, H.; Zhang, Y.; Gorocs, Z.; Liang, K.; Tseng, D. Deep learning enhanced mobile-phone microscopy. *ACS Photonics* **2018**, *5*, 2354–2364. [CrossRef]

Article

Stearic Acid Coated MgO Nanoplate Arrays as Effective Hydrophobic Films for Improving Corrosion Resistance of Mg-Based Metallic Glasses

Yonghui Yan [1], Xiaoli Liu [2], Hanqing Xiong [3], Jun Zhou [1], Hui Yu [1], Chunling Qin [1] and Zhifeng Wang [1,4,*]

[1] School of Materials Science and Engineering, Hebei University of Technology, Tianjin 300401, China; 201821803008@stu.hebut.edu.cn (Y.Y.); zhoujun@hebut.edu.cn (J.Z.); yuhuidavid@hebut.edu.cn (H.Y.); clqin@hebut.edu.cn (C.Q.)

[2] School of Materials Science and Engineering, Hebei University of Science & Technology, Shijiazhuang 050018, China; iven308@126.com

[3] Department of Mechanical and Electronic Engineering, Changsha University, Changsha 410022, China; xhanqing@ccsu.edu.cn

[4] Key Laboratory for New Type of Functional Materials in Hebei Province, Hebei University of Technology, Tianjin 300401, China

* Correspondence: wangzf@hebut.edu.cn; Tel.: +86-22-6020-2006

Received: 29 April 2020; Accepted: 12 May 2020; Published: 15 May 2020

Abstract: Mg-based metallic glasses (MGs) are widely studied due to their high elasticity and high strength originating from their amorphous nature. However, their further application in many potential fields is limited by poor corrosion resistance. In order to improve this property, an MgO nanoplate array layer is first constructed on the surface of Mg-based MGs by cyclic voltammetry (CV) treatments. In this situation, the corrosion resistance and hydrophilicity of the material are enhanced. Then, stearic acid (SA) can effectively adhere onto the surface of the MgO layer to form a superficial hydrophobic film with a water contact angle (WCA) of 131°. As a result, the SA coated MgO hydrophobic film improves the corrosion resistance of Mg-based MGs in 3.5 wt.% NaCl solution obviously. In addition, the effects of four technological parameters (solution concentration, sweep rate, cycle number, and reaction temperature) in the CV process on the morphology and size of nano-products are investigated in detail. The work proposes a new method for the creation of nanostructures on the surface of materials and provides a new idea to increase the corrosion resistance of MGs. The related method is expected to be applied in wider fields in future.

Keywords: Mg; metallic glass; hydrophobic film; corrosion resistance

---

1. Introduction

Magnesium alloys are the lightest metal structural materials, which possess the characteristics of high specific stiffness, high specific strength, good shock absorption, good electromagnetic shielding, as well as easy processing and recovery. Magnesium alloys receive broad research interest across many fields including the automotive industry, aerospace, electronic communications, and biomedicine [1–5]. Unfortunately, the poor corrosion resistance of magnesium alloys limits their further applications, especially in coastal areas. That is because elemental Mg presents high chemical activity and low electrochemical potential in liquid environments containing chloride ions [6,7], leading to an etch towards Mg from the alloy. Mg-TM-RE (TM: transition metal such as Zn, Cu, Ni, and RE: rare-earth metals such as Nd, Yb, or Y) metallic glasses (MGs) exhibit better compressive strength and corrosion resistance than crystalline state alloys with the same composition due to their uniform and long range disordered structure [8–12]. In recent years, many studies related to Mg-TM-RE MGs have been carried out, but the corrosion resistance of Mg-based MGs still represents a key issue to be addressed and improved in order to meet the demand of the market [13].

The existing works show some effective strategies to improve the corrosion resistance of Mg-based MGs. Adjusting the element composition and proportion of the MGs [14–16] is one of the simple strategies. By introducing iron particles into Mg-based MGs [17] and by forming a micro-arc oxidation (MAO) layer containing Si on the surface of Mg-based MGs [18], the corrosion resistance of the MGs can also be improved. Moreover, the preparation of hydrophobic film (a water contact angle of more than 90°) on the surface of materials is also an effective method to enhance the corrosion resistance [19–21]. The hydrophobic surface is constructed by introducing low surface energy materials onto the sample surface to prevent the corrosive ions from etching the alloy. In this way, the anti-corrosion ability of the sample can be improved. The reported methods for synthesizing hydrophobic films include the hydrothermal method [22], ball milling [23], flashlight irradiation [24], multi-arc ion plating [25], hydrolysis co-precipitation method [26], etc. Many existing methods are more costly and complicated [27,28], so it is necessary to develop a new method with lower cost and more convenient operation. In addition, both MgO protection layers and stearic acid layers are reported to effective in improving corrosion resistance of Mg alloys [29–31]. The research on the combination of the two protective layers is still lacking.

In this study, the MgO nanoplate array is fabricated on the surface of $Mg_{66}Zn_{30}Yb_4$ MGs by cyclic voltammetry (CV) and dehydration treatment for the first time. The stearic acid (SA) coated MgO composite hydrophobic film is then synthesized by immersing the sample into SA to enhance the corrosion resistance of the $Mg_{66}Zn_{30}Yb_4$ MG significantly. The treatment technology developed in this work provides a new idea to synthesize various nanostructures on the surface of materials, and also provides a simple method for improving the corrosion resistance of MGs, establishing a research foundation for the further application of Mg-based MGs in corrosive environments.

2. Materials and Methods

2.1. Preparation of Glassy Ribbons

Mg, Zn, and Yb metallic ingots (>99.9 wt.%) were melted into a button-type master alloy ingot with nominal composition of $Mg_{66}Zn_{30}Yb_4$ (at %) by arc-melting method [32]. The ingot was melted three times to ensure a uniform composition. Melt-spinning method [33] was adopted to spray the remelted master alloy melts from the small hole at the tip of a quartz tube onto the high-speed rotated copper roller (1500 r/min). Finally, the $Mg_{66}Zn_{30}Yb_4$ glassy ribbon (GR) [34] 2 mm wide, 30 μm thick, and dozens of centimeters long was produced.

2.2. Preparation of Hydrophobic Surface

Firstly, CV treatment was performed through the electrochemical workstation to construct the $Mg(OH)_2$ micro/nanostructure on the surface of the $Mg_{66}Zn_{30}Yb_4$ GR. The GR, platinum plate, and Ag/AgCl electrode were used as working electrode, counter electrode, and reference electrode, respectively. The CV cycling was carried out in the voltage range of 1~3 V with a scan rate of 0.01 Vs^{-1} in 1 M KOH electrolytes at 25 °C. Several experiments were carried out in order to investigate the influence of electrolyte concentration, scanning rate, cycle number and experimental temperature on the morphology and size of the $Mg(OH)_2$ products. These experiments were made to individuate the more appropriate process parameters before the synthesis of the SA layer. After the construction of the $Mg(OH)_2$ layer, the sample was cleaned with deionized water and dried at 60 °C in an oven for 1 h. In this situation, MgO can be formed by dehydration of $Mg(OH)_2$. Then MgO coated glassy ribbon was immersed into a 0.01 mol L^{-1} stearate ethanol solution at 50 °C for 5 min. Finally, the sample was removed and dried in an oven at 60 °C for 1 h to obtain a SA coated MgO composite hydrophobic film on the surface of the $Mg_{66}Zn_{30}Yb_4$ GR. The preparation process of the film is shown in Figure 1.

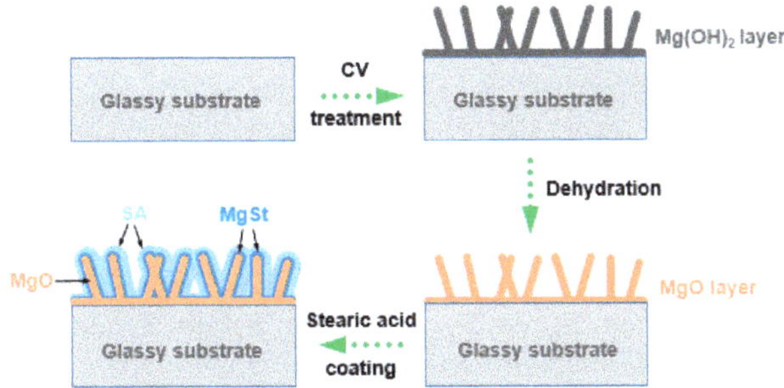

Figure 1. Schematic illustration showing the synthesis procedure of the stearic acid coated MgO hydrophobic film on the $Mg_{66}Zn_{30}Yb_4$ glassy ribbon (GR).

2.3. Characterization

Morphological characterization of samples produced with different process parameters were observed by scanning electron microscope (SEM, Nova NanoSEM 450, Nebraska Center for Materials and Nanoscience, Lincoln, NE, USA) with back-scattered electron (BSE) mode (1.00 kV). The sizes of a large number of nanoplates obtained under different process parameters were measured by Nano Measurer V1.2, software developed by Fudan University, Shanghai, China.

The phase composition, the variation in elemental contents with the sample depth, the magnified microstructure, the element composition and valence state of surface MgO layers were characterized by X-ray diffraction (XRD, Bruker, Billerica, MA, USA, D8-advance), SEM-EDS (energy disperse spectroscopy) with line scanning mode, transmission electron microscope (TEM, JEOL JEM-2010FEF, JEOL, Boston, MA, USA) and X-ray photoelectron spectroscopy (XPS, Microlab 350, MMT, LLC., Denver, CO, USA).

Characterizations of the SA coated specimen were revealed by SEM (BSE mode), TEM and the water contact angles (WCA) tests. The WCA were measured at room temperature through a contact angle measuring instrument (Dataphysics OCA20, Filderstadt, Germany). The volume of one drop of deionized water used in the measurement was 4 µL, and the measurement started after the water drop staying on the surface of the film for 1 min.

The superficial morphologies of the experimental samples after immersion were observed by SEM (BSE mode for SA treated specimen, SE mode for original and crystalline state $Mg_{66}Zn_{30}Yb_4$ sample) and EDS (line scanning mode).

2.4. Corrosion Resistance and Immersing Test

Polarization curve and open circuit potential tests were carried out to compare the corrosion resistance of different materials. A three-electrode system was adopted in the tests. The CV treated samples, platinum plate, and Ag/AgCl electrode were used as working electrode, counter electrode, and reference electrode, respectively. The potentiodynamic polarization curves were measured in the 3.5 wt.% NaCl solution at a scanning rate of 0.01 V/s. In order to ensure the accuracy of the experiment, the test area of the working electrode was limited in the defined range of 0.4 cm^2 (0.2 cm × 2 cm). The samples were put into the solution to stabilize for 5 min before testing. For immersing tests, the original GR, crystalline state $Mg_{66}Zn_{30}Yb_4$ alloy and SA treated GR were immersed into 3.5 wt.% NaCl solution at 25 °C. After immersing for 10 min, the surface morphology, cross-sectional morphology, and element distribution of the film were detected.

3. Results and Discussion

Figure 2a shows the XRD patterns of the original $Mg_{66}Zn_{30}Yb_4$ GR and the GR after CV and dehydration treatment. The original GR shows typical diffuse-scattering peaks (broad peaks centered at about 36°) [35], indicating an amorphous nature [36]. The amorphous structure of the ribbon after CV treatment is partially preserved, accompanied by additional clear crystalline peaks which corresponds to the (200) and (220) diffraction peaks of MgO (JCPDS No. 65-0476). Figure 2b shows the CV process of the GR with a scan rate of 0.01 Vs^{-1} in 1 M KOH solution at 25 °C (5th lap). During the CV process, the potential changes in the range of 1~3 V, playing a similar role of anodic oxidation. For macro-morphology and bendability testing, GRs in 3~5 cm long were used. The surface color of the ribbon changes from bright silver-white to light gray after CV treatment (Figure 2c), indicating the GR is covered with a layer of products. Moreover, the insert in Figure 2b shows that the Mg$_{66}$Zn$_{30}$Yb$_4$ GR still maintains good flexibility after CV treatment, and can be bent continuously for 180°. This provides more possibilities for the application of the material in broader fields.

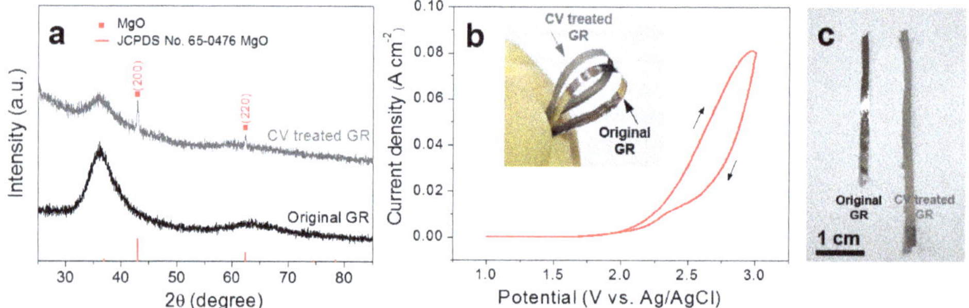

Figure 2. (**a**) X-ray diffraction (XRD) pattern; (**b**) cyclic voltammogram (CV) treatment towards a GR; the inserts: digital photo of GRs with good bendability; (**c**) digital photo of GRs before and after CV treatment.

Figure 3 shows BSE images towards GR surface after CV and dehydration treatment under different process parameters, including solution concentration, scanning rate, cycle number and reaction temperature. It can be seen that the sample surface presents some typical morphologies such as nanoplate and nanosphere. When the concentration of KOH solutions is low, the MgO nanoplates distribute evenly. With the increase in solution concentration, the nanoplates gradually aggregate into nanospheres (Figure 3a–d). When the scanning rate increases from 0.003 V/s to 0.01 V/s, the size of nanoplates enhances. At the same time, the distribution of nanoplates changes from aggregation to dispersion. When the scanning rate is more than 0.03 V/s, the number of nanoplates is greatly reduced and they become sparser with each other (Figure 3e–h). As the cycle number increases from 3 to 30, the surface nanoplates become more and more dense. However, when the cycle number reaches to 100, the surface nanoplates fall off from the matrix and only a small amount remains on the surface of the material (Figure 3i–l). When the temperature is relatively low (5 °C), a very small number of nanoplates are formed on the sample surface. When the temperature rises to 25 °C or above, the dynamic process of the nanoplate formation is accelerated, and the substrate surface is filled with nanoplates (Figure 3m–p). The possible reasons for the above phenomenon are explained as follows. The higher solution concentration increases the yield of the product, causing aggregating of the nanoplates into nanospheres. The faster scanning speed reduces the reaction time, resulting in a sharp decrease in the yield of nanoplates. Too many reaction cycles increase the yield of the product, but may result in the separation of nanoplates from the matrix. Too low a reaction temperature reduces the velocity of ions and impedes the kinetic process of the reaction, which is not conducive to the synthesis of the products. Based on the above results, it can be seen that the process parameters of the CV treatment demonstrate a great influence on the surface morphology and yield of nanoplates.

Figure 4 shows the size and proportion statistics of the nanoplate generated under different experimental parameters. It is found that the scanning rate and reaction temperature has a greater influence on the size of nanoplates. For different solution concentrations and cycle numbers, the size differentiation towards the nanoplates is low. In addition to individual process conditions, the nanoplates with a length of 120–180 nm occupy a major proportion in the size statistics. After a number of experiments and data analysis, the stable nanoplate array structure, obtained under the process parameter of 25 °C, 1 M KOH, 0.01 V/s sweep speed for 10 cycles, is selected for subsequent experiments. That is because under these experimental conditions, the uniform nanoplate array structure can be obtained, which is in favor of improving hydrophilicity to further reaction with SA.

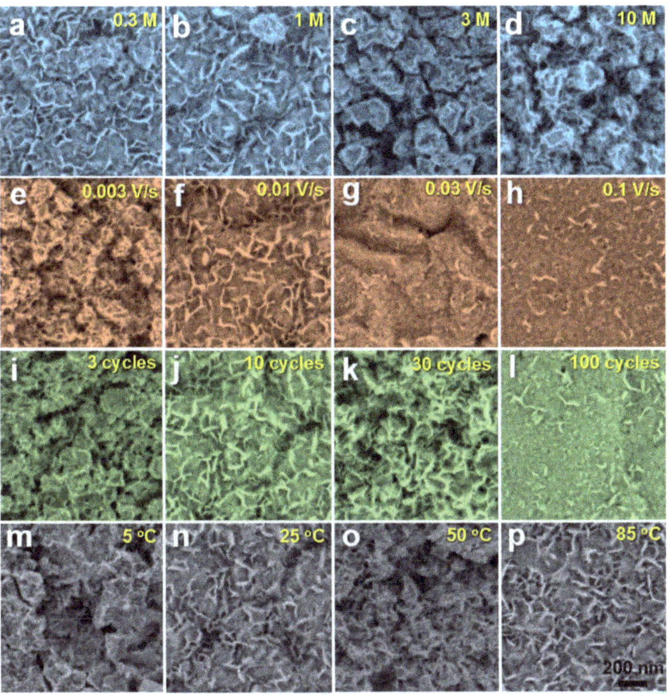

Figure 3. Back-scattered electron (BSE) images of GRs treated by CV under different technological conditions: (**a**–**d**) different KOH solution concentrations, with a scan rate of 0.01 Vs^{-1} at 25 °C for 10 cycles; (**e**–**h**) different sweep rates, in 1 M KOH solution at 25 °C for 10 cycles; (**i**–**l**) different cycle numbers, with a scan rate of 0.01 Vs^{-1} in 1 M KOH solution at 25 °C; (**m**–**p**) different reaction temperatures, with a scan rate of 0.01 Vs^{-1} in 1 M KOH solution for 10 cycles.

The SEM image of cross section of the sample prepared under selected parameters is shown in Figure 5. The nanoplate layer is about 5 μm in thickness with a porous and loose structure, which is conducive to the infiltration and attachment of SA. The line scanning result from the substrate to the nanoplate layer shows that the content of O element in the substrate is almost zero, and it increases dramatically after the line scanning reaches to the nanoplate layer. However, the contents of Zn and Yb elements decrease rapidly. The closer to the sample surface, the higher the O content and the lower the Zn and Yb content. Compared with the substrate, the content of Mg element in the film increases slightly, while the contents of Zn and Yb element decrease greatly. The above experimental results further confirm that the material surface is mainly composed of magnesium oxides.

In order to better understand the morphology and structure of the nanoplate, the sample is characterized by TEM as shown in Figure 6. Figure 6a,b show the morphology of the nanoplate, which is consistent with the SEM result. Figure 6c is a high-resolution transmission electron microscopy (HRTEM) image showing the measured lattice fringe spacing of 0.210 nm, corresponding to the (200) plane of MgO. The selected area electron diffraction (SAED) diagram of the nanoplate (Figure 6d) shows a polycrystalline concentric ring feature, corresponding to the (111), (200), (220), (222), (400), (420), and (422) crystal planes of MgO. TEM detection further confirms the successful synthesis of MgO.

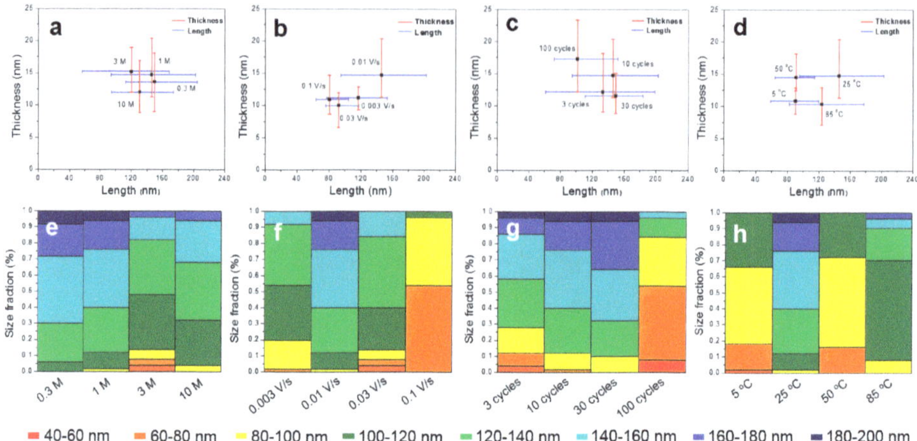

Figure 4. (a–d) The size statistics of MgO nanoplates obtained under different experimental parameters; (e–h) fractions of MgO nanoplates with different sizes obtained under different experimental parameters; (a,e): solution concentrations; (b,f): sweep rates; (c,g): cycle numbers; (d,h): reaction temperatures.

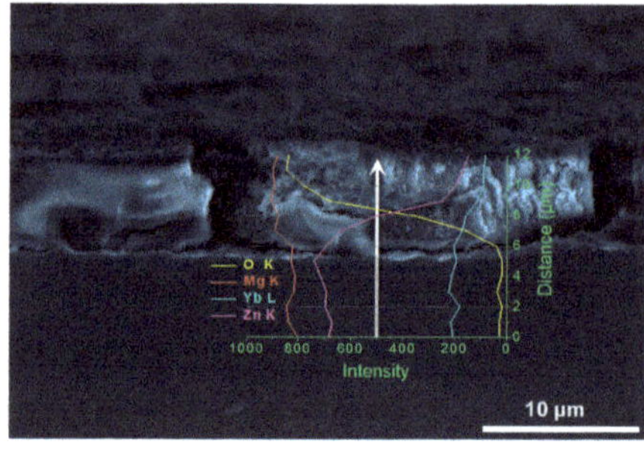

Figure 5. SEM image of cross section of the sample and the corresponding EDS line scanning result.

XPS is used to detect the elemental composition and valence state on the surface of nanoplate films. The fully scanned spectra (Figure 7a) shows that Mg, Zn, Yb and O peaks can be detected on the sample surfaces without other impurities. The high resolution XPS spectrum of Mg 1s is analyzed in Figure 7b. The Mg 1s peak position shifts from 1305.2 eV to about 1304.5 eV after the CV treatment, which is consistent with the previously reported peak position of MgO [37]. In Figure 7c, there are two peaks in the Zn 2p spectra. The difference in binding energy between Zn 2p$^{3/2}$ (1021.8 eV) and Zn 2p$^{1/2}$ (1044.9 eV) is about 23.1 eV, which is in accord with the reported value of ZnO [38], indicating that the nanoplate layer also contains a small amount of ZnO. In addition, the major peak of the O 1s (Figure 7d) can be decomposed into three peaks at 529.8 eV, 531.3 eV, and 532.1 eV, corresponding to OM, OH, and OH$_2$ peaks respectively [39,40], which are derived from metal oxides (MgO and ZnO), hydroxides and surface water absorbed in the environment, respectively. The XPS detection demonstrates that the material surface is mainly composed of MgO and a small amount of ZnO as well as metal hydroxides (un-dehydrated hydroxides or formed by the reaction between oxides and water).

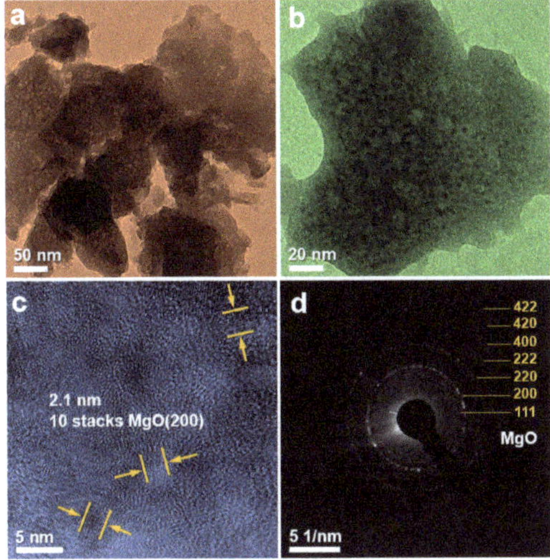

Figure 6. (**a–c**) Transmission electron microscope (TEM) images of MgO nanoplates; (**d**) selected area electron diffraction (SAED) pattern of a nanoplate.

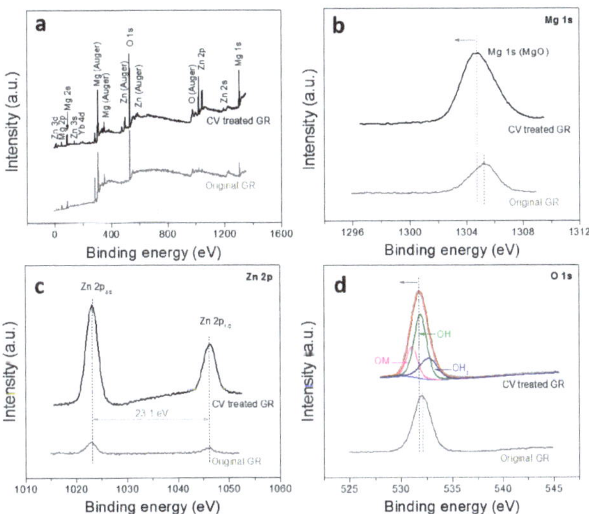

Figure 7. XPS spectra of the nanoplate film: (**a**) fully scanned spectra; (**b**) Mg 1s spectra; (**c**) Zn 2p spectra; (**d**) O 1s spectra.

The effect of MgO nanoplate layers on the corrosion resistance of Mg-based MGs is studied by potentiodynamic polarization curve in 3.5 wt.% NaCl solutions as shown in Figure 8. The original GR shows a certain range of passivation region. When the MgO nanoplate array layer is coated on the surface of GRs, the corrosion potential and the self-corrosion current of the materials are reduced under most of testing conditions in this paper. In particular, the self-corrosion current can decrease by about two orders of magnitude, indicating that the construction of MgO nanoplate layers can effectively improve the corrosion resistance of Mg-based MGs in NaCl solutions. By comparing the polarization curve positions under different conditions, the technological parameter performed by the blue curve (25 °C, 1 M KOH, 0.01 V/s sweep speed for 10 cycles) is selected in this paper for follow-up experiments.

Compared with samples with a greater improvement in corrosion resistance, the material treated by selected technological parameter exhibits a wider passivation area and makes the experimental process more time-saving and easier to implement (no need for high temperature, high solution concentration and CV treatment for multiple cycles).

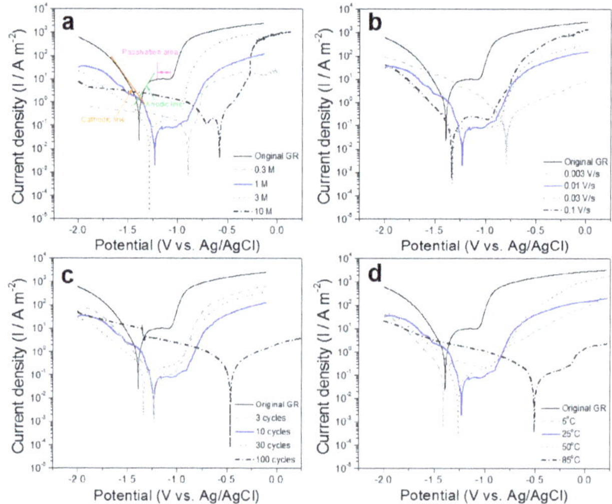

Figure 8. Potentiodynamic polarization curves of samples tested in 3.5 wt.% NaCl solution under different experimental parameters: (**a**) solution concentrations; (**b**) sweep rates; (**c**) cycle numbers; (**d**) reaction temperatures.

Figure 9a shows the surface SEM morphology of the $Mg_{66}Zn_{30}Yb_4$ GR after immersing in SA. The nanoplate array maintains the original morphology after immersion, covered by some flocculent SA. The TEM image (Figure 9b) shows that the surface of the nanoplate is covered with a layer of magnesium stearate (MgSt) and the outermost layer is covered with porous SA. The related reaction is shown in Equations (1) and (2) [41,42], revealing that the reaction between SA and MgO as well as the reaction between SA and $Mg(OH)_2$ will form a $[CH_3(CH_2)_{16}COO]_2Mg$ (MgSt) interlayer between the two reactants. The nanoplates with good hydrophilcity, distributed in a form of porous networks, are very conducive to the diffusion of SA onto their surfaces, resulting in the formation of a composite hydrophobic layer.

$$MgO + 2CH_3(CH_2)_{16}COOH \rightarrow [CH_3(CH_2)_{16}COO]_2Mg + H_2O \tag{1}$$

$$Mg(OH)_2 + 2CH_3(CH_2)_{16}COOH \rightarrow [CH_3(CH_2)_{16}COO]_2Mg + 2H_2O \tag{2}$$

Figure 9c shows the potentiodynamic polarization curves of the original GR, CV treated GR, and SA treated GR in 3.5 wt.% NaCl solutions. The SA treated GR shows the lowest self-corrosion current and the highest corrosion potential among the three samples, indicating that the corrosion resistance of the CV treated GR is further improved by the coverage of SA. It can be seen from Figure 8d that the open circuit potential (OCP) of the GR increases from -1396 mV to -1246 mV after the CV treatment, while it further increases to -1197 mV after the SA treatment. The increase in OCP (Figure 8d) and the decrease in self-corrosion current (Figure 8c) mutually support that the corrosion resistance of the SA treated GR improves.

The wettability of different materials can be compared by measuring the water contact angle (WCA). If the contact angle is more than 90°, the surface is regarded as hydrophobic, while if the contact angle is less than 90°, the surface is regarded as hydrophilic [42]. As shown in Figure 10a, the WCA of the original GR surface is 96°, indicating that the original material possesses certain hydrophobicity. After CV treatment, the WCA of GR changes to 30° (Figure 10b), indicating that the material presents good hydrophilicity in this situation, which is ready for the good adhesion of SA on the MgO surface. After SA immersion, the WCA of GR increase to 131° (Figure 10c), showing significantly improved hydrophobicity. Moreover, after immersing in 3.5 wt.% NaCl solution for 24 h, the WCA can still be kept at about 123°, showing good hydrophobic characteristic and good maintenance feature. The above phenomena confirm that the SA treated MgO film has good hydrophobicity and can protect the internal sample effectively for a long time.

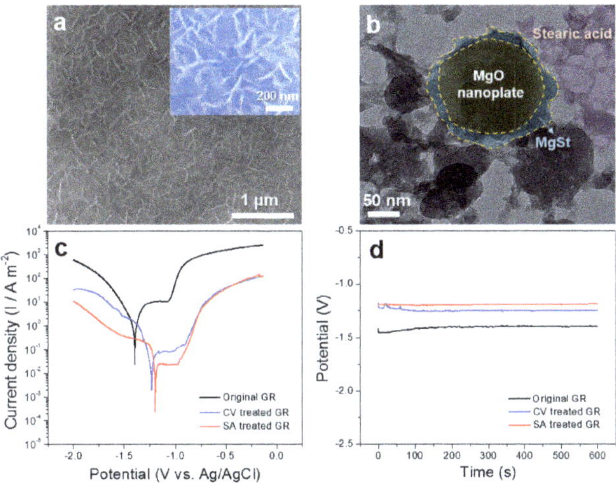

Figure 9. (**a**) BSE image and (**b**) TEM image of the hydrophobic film; (**c**) polarization curves and (**d**) open circuit potential of samples treated in different ways.

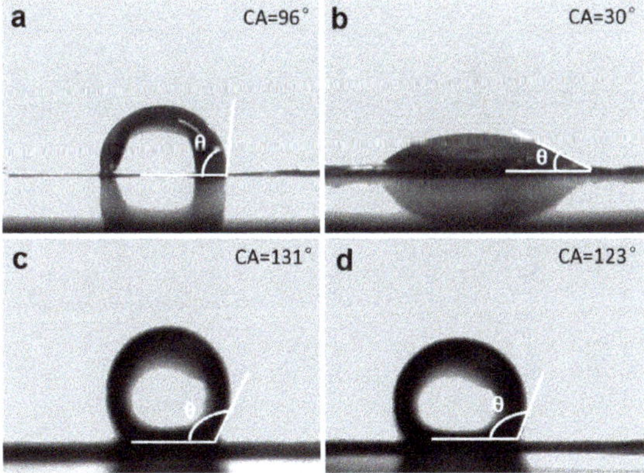

Figure 10. Images of water contact drops for different samples: (**a**) original GR; (**b**) CV treated GR; (**c**) SA treated GR (**d**) SA treated GR after immersion in 3.5 wt.% NaCl for 24 h.

Figure 11 presents superficial SEM morphologies of the experimental samples after immersion in 3.5 wt.% NaCl solution at 25 °C for 10 min. Figure 11a shows that the nanoplate array structure on the SA treated GR remains good after corrosion, with no obvious change in morphology (Figure 11e), showing the good effect of the film in improving corrosion resistance. Cross-sectional SEM image (Figure 11b) shows that the thickness of the hydrophobic film after corrosion does not change much and remains in about 5~6 μm. The linear scanning spectrum shows that the content of Cl element in the film is very low, indicating that the film plays an effect role to protect the substrate from corrosion. Figure 11c,f reveals that the obvious surface pitting of the original GR takes place during corrosion, indicating that, although the original GR presents relatively good corrosion resistance, local corrosion still exists. When the crystalline $Mg_{66}Zn_{30}Yb_4$ material is corroded, a certain degree of corrosion occurs inside the grain, forming a porous network. At the same time, a large number of hexagonal nanoplates generate at the grain boundary (Figure 11g), indicating that the crystalline state $Mg_{66}Zn_{30}Yb_4$ material possesses a very general corrosion resistance and is very easy to be corroded. Supplementary Video S1 shows a video comparing the macroscopic corrosion phenomenon of the original GR and the SA treated GR during the corrosion process in a more extreme corrosion environment (a mixture of 20 mL 3.5 wt.% NaCl solution and 0.5 mL 1 M HCl solution). It is found that in such a corrosive environment, a large number of bubbles appear immediately on the surface of amorphous GR when it is put into the corrosive solution, indicating that a drastic reaction occurs on the surface of the sample. However, after the SA treated GR is added to the corrosive solution, there is no obvious corrosion visible to the naked eye on the material surface. Based on the above analysis, the hydrophobic film synthesized in this paper presents a good protective effect to the glassy substrate and can effectively improve the corrosion resistance of Mg-based MGs in Cl ion containing solutions.

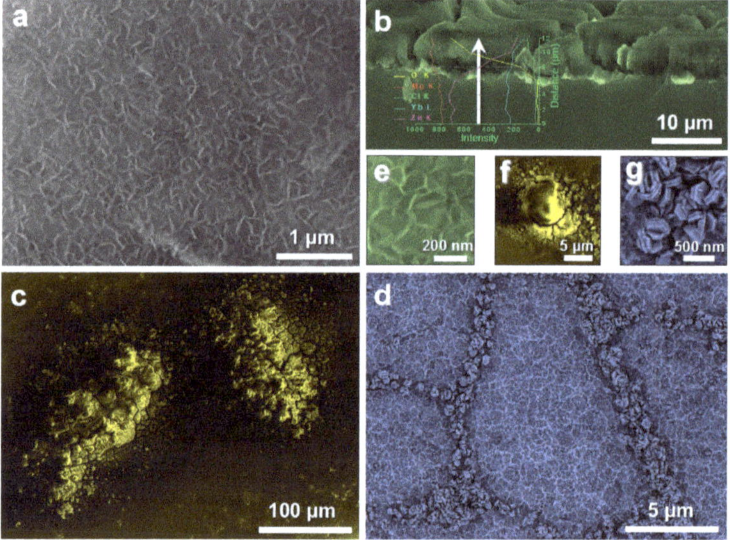

Figure 11. SEM images showing superficial morphologies of the experimental samples after immersion in 3.5 wt.% NaCl solution for 10 min: (**a**) SA treated $Mg_{66}Zn_{30}Yb_4$ GR; (**b**) cross section of the hydrophobic film and the EDS line scanning; (**c**) original $Mg_{66}Zn_{30}Yb_4$ GR; (**d**) crystalline state $Mg_{66}Zn_{30}Yb_4$ sample; (**e–g**) enlarged images of (**a,c,d**) respectively.

4. Conclusions

By utilizing the CV treatment method, the MgO nanoplate array films are synthesized on the surface of $Mg_{66}Zn_{30}Yb_4$ GRs successfully. Four CV treated parameters, including solution concentration, scanning rate, cycle number, and reaction temperature, are examined to analyse their effects on the sizes and shapes of the nano-products. It is found that the scanning rate and reaction temperature have a greater influence on the size of nanoplates. In addition, the nanoplates present a main length of 120~180 nm under the experimental conditions. The stable nanoplate array structure, obtained under the process parameter of 25 °C, 1 M KOH, 0.01 V/s sweep speed for 10 cycles, is selected for surface SA modification. As a result, the SA coated MgO composite hydrophobic film with a WCA of 131° is prepared on the surface of the $Mg_{66}Zn_{30}Yb_4$ GR. The composite shows improved corrosion resistance compared with the original amorphous alloy and the crystal material with the same composition. The work provides us a simple, low-cost, energy-saving, and environmentally friendly CV treatment method to create surface nanostructures and improve the corrosion resistance of the $Mg_{66}Zn_{30}Yb_4$ GR, which may promote the development and application of the CV treatment method and is expected to lead to the synthesis of different films with various nanostructures.

Supplementary Materials: The following are available online at http://www.mdpi.com/2079-4991/10/5/947/s1, Video S1: Macroscopic corrosion phenomenon of the GRs.

Author Contributions: Data curation, Y.Y. and J.Z.; Formal analysis, Y.Y., X.L. and H.X.; Funding acquisition, Z.W.; Investigation, Y.Y. and X.L.; Methodology, H.Y., C.Q. and Z.W.; Project administration, C.Q. and Z.W.; Resources, H.X. and H.Y.; Supervision, J.Z. and C.Q.; Writing—original draft, J.Z.; Writing—review & editing, Z.W. All authors have red and agreed to the published version of the manuscript.

Funding: The authors would like to acknowledge the financial support from Key Project of Science & Technology Research of Higher Education Institutions of Hebei Province, China (ZD2018059).

Conflicts of Interest: The authors declare no conflict of interest.

References

1. Wei, J.F.; Li, B.C.; Jing, L.Y.; Tian, N.; Zhao, X.; Zhang, J.P. Efficient protection of Mg alloy enabled by combination of a conventional anti-corrosion coating and a superamphiphobic coating. *Chem. Eng. J.* **2020**, *390*, 124562. [CrossRef]

2. Munir, K.; Lin, J.X.; Wen, C.E.; Wright, P.F.A.; Li, Y.C. Mechanical, corrosion, and biocompatibility properties of Mg-Zr-Sr-Sc alloys for biodegradable implant applications. *Acta Biomater.* **2020**, *102*, 493–507. [CrossRef] [PubMed]

3. Wang, Z.F.; Zhao, W.M.; Li, H.P.; Ding, J.; Li, Y.Y.; Liang, C.Y. Effect of titanium, antimony, cerium and carbon nanotubes on the morphology and microhardness of Mg-based icosahedral quasicrystal phase. *J. Mater. Sci. Technol.* **2010**, *26*, 27–32. [CrossRef]

4. Wu, Y.P.; Wang, Z.F.; Liu, Y.; Li, G.F.; Xie, S.H.; Yu, H.; Xiong, H.Q. AZ61 and AZ61-La alloys as anodes for Mg-air battery. *J. Mater. Eng. Perform.* **2019**, *28*, 2006–2016. [CrossRef]

5. Xiong, H.Q.; Liang, Z.F.; Wang, Z.F.; Qin, C.L.; Zhao, W.M.; Yu, H. Mechanical properties and degradation behavior of $Mg_{(100-7x)}Zn_{6x}Y_x$ (x = 0.2, 0.4, 0.6, 0.8) alloys. *Metals* **2018**, *8*, 261. [CrossRef]

6. Cheng, S.M.; Cheng, W.L.; Gu, X.J.; Yu, H.; Wang, Z.F.; Wang, H.X.; Wang, L.F. Discharge properties of low-alloyed Mg-Bi-Ca alloys as anode materials for Mg-air batteries: Influence of Ca alloying. *J. Alloys Compd.* **2020**, *823*, 153779. [CrossRef]

7. Wang, X.J.; Chen, Z.N.; Ren, J.; Kang, H.J.; Guo, E.Y.; Li, J.H.; Wang, T.M. Corrosion behavior of as-cast Mg–5Sn based alloys with in additions in 3.5 wt.% NaCl solution. *Corros. Sci.* **2020**, *164*, 108318. [CrossRef]

8. Xu, Y.K.; Ma, H.; Xu, J.; Ma, E. Mg-based bulk metallic glass composites with plasticity and gigapascal strength. *Acta. Mater.* **2005**, *53*, 1857–1866. [CrossRef]

9. Eckert, J.; Das, J. Mechanical properties of bulk metallic glasses and composites. *J. Mater. Res.* **2007**, *22*, 285–301. [CrossRef]

10. Gao, R.; Hui, X.; Fang, H.Z.; Liu, X.J.; Chen, G.L.; Liu, Z.K. Structural characterization of $Mg_{65}Cu_{25}Y_{10}$ metallic glass from ab initio molecular dynamics. *Comput. Master. Sci.* **2008**, *44*, 802–806. [CrossRef]

11. Men, H.; Hu, Z.Q.; Xu, J. Bulk metallic glass formation in the Mg–Cu–Zn–Y system. *Scr. Mater.* **2002**, *46*, 699–703. [CrossRef]

12. Yao, H.B.; Li, Y.; Wee, A.T.S. Corrosion behavior of melt-spun $Mg_{65}Ni_{20}Nd_{15}$ and $Mg_{65}Cu_{25}Y_{10}$ metallic glasses. *Electrochim. Acta* **2003**, *48*, 2641–2650. [CrossRef]

13. Babilas, R.; Bajorek, A.; Simka, W.; Babilas, D. Study on corrosion behavior of Mg-based bulk metallic glasses in NaCl solution. *Electrochim. Acta* **2016**, *209*, 632–642. [CrossRef]

14. Guo, S.F.; Chan, K.C.; Jiang, X.Q.; Zhang, H.J.; Zhang, D.F.; Wang, J.F.; Jiang, B.; Pan, F.S. Atmospheric RE-free Mg-based bulk metallic glass with high bio-corrosion resistance. *J. Non-Cryst. Solids.* **2013**, *379*, 107–111. [CrossRef]

15. Babilas, R.; Bajorek, A.; Sakiewicz, P.; Kania, A.; Szyba, D. Corrosion resistance of resorbable Ca-Mg-Zn-Yb metallic glasses in Ringer's solution. *J. Non-Cryst. Solids* **2018**, *488*, 69–78. [CrossRef]
16. Zhang, X.L.; Chen, G.; Bauer, T. Mg-based bulk metallic glass composite with high bio-corrosion resistance and excellent mechanical properties. *Intermetallics* **2012**, *29*, 56–60. [CrossRef]
17. Wang, J.F.; Huang, S.; Wei, Y.Y.; Guo, S.F.; Pan, F.S. Enhanced mechanical properties and corrosion resistance of a Mg-Zn-Ca bulk metallic glass composite by Fe particle addition. *Mater. Lett.* **2013**, *91*, 311–314. [CrossRef]
18. Chen, S.S.; Tu, J.X.; Hu, Q.; Xiong, X.B.; Wu, J.J.; Zou, J.Z.; Zeng, X.R. Corrosion resistance and in vitro bioactivity of Si-containing coating prepared on a biodegradable Mg-Zn-Ca bulk metallic glass by micro-arc oxidation. *J. Non-Cryst. Solids* **2017**, *456*, 125–131. [CrossRef]
19. Yang, J.X.; Cui, F.Z.; Lee, I.S.; Wang, X.M. Plasma surface modification of magnesium alloy for biomedical application. *Surf. Coat. Technol.* **2010**, *205*, S182–S187. [CrossRef]
20. Wu, G.S.; Ibrahim, J.M.; Chu, P.K. Surface design of biodegradable magnesium alloys—A review. *Surf. Coat. Technol.* **2013**, *233*, 2–12. [CrossRef]
21. Xie, J.; Hu, J.; Fang, L.; Liao, X.L.; Du, R.L.; Wu, F.; Wu, L. Facile fabrication and biological properties of super-hydrophobic coating on magnesium alloy used as potential implant materials. *Surf. Coat. Technol.* **2020**, *384*, 125223. [CrossRef]
22. Zhou, J.; Li, K.; Wang, B.; Ai, F.R. Nano-hydroxyapatite/ZnO coating prepared on a biodegradable Mg–Zn–Ca bulk metallic glass by one-step hydrothermal method in acid situation. *Ceram. Int.* **2020**, *46*, 6958–6964. [CrossRef]
23. Bhushan, B.; Katiyar, P.K.; Murty, B.S.; Mondal, K. Synthesis of hydrophobic Ni-VN alloy powder by ball milling. *Adv. Powder. Technol.* **2019**, *30*, 1600–1610. [CrossRef]
24. Wang, K.; Pang, J.B.; Li, L.W.; Zhou, S.Z.; Li, Y.H.; Zhang, T.Z. Synthesis of hydrophobic carbon nanotubes/reduced graphene oxide composite films by flash light irradiation. *Front. Chem. Sci. Eng.* **2018**, *12*, 376–382. [CrossRef]
25. Du, X.Y.; Gao, B.; Li, Y.H.; Song, Z.X. Super-robust and anti-corrosive NiCrN hydrophobic coating fabricated by multi-arc ion plating. *Appl. Surf. Sci.* **2020**, *511*, 145653. [CrossRef]
26. Fang, L.; Hou, L.X.; Zhang, Y.H.; Wang, Y.K.; Yan, G.H. Synthesis of highly hydrophobic rutile titania-silica nanocomposites by an improved hydrolysis co-precipitation method. *Ceram. Int.* **2017**, *43*, 5592–5598. [CrossRef]
27. Cui, X.J.; Lin, X.Z.; Lin, C.H.; Yang, R.S.; Zheng, X.W.; Gong, M. Fabrication and corrosion resistance of a hydrophobic micro-arc oxidation coating on AZ31 Mg alloy. *Corros. Sci.* **2015**, *90*, 402–412. [CrossRef]
28. Hu, R.G.; Zhang, S.; Bu, J.F.; Lin, C.J.; Song, G.L. Recent progress in corrosion protection of magnesium alloys by organic coatings. *Prog. Org. Coat.* **2012**, *73*, 129–141. [CrossRef]
29. Zhu, J.Y.; Dai, X.J.; Xiong, J.W. Study on the preparation and corrosion resistance of hydroxyapatite/stearic acid superhydrophobic composite coating on magnesium alloy surface. *AIP Adv.* **2019**, *9*, 045314. [CrossRef]
30. Khalifeh, S.; Burleigh, T.D. Super-hydrophobic stearic acid layer formed on anodized high purified magnesium for improving corrosion resistance of bioabsorbable implants. *J. Magnes. Alloy.* **2018**, *6*, 327–336. [CrossRef]
31. Qiao, L.Y.; Wang, Y.; Wang, W.L.; Mohedano, M.; Gong, C.H.; Gao, J.C. The preparation and corrosion performance of self-assembled monolayers of stearic acid and MgO layer on pure magnesium. *Mater. Trans.* **2014**, *55*, 1337–1343. [CrossRef]
32. Wang, Z.F.; Zhang, X.M.; Liu, X.L.; Wang, Y.C.; Zhang, Y.G.; Li, Y.Y.; Zhao, W.M.; Qin, C.L.; Mukanova, A.; Bakenov, Z. Bimodal nanoporous NiO@Ni–Si network prepared by dealloying method for stable Li-ion storage. *J. Power Sources* **2020**, *449*, 227550. [CrossRef]
33. Wang, Z.F.; Zhang, X.M.; Liu, X.L.; Zhang, W.Q.; Zhang, Y.G.; Li, Y.Y.; Qin, C.L.; Zhao, W.M.; Bakenov, Z. Dual-network nanoporous NiFe$_2$O$_4$/NiO composites for high performance Li-ion battery anodes. *Chem. Eng. J.* **2020**, *388*, 124207. [CrossRef]
34. Qin, C.L.; Zheng, D.H.; Hu, Q.F.; Zhang, X.M.; Wang, Z.F.; Li, Y.Y.; Zhu, J.S.; Ou, J.Z.; Yang, C.H.; Wang, Y.C. Flexible integrated metallic glass-based sandwich electrodes for high-performance wearable all-solid-state supercapacitors. *Appl. Mater. Today* **2020**, *19*, 100539. [CrossRef]
35. Li, Y.Y.; Liang, Z.F.; Yang, L.Z.; Zhao, W.M.; Wang, Y.L.; Yu, H.; Qin, C.L.; Wang, Z.F. Surface morphologies and mechanical properties of Mg-Zn-Ca amorphous alloys under chemistry-mechanics interactive environments. *Metals* **2019**, *9*, 327. [CrossRef]

36. Wang, Z.F.; Liu, J.Y.; Qin, C.L.; Yu, H.; Xia, X.C.; Wang, C.Y.; Zhang, Y.S.; Hu, Q.F.; Zhao, W.M. Dealloying of Cu-based metallic glasses in acidic solutions: Products and energy storage applications. *Nanomaterials* **2015**, *5*, 697–721. [CrossRef]

37. Yang, C.; Wang, Y.S.; Fan, H.L.; Falco, G.; Yang, S.; Shangguan, J.; Bandosz, T.J. Bifunctional ZnO-MgO/activated carbon adsorbents boost H_2S room temperature adsorption and catalytic oxidation. *Appl. Catal. B-Environ.* **2020**, *266*, 118674. [CrossRef]

38. Wang, H.; Li, Q.; Zheng, X.K.; Wang, C.; Ma, J.W.; Yan, B.B.; Du, Z.N.; Li, M.Y.; Wang, W.J.; Fan, H.Q. 3D porous flower-like ZnO microstructures loaded by large-size Ag and their ultrahigh sensitivity to ethanol. *J. Alloys Compd.* **2020**, *829*, 154453. [CrossRef]

39. Wang, Z.F.; Fei, P.Y.; Xiong, H.Q.; Qin, C.L.; Zhao, W.M.; Liu, X.Z. $CoFe_2O_4$ nanoplates synthesized by dealloying method as high performance Li-ion battery anodes. *Electrochim. Acta* **2017**, *252*, 295–305. [CrossRef]

40. Zhang, Q.; Li, M.; Wang, Z.F.; Qin, C.L.; Zhang, M.M.; Li, Y.Y. Porous Cu_xO/Ag_2O (x = 1, 2) nanowires anodized on nanoporous Cu-Ag bimetal network as a self-supported flexible electrode for glucose sensing. *Appl. Surf. Sci.* **2020**, *515*, 146062. [CrossRef]

41. Liu, A.H.; Xu, J.L. Preparation and corrosion resistance of superhydrophobic coatings on AZ31 magnesium alloy. *Trans. Nonferrous Met. Soc. China* **2018**, *28*, 2287–2293. [CrossRef]

42. Wu, L.; Wu, J.H.; Zhang, Z.Y.; Zhang, C.; Zhang, Y.H.; Tang, A.T.; Li, L.J.; Zhang, G.; Zheng, Z.C.; Atrens, A.; et al. Corrosion resistance of fatty acid and fluoroalkylsilane-modified hydrophobic Mg-Al LDH films on anodized magnesium alloy. *Appl. Surf. Sci.* **2019**, *487*, 569–580. [CrossRef]

 *nanomaterials*

Article

Polyvinyl Alcohol-Few Layer Graphene Composite Films Prepared from Aqueous Colloids. Investigations of Mechanical, Conductive and Gas Barrier Properties

Benoit Van der Schueren [1], Hamza El Marouazi [1], Anurag Mohanty [1], Patrick Lévêque [2], Christophe Sutter [1], Thierry Romero [1] and Izabela Janowska [1,*]

[1] Institut de Chimie et Procédés pour l'Énergie, l'Environnement et la Santé (ICPEES), CNRS UMR 7515-University of Strasbourg, 25 rue Becquerel, 67087 Strasbourg, France; benelux88@gmail.com (B.V.d.S.); hamza.el-marouazi@etu.unistra.fr (H.E.M.); anurag.mohanty@etu.unistra.fr (A.M.); christophe.sutter@unistra.fr (C.S.); thierry.romero@unistra.fr (T.R.)
[2] Laboratoire des sciences de l'Ingénieur, de l'Informatique et de l'Imagerie (ICube), UMR 7357, CNRS, Université de Strasbourg, 67400 Strasbourg, France; patrick.leveque@unistra.fr
*   Correspondence: janowskai@unistra.fr

Received: 11 February 2020; Accepted: 26 April 2020; Published: 29 April 2020

Abstract: Quasi all water soluble composites use graphene oxide (GO) or reduced graphene oxide (rGO) as graphene based additives despite the long and harsh conditions required for their preparation. Herein, polyvinyl alcohol (PVA) films containing few layer graphene (FLG) are prepared by the co-mixing of aqueous colloids and casting, where the FLG colloid is first obtained via an efficient, rapid, simple, and bio-compatible exfoliation method providing access to relatively large FLG flakes. The enhanced mechanical, electrical conductivity, and O_2 barrier properties of the films are investigated and discussed together with the structure of the films. In four different series of the composites, the best Young's modulus is measured for the films containing around 1% of FLG. The most significant enhancement is obtained for the series with the largest FLG sheets contrary to the elongation at break which is well improved for the series with the lowest FLG sheets. Relatively high one-side electrical conductivity and low percolation threshold are achieved when compared to GO/rGO composites (almost 10^{-3} S/cm for 3% of FLG and transport at 0.5% FLG), while the conductivity is affected by the formation of a macroscopic branched FLG network. The composites demonstrate a reduction of O_2 transmission rate up to 60%.

Keywords: nano composites; polymer-matrix composites; electrical properties; mechanical properties; few layer graphene composites

---

1. Introduction

Most of the polymer-graphene composites contain graphene oxide (GO) or graphene platelets, usually prepared by exfoliation in organic solvent. In case of water-soluble polymers such as PVA, GO would be the first choice due to the high content of hydrophilic/oxygenated groups and related ability to be well dispersed in water. The second advantage is usually a high aspect ratio, i.e., high lateral size of sheets vs. low thickness. However, the required reduction to restore conjugated C=C lattice and harsh synthesis conditions are important drawbacks of GO [1,2]. Finally, the reduced GO (rGO) is much less dispersible in water and some reports include additional functionalization of rGO to mix and get higher interface with PVA [3–6]. High strength and ductility composites were obtained for instance using rhodamine to functionalize rGO by cation-π and π-π interactions with an rGO surface [5]. The interesting mechanical and conductive properties were achieved for the composites with self-assemblies of microscale PVA lamellas, dendrites, and rods containing rGO functionalized by sulfonate groups [6].

The common way to obtain GO/rGO involves harsh conditions via the Hummer method, and reduction with hydrazine followed by preparation of the composite [7–11]. Few other reports deal for instance with the use of stress induced exfoliation to graphene based on direct ultrasonication of graphite with PVA. However, this latter method is long as well (48 h of sonication), and the properties of the final composites are not spectacular, as eventually, additional drawings to aligned graphene are applied [12].

Since GO is much more dispersible in water, graphene/rGO based composites show in general superior mechanical and conductive properties. The comparative studies between GO and rGO have been done for instance by Bao et al. [7]. The work clearly shows the advantage of graphene (rGO) over GO in PVA composites with related mechanisms in term of mechanical, conductivity and thermal stability properties [7]. On the other hand, a study demonstrated that, due to the presence of the basal oxygen containing groups and consequently adequate interfacial interactions with PVA, GO containing composite (0.3%) showed higher tensile strength than the one containing rGO [9].

Most reported works deal with low graphene loading and an interesting enhancement of mechanical, electrical or barrier properties were obtained at low graphene (rGO) content: 0.5–0.8% [7,11,13,14]. Some improvements were observed at much lower concentrations [4,15]. The best results in term of tensile strength were obtained by Zhao for the composite containing rGO, obtained in the presence of surfactant Sodium Dodecyl Benzene Sulphonate (SDBS), showing a 1000% enhancement of Young's modulus for 1.8% rGO content. However, no conductivity properties were demonstrated [8].

Despite a number of studies reporting on GO and rGO additives, the use of FLG/multi-layer graphene (MLG) for the same purpose is rare [16,17]. This is probably due to the lack of FLG that would fulfil the requirement of efficient and simple synthesis with a relatively high aspect ratio of the product. Most recently synthetized bio-compatible FLGs, or graphene platelets differently speaking, do not exceed a few hundred nm in size [18]. According to the experimental and theoretical evaluations, FLG/MLG has a similar Young's modulus, i.e., around 1TPa, to monolayer graphene [19], while the conductivity is affected by the perpendicular component that is three orders of magnitude lower than in the planar direction [20].

The aim of this work is to investigate FLG-containing PVA films obtained by casting from the co-mixed aqueous colloids, where FLG colloid are previously prepared by rapid, efficient and simple mixing assisted-exfoliation of expanded graphite in the presence of bovine serum albumin: a natural system possessing a high hydrophilic/lipophilic balance (HLB) [21,22]. The resulting FLG has a relatively high aspect ratio due to the large size of the sheets that is comparable with some reported GO/rGO sheets with lower size, while the simplicity of its preparation is a significant improvement on the harsh conditions typically applied for GO/rGO synthesis. On the other hand, the number of sheets is clearly more important. It is then of interest to check the possibility to use such FLGs and their water processable colloid, and compare the properties of the obtained composites with those known in the literature.

2. Material and Methods

2.1. Materials

Few layer graphene (FLG) was synthesized according to the exfoliation method described earlier [21]: expanded graphite (EG) was submitted to the mixing-assisted ultrasonication in distilled water in the presence bovine serum albumin (BSA) with 5–10 wt. % vs. FLG. In this approach, three different durations of the sonication were applied, i.e., 2, 4, or 6 h. The aqueous colloids obtained after sonication were left for decantation of 4 h and then the stable colloidal supernatants were separated and used later for the preparation of the composites.

EG was purchased from Mersen France Gennevilliers (Gennevilliers, France).

BSA was purchased from Merck (Sigma-Aldrich) company, St. Quentin Fallavier, France.

PVA powder (M.W. 57,000–66,000) was purchased from Alfa Aesar (Kandel, Germany).

For the preparation of composite films (PVA-FLG I, PVA-II, PVA-III, PVA-IV), PVA powder was first dissolved in distilled water by mixing and heating of the solution at 90 °C for 3–6 h (6.25 g/150 mL). Then, the given volume of aqueous suspension containing appropriate weight of FLG were added to the solution. All was next mixed for additional 0.5h and casted in Petri dishes. The casted colloids were dried under ambient conditions for 24 h and next under vacuum at 40–60 °C (or directly in oven at 60 °C in the case of PVA-FLG I).

Additional drying for 48 h at 60 °C was applied prior to the mechanical, conductive, and barrier properties measurements.

2.2. Characterization Tools

Thermogravimetric analysis (TGA) was performed on TA instrument SDT Q600 (Luken's drive, New Castle, DE, USA) under N_2 and air flow with the heating rate of 15°/min.

Differential scanning calorimetry (DSC) was carried out on Q200 V24.11 Build 124 (USA) for two cycles under N_2 with ramp rate of heating and cooling of 10 and 5 °C/min respectively, and isothermal steps of 2 min.

Scanning electron microscopy (SEM) was performed on Jeol JSM-6700 F (Tokyo, Japan) working at 3 kV accelerated voltage. In order to analyze the cross-section some films were broken (or cut) after freezing in liquid N_2.

X-ray photoelectron spectroscopy (XPS) was performed on a MULTILAB 2000 Thermo Fisher Scientific (Dreieich, Germany) spectrometer equipped with Al K anode (hv $\frac{1}{4}$ 1486.6 eV). CASA XPS program with a Gaussian-Lorentzian mix function and Shirley background subtraction were employed for processing of spectra.

Transmission electron microscopy (TEM) was performed on 2100 F Jeol microscope (Tokyo, Japan).

Tensile testing was carried out on INSTRON ElectroPuls E3000 (Norwood, MA, USA). Prior to the measurements the composites films were cut into specimens with appropriate dimensions for using in serrated grips.

Conductivity measurements (sheet resistance) of the films were carried out using four points probes method (FPPs) under N_2 atmosphere (in glove-box) on KEITHLEY 4200-SCS (Beaverton, OR, USA).

Oxygen transmission analysis was performed on apparatus module Ox-Tran 2/10 (Mocon, USA). For each film two measurements were carried out in accordance with standard ASTM D 3985–05. The samples were first conditioned and next measured for at least 24 h at t = 23 °C and RH = 0%.

3. Results and Discussion

3.1. Synthesis and Structures of FLG and PVA-FLG

Four series of PVA-FLG composites containing up to 3% of FLG are prepared and denoted as PVA-FLG I, PVA-FLG II, PVA-FLG III, and PVA-FLG IV. The series I and III include FLG that was exfoliated for 2 h (FLG-2h), series II includes FLG obtained after 4 h of exfoliation (FLG-4h), while series IV contains FLG after 6 h of exfoliation (FLG-6h). According to our earlier observations, the duration of exfoliation/ultrasonication of expanded graphite influences the size of FLG flakes which successively decreases with the increase of sonication duration [21]. Additionally, the final drying of PVA-FLG I film was different from the one applied in the three others. Different average size of FLG sheets in series could be observed through SEM microscopy, as shown in Figure 1a,b. Figure 1c shows an additional representative micrograph of a global view of FLG-2h flakes with average size of 5 μm. The prolonged time of sonication also causes the introduction of defects that are mostly linear defects, i.e., edges (due to the lower size of flakes) which is reflected by a lower combustion temperature, as shown in

Figure 1d, and enlarged C1s XPS spectra (full width at half-maximum, FWHM), shown in Figure 1e. It is known that, contrary to well crystallized planar conjugated sp^2 C=C lattice, graphene/FLG edges contain unsaturated carbon atoms and consequently are more sensitive towards oxidation and greedy for more electronegative oxygen, nitrogen-rich groups. The relative content of O and N, Figure 1e, increases with sonication time, where its eventual negligible variation possibly has origins in the variable amount of stabilizing BSA. A different stabilization behavior in the FLG-BSA suspensions during the decantation step and so different amount of BSA remaining in the suspensions would be a consequence of vary FLG flake size. Figure 1f illustrates additionally deconvoluted C1s XPS spectra of FLG-6h, confirming the presence of very few commonly recognized oxygen-containing groups and sp^3 type defect-linked C [23]. According to our current and earlier observations, these types of groups are mostly localized at the edges of FLG and not at the surface as is the case of GO or weakly reduced GO with a number of hydroxyl and epoxy surface-localized groups [21].

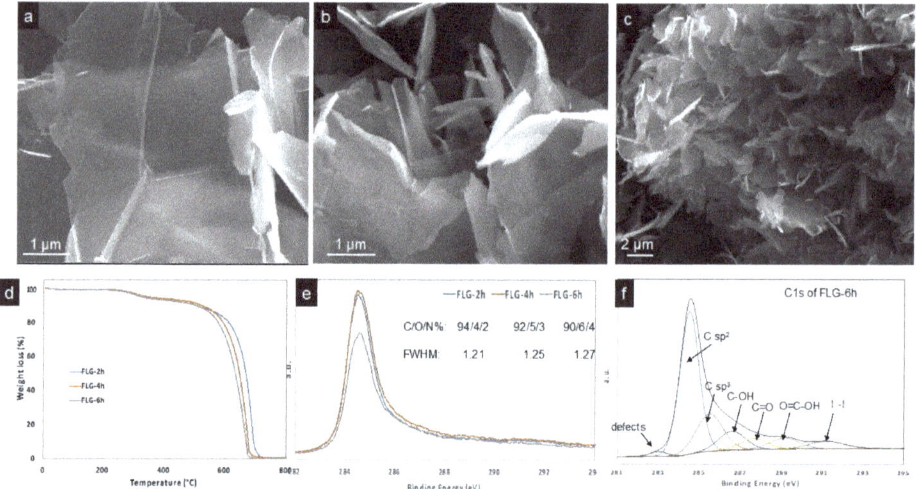

Figure 1. Representative SEM micrographs of (**a,c**) FLG-2h (**b**) FLG-6 h. (**d,e**) TGA curves and XPS spectra of FLG-2h, FLG-4h, FLG-6h, (**f**) deconvoluted XPS spectra of FLG-6h.

The prepared films are the composites of PVA, FLG and very small quantities of BSA present in the amount of 5–10% of FLG. The content of BSA is then very low but sufficient to get a homogenous aqueous dispersion of FLG and next of PVA-FLG (for 1% of FLG vs. weight of PVA, it represents only 0.05–0.1%). BSA is a large system consisting of transport type proteins containing almost 600 amino acid residues, i.e., a huge number of groups able to support hydrogen bonding [24]. Some of these groups are basically occupied to form folded quaternary structures and other ones are responsible for the very good solubility in water. Once submitted to the strong ultrasonication in water, the sonolysis process occurs and BSA form large microspheres through the interactions of disulphide from cysteine residue [21]. According to our observations, during the sonolysis of BSA in the presence of FLG smaller BSA spheres were formed from the excess of BSA [21], while unfolded linear structures of BSA were adsorbed on FLG surface [25]. Concerning the interactions with graphene surfaces, it was somehow reported that the highest affinity in BSA towards graphite surfaces have peptide groups [26]. This time, we could have localized adsorbed BSA and show that the chains of BSA are adsorbed at the surface of the FLG flakes as well as locally at the FLG edges (Supplementary Materials, Figure S1). The adsorption on the edges is not surprising taking into account the existence of reactive groups from the edges that can easily form hydrogen bonding with BSA. It seems also that BSA partially exists in aggregate form and partially as a thin, more planar structure adsorbed at FLG surface, but detailed investigations are necessary for the determination of this structure, which is out of scope of this work, especially given that BSA did not affect the properties studied below (PVA + BSA).

Hydrogen bonds are indeed the central interactions to discuss. They allow first to overcome van der Waals interactions between the sheets in graphite to produce FLG flakes and keep the hydrophobic FLG flakes as a stable aqueous suspension, and further, to make aqueous suspension with PVA. Concerning PVA-FLG interactions, most hydrogen bonding is established due to the presence of BSA adsorbed over FLG and not by oxygen groups covalently bonded to the surface as the case of GO or rGO sheets. Still, other interactions can occur with naked, BSA-free edges of FLG flakes. The role of water in the formation of intermediate hydrogen bonds cannot be forgotten, since possibly affect the properties of the composites. The importance of water in hydrogen bonding formation was demonstrated clearly by Brinson et al., who proposed a gallery-bridging hydrogen bond network associated with rotational and translational degrees of freedom thanks to water molecules in filtered GO paper and its PVA composite. The gallery played a role in the transfer of stress and modification of the stiffness [27].

The common feature of all films is the formation a macroscopic branched network of FLG with dendritic features at the periphery, visible from the upper side. The network could be observed during the drying step of casted colloids, Figure 2a. We associate this phenomenon with the diffusion limited aggregation (DLA) process induced by drying, quite often present in colloid science. A similar kind of branched pattern was also previously obtained upon the drying of pure FLG dispersed in organic solvents [28]. The DLA process and resulting patterns depend on many parameters related to the chemical potential of the matter, its concentration, overall interactions between the matter, matter-solvent and matter-support, matter-air, and drying conditions. It seems then that the present system including PVA, FLG covered by BSA, and water under casting/drying conditions is favorable for DLA to occur (for the reason of comparison FLG-BSA-PLA film prepared in parallel under the same drying conditions did not reveal the formation of such network). Some tendency to form a kind of wavy network is observed in ref. PVA film (Supplementary Materials Figure S2), while in the composites, the branches of the network are full of arranged FLG flakes as we could have confirmed by low resolution SEM analysis (Supplementary Materials Figure S3). Figure 2b–e shows the representative SEM micrographs of PVA-FLG films. Two sides of films can be distinguished after their drying, the one having an opaque optical aspect, that being upper side, seen in Figure 2b, and the one shining more, being in contact with glass (or plastic) Petri dish under drying, shown in Figure 2c. In Figure 2d,e, the transversal cross-section of the films shows the FLG flakes embedded in, and locally protruded, from polymer matrix. We can see clearly the parallel to the film surface (planar) orientation of the FLG flakes within the composite, the tendency observed in the literature rather for graphene (rGO) than for GO [9]. It was indeed observed that, due to the presence of the basal oxygen containing groups, and consequently adequate interfacial interactions with PVA, GO was randomly arranged within PVA matrix contrary to rGO that demonstrate more aligned arrangement [9]. According to this statement, the planar arrangement of FLG flakes suggest that the interactions between FLG edges (free or BSA decorated) and PVA occur and are relatively significant. Some deviation from planar arrangement was observed only close to the upper side of the film.

The TEM micrographs in Figure 3 confirm the existence of few micrometer size FLG flakes having good interface with the polymer, as well as few sheets in the flakes, i.e., nine in the presented micrograph of Figure 3c. According to our earlier studies the estimation of the average number of the sheets in FLG used here is around seven, while the presence of up to 15 as well as 2, 3 sheets are observed by TEM microscopy. Such number of sheets in FLG flakes seems to be high compared to GO/rGO materials but the thickness of the flakes is counterbalanced by relatively large size of the flakes, i.e., 3–5 μm on average.

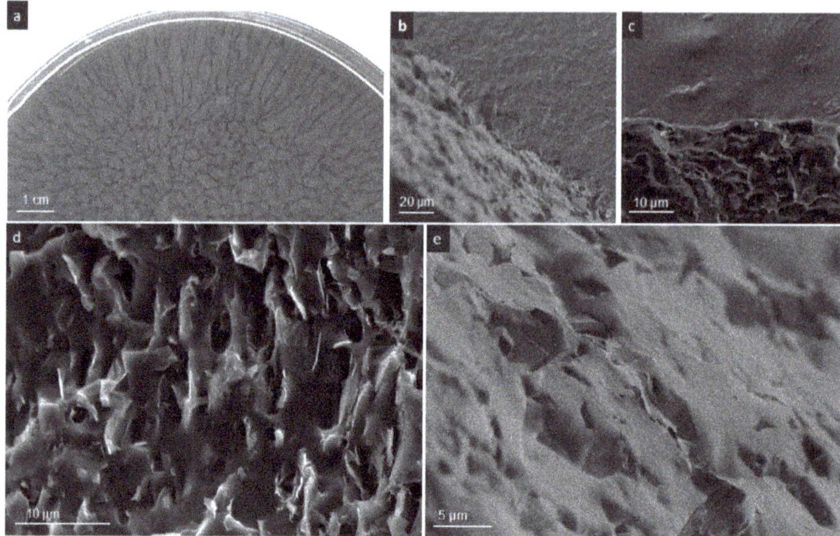

Figure 2. Representative PVA-FLG film structure, (**a**) optical photo of the drying film with formation of macroscopic branched network, (**b**,**c**) SEM micrographs of two sides of the films: opaque/upper and shining/down ones, (**d**,**e**) cross-sectional micrographs of the films: (**d**) closer to upper side, (**e**) closer to down side.

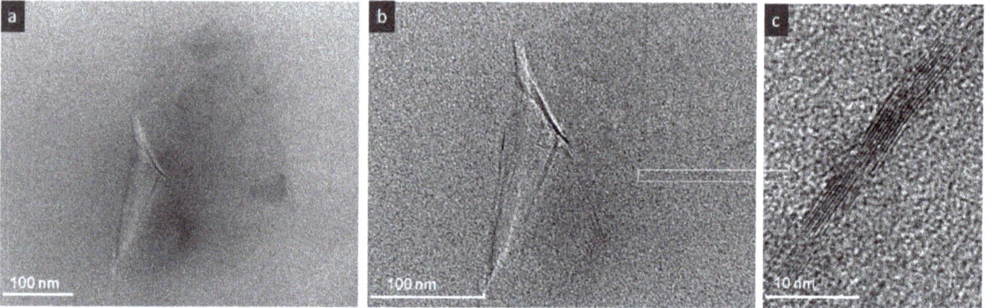

Figure 3. TEM micrographs of FLG flakes embedded in PVA matrix, (**a**) focus on the FLG morphology/size, (**b**) focus on the FLG edges (**c**) high resolution TEM of the edge of the flake demonstrating to have 9 sheets.

3.2. Thermal Properties of PVA-FLG

The modification of the thermal stability of PVA films induced by the addition of FLG flakes is reflected in TGA analysis, series III, Figure 4. It seems that the weight loss at lower temperatures is more significant for the composites compared to PVA, while at higher temperatures, this tendency in the composites is less clear. The faster weight loss at the beginning in the composites can be first of all related to the presence of BSA, which as mentioned above has high hydrophilic lipophilic balance properties and can enhance the ability to lock up water molecules. In the last region of combustion, i.e., below 500 °C, the interface between FLG and remaining PVA slightly slows down the process. For more precise investigations, the combustion temperature at 10% and 50% weight loss ($T_{10\%}$, $T_{50\%}$) are shown in Table 1, Figure 4. As can be seen, along with the increase of FLG loading, $T_{10\%}$ decreases in a significant and consecutive manner. This tendency changes once all water and large part of BSA are gone, and at $T_{50\%}$, the most robust is the composite with 1% of FLG, then PVA and PVA-FLG 2%

has similar thermal resistance, and only PVA-FLG 3% shows lower $T_{50\%}$. This random variation is probably linked to two phenomena, initially to the formation of a physical barrier from well dispersed FLG at 1% (decrease of the diffusion of degraded PVA to the gas phase) [29,30]. Next, at 3% FLG, to good thermal conductivity and related high heat transfer of FLG that accelerates a degradation of PVA.

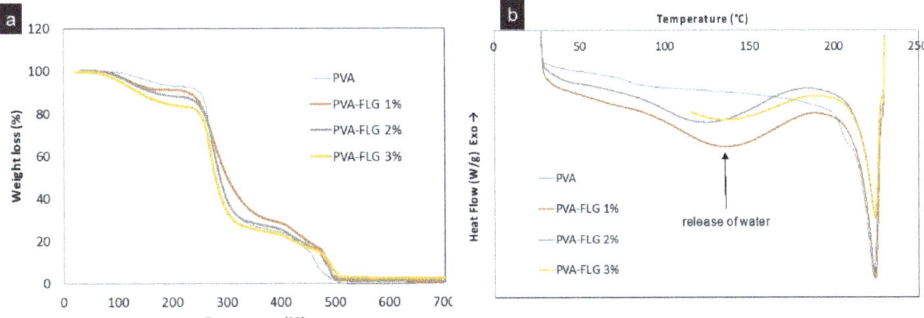

Figure 4. Thermal stability properties of the PVA-composites: (**a**) representative TGA (series III), (**b**) representative DSC (series IV).

Table 1. Temperatures for weight loss of 10 and 50% (series IV).

Sample (Series IV)	$T_{10\%}$	$T_{50\%}$
PVA	250	286
PVA-FLG 1%	228	300
PVA-FLG 2%	165	287
PVA-FLG 3%	141	278

The release of bound water, especially from the composites, is also clearly seen on DSC curves for the first heating, Figure 4b (series IV). According to DSC analysis, the melting temperatures (T_m) of the composites are in general slightly higher than T_m of PVA (Table 2) but the calculated crystallinity degrees (%C) clearly decrease with addition of FLG. The crystallinities were obtained from melting enthalpies, ΔH_m, values vs. ΔH_m of 100% crystalized PVA reference, i.e., 138.6 J/g [31], as shown in Table 2. The decrease of %C in the composites indicates disturbing effect of FLG additive on the ordering of the polymer chains and could be associated with so called hydrogen bond barrier blocking the hydrogen bonding within PVA. Such a hydrogen bond barrier was observed also in GO composites [7].

Likewise, the reduced melting enthalpy of the composites compared to the PVA in the first heat indicates loosely co-interacting species in the formers. This tendency is kept once water is released in the second heating only for certain samples. In few cases, the difference between relative maximum ΔH_{mII} obtained for the composites and reference PVA samples is much smaller, which can be attributed to the reduced mobility of the polymer chains interacting with BSA covered FLG. Some discrepancies in the modification of the relative T_m and ΔH_m between the series are induced probably by coupled factors related to quantitative variation of the interactions between PVA, water, BSA-FLG and FLG edges considering variable size of FLG flakes. Despite these discrepancies, it can be observed that the crystallinity (and ΔH_{mII}) in the second heating is relatively high for FLG at 1% and 2% loading compared to other composites and the reference samples.

Due to the locked-up water molecules the composites were additionally dried prior to the mechanical, conductive and barrier properties measurements.

Table 2. Melting temperatures and crystallinity degrees of PVA-FLG composites obtained from DSC analysis.

		T_m [°C]	ΔH_m [J/g]	%C	$\Delta H_{m\ II}$ [J/g]
Series I	PVA	224.1	66.8	48.8	93.3
	PVA-FLG 0.5%	224.6	63.3	45.7	75.6
	PVA-FLG 1%	224.4	66.2	47.7	78.7
	PVA-FLG 3%	224.68	58.9	42.5	73.2
Series II	PVA	220.6	49.9	36.0	87.5
	PVA-FLG 0.3%	224.6	54.5	39.3	79.6
	PVA-FLG 0.5%	220.6	47.1	34.0	63.1
	PVA-FLG 1%	224.1	61.9	44.7	71.4
	PVA-FLG 2%	224.8	67.6	48.8	87.4
Series IV	PVA	223.7	86.6	62.5	62.3
	PVA-FLG 1%	224.3	70.1	50.6	68.8
	PVA-FLG 2%	224.5	71.8	51.8	67.1
	PVA-FLG 3%	224.2	57.1	41.2	54.5

3.3. Mechanical Properties

The investigated mechanical properties of the four composites series are presented in Figures 5 and 6a and Table 3. They include the representative stress-strain curves, the modification of tensile modulus as a function of FLG loading and enhancement of tensile modulus, strength and elongation at break. One can see in Figure 6a that the tensile modulus (E) varies in more or less significant manner between the series, while the optimum and critical loading of FLG in all series are the same, i.e., closed to 1%. The presented points are the average values while, taking into account the margins of error, we see that, for few series, optimum loading can be obtained at lower loading (around 0.5). Already, the addition of 0.3% wt. of FLG induces the enhancement of the properties, then the tensile modulus in each series increases with the increase of FLG amount until 1% of FLG and next it is strongly reduced demonstrating even poorer properties compared to the neat PVA. A similar tendency is maintained for the tensile strength. It seems that the improvement of these properties depends on the size of FLG flakes being the most significant in series III (the largest flakes), weaker in series II (the average size) with the smallest in series IV (the smallest flakes). The greatest improvement of E and tensile strength is measured for series III for 1% of FLG loading, 114% and 60% respectively. Series I, containing also large flakes, shows the same tendency but the enhancement is somehow weaker. (Let recall that the drying step of films in series I was much faster, occurring at higher temperature, then drying of other series.) The elongation at break diminishes with the addition of FLG and this behavior is also measured in series III. On the contrary, the elongation at break increases with addition of FLG in series II and especially in series IV, so, inversely to FLG size. The superior elongation at break properties of series II and IV containing smaller sheets, FLG-4h and FLG-6h, indicate that the interactions between FLG and PVA occur mostly via the FLG edges and the impact of the adsorbed over FLG surface albumin on the overall interactions is lower. Consequently, the gliding of the FLG sheets is easier compared to the other series. On the contrary, lower ductility along with superior tensile modulus and strength in series with bigger FLG sheets suggests a relatively higher interaction through the bonding between PVA and BSA adsorbed over FLG surface. The potential impact of the FLG size on the formed branched patterns and consequently on the mechanical properties cannot indeed be excluded and would require separate investigations, while initial observations suggest no remarkable modification of the pattern between the series.

Our results are to a certain extent in agreement with the ones reported earlier where the best results were obtained for the PVA composites containing near 1% of GO and rGO (0.8%). Likewise, the enhancement of GO or rGO content to 1.6 resulted in the reduction of mechanical properties [7]. The composites prepared by Zhao behaved in a similar way, the mechanical percolation was obtained for 1.8% [8]. We can say then, that in this work, despite the formation of branched FLG pattern, the optimum percolation is obtained at similar FLG content range to the one obtained with GO/rGO PVA films.

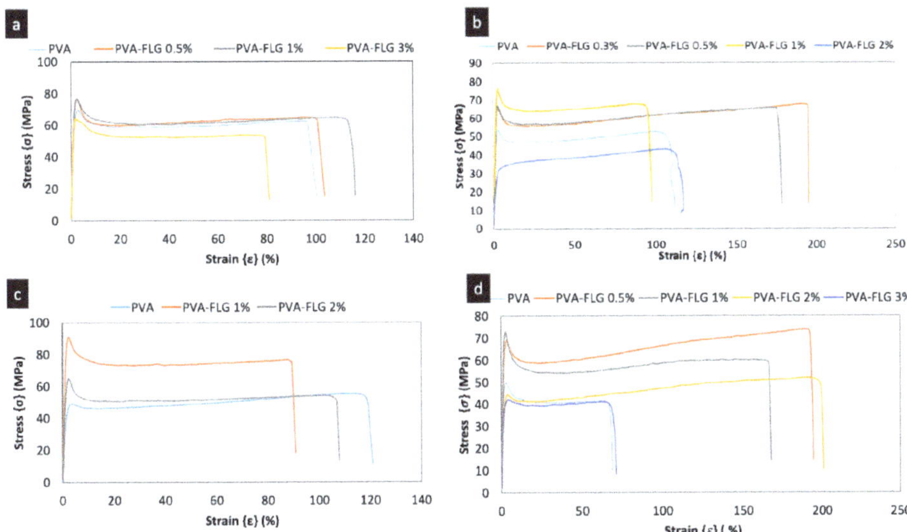

Figure 5. The representatives stress-strain curves of PVA-FLG composites: (**a**) series I, (**b**) series II, (**c**) series III, (**d**) series IV.

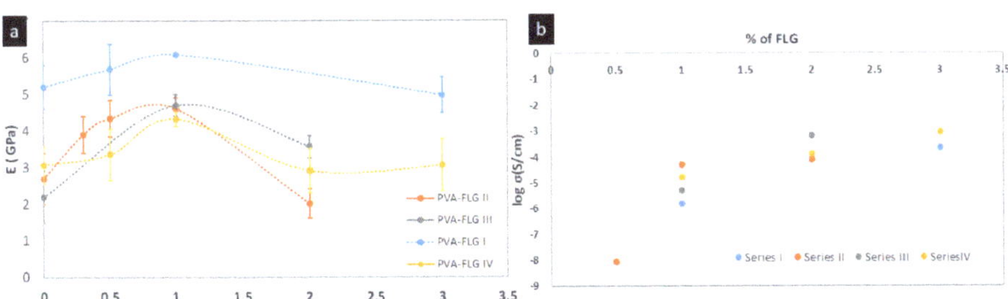

Figure 6. (**a**) Young modulus of PVA-FLG composites as function of FLG %, (**b**) conductivity of PVA-FLG composites as a function of FLG%.

Table 3. Enhancements of mechanical properties in PVA-FLG composites.

	PVA-FLG (%)	Enhancement of Tensile Modulus (GPa) (%)	Enhancement of Tensile Strength (MPa) (%)	Enhancement of Elongation at Break (%) (%)
	0.5	9	2	−6
Series I	1	**22**	**2**	17
	3	−4	−16	−20
	0.3	44	18	52
Series II	0.5	61	17	9
	1	**70**	**29**	22
	2	−25	−17	9
Series III	1	**114**	**60**	−36
	2	62	41	−16
	0.5	9	43	73
Series IV	1	**42**	39	75
	2	−5	24	**67**
	3	−0.3	0	−30

Since PVA is a semicrystaline polymer, its mechanical properties should also be affected by crystallinity features. Indeed, in series IV, the melting temperature slightly increases in the nanocomposites compared to PVA, while the crystallinity (from the second DSC heat rate) in the composites with 1% and 2% FLG loading is even higher than for the reference PVA.

Taking all these into account, the improvement of the mechanical properties in the composites at FLG loading around 1% wt. can then be attributed to the enhanced crystallinity, interactions between FLG-BSA/FLG-edges and PVA chains through hydrogen bonding, and van der Waals interactions as described above.

3.4. Electrical Conductivity

According to the FPPs measurements, the films are conductive, starting from 0.5% of FLG content with conductivity order of 10^{-9} S/cm, as shown in Figure 6b. The conductivity strongly increases, reaching values in the order of 10^{-6}–10^{-5} S/cm at 1% FLG loading and next increases slowly to be the order of 10^{-5}–10^{-4} S/cm at 2% FLG loading, being finally stabilized with the order of 10^{-4} S/m for FLG content of 3%. The highest value, almost 10^{-3} S/cm of order (8.5×10^{-4} S/cm), was achieved for series IV with 3% of FLG content.

A similar conductivity enhancement behavior was observed for rGO based composite, where the conductivity clearly jumped going from 0.8% to 1.6% of rGO and then stabilized at higher rGO concentration [7].

We assume that the inhibited increase of conductivity at certain FLG concentration is related to the increasing face-to-face arrangement of the FLG flakes. Such overlapping of the flakes in z-direction increases more-and-more along with FLG content. The conductivity in the direction perpendicular to the plane is three orders of magnitude lower than the planar one in FLG flake as it is in graphite [20], and an additional resistance in the films occurs at the connections of the flakes. Likewise, some extra overlapping of the flakes at higher FLG concentration can be induced by the formation of branched network, instead of homogenous distribution of add-FLG in the plane. Taking into account the limit of pattern formation (or random distribution) of FLG in the x direction, the obtained conductivities are relatively high. The conductivities and percolation thresholds of the composites are within the highest and the lowest ones, respectively, known in the literature for PVA-graphene based films [10,14]. Comparing our results with the ones achieved for PVA-rGO film, the PVA-FLG composites show around 5 orders of magnitudes higher conductivities for the same content of carbon [7]. The measured here conductivities are also comparable with the ones obtained for the composites containing large rGO of few tens of macrons in majority [10]. The effect of the aspect ratio on the conductivity properties and more precisely on the percolation threshold has been clearly shown for rGO flakes [10]. Indeed, we highlight that four main factors impact the transport properties in the films: carbon crystallinity, carbon morphology, carbon arrangement/interactions, and carbon-PVA interactions.

We did not find any specific correlation between the size of FLG sheets and conductivity percolation or behavior in general, but the differences in FLG size and % loading are possibly too small to observe a conductivity–size relationship. Likewise, the formation of the branched network would modify this simple reliance.

The good conductive characteristic despite a relatively significant thickness of the flakes is first of all ensured by the well crystalized carbon lattice of FLG with a low amount of defects and relatively large lateral size of the flakes compared to other FLG materials obtained by the liquid exfoliation of graphite [21]. In terms of FLG arrangement, there are two main factors here that play a role, i.e., the formation of connected FLG branched macroscopic patterns and the flat arrangement of the flakes at the conductive side as discussed next. Having seven graphene sheets on average and comparable or lower to rGO lateral size of the sheets, the FLG used here would make percolation a few times higher in terms of concentration than the thin graphene (rGO) if the advantages of the arrangement are not present. It was shown previously that the formation of fractal-like networks formed by pure FLG flakes over glass substrate enhanced the transport properties over macroscopic substrate compared to random distribution, but this effect in herein composites is for now difficult to determine [28]. Regardless the macroscopic network, we need to note the fact that, in all measured films, only the opaque side of the films is conductive—the one that initially was in the contact with glass substrate during the casting of the films, as shown in Figure 2c. This was quite surprising taking into account that, according to the first SEM observations, the presence of FLG flakes is clearly pronounced in the upper side of films. Despite relatively rapid drying and quite viscous colloids, we could suppose that some low difference in % loading exists between the two sides. For this purpose, the XPS analysis on the surface of both sides of the film was performed and did not reveal any chemical difference (the relative C to O ratio would show this difference if existing). This means that the concentration in FLG of both sides is very close while morphology can be side-dependent. This latter was next confirmed by applying high electron intensity mode under SEM analysis.

The low and higher magnifications micrographs on Figure 7 demonstrate two sides of the film where, due to the charging effect the FLG, flakes and PVA domains can be clearly distinguished as black and white, respectively. It is clear that, at the conductive side, we see a quasi-totally flat arrangement of the flakes that results in their good connectivity making impression of FLG excess, Figure 7a,b. Such an arrangement is disturbed at the opposite, upper side, through which the evaporated water molecules and air are evacuated upon drying living space for FLG flakes collapsing. On this side, we see more protruded flakes with their edges then the flakes surface as it is the case of the conductive side, this time giving impression of PVA excess. This side view is in agreement with the above cross-sectional observations, Figure 2d. It results in a lack of FLG connection and conductivity.

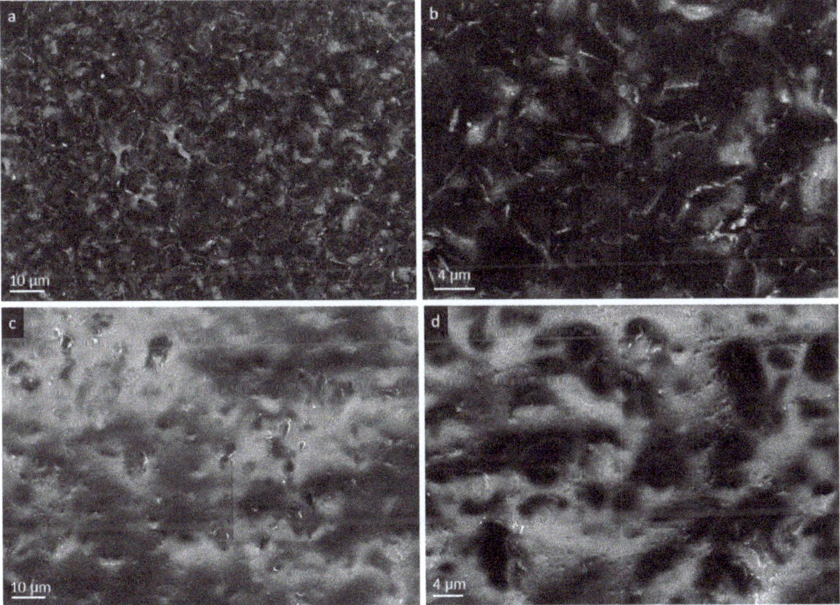

Figure 7. SEM micrographs of PVA-FLG films, series IV, 3% of FLG (high electron intensity): (**a,b**) conductive side, (**c,d**) non-conductive side.

3.5. Oxygen Barrier Properties

Few of prepared composites from series I and II with different FLG % were submitted to the oxygen barrier properties measurements. The oxygen transmission decreases up to 60%, from 10.7 for pure PVA to 4.3 cc/m^2 day for series I-0.5%FLG, and close to this, to 4.5 and 4.7 cc/m^2 for series II-0.3%FLG and series I-0.5%FLG. These average barrier properties are in agreement with the non-homogenous distribution of and specific arrangement of FLG into connected macroscopic network. It seems also that, for higher concentration of FLG, the composites show slightly higher transmission (4.9–6.1 cc/m^2), which would also suggest that the successive addition of FLG allows added overlapping of FLG in the z direction within the network than the extension of the network in the plane.

4. Conclusions

PVA composites containing FLG additive show enhanced mechanical, conductive, and gas barrier properties. The best mechanical properties are observed for PVA containing 1% wt. of FLG. The most significant enhancements of tensile modulus and strength are measured for the composites containing larger FLG flakes contrary to the elongation at break for which the best improvement is noticed for PVA containing lower size flakes. The rapid, simple, bio-compatible, and efficient synthesis of FLG together with the resulting large size of FLG sheets make this FLG material a suitable choice for further applications and their optimization in polymer composites. Preliminary studies show that highly conductive polymer-FLG films can be obtained, while at present, the formation of a macroscopic branched FLG network and one-side conductivity are achieved.

Supplementary Materials: The following are available online at http://www.mdpi.com/2079-4991/10/5/858/s1, Figure S1: SEM micrograph of FLG flakes with adsorbed BSA, Figure S2: Optical image of PVA film, Figure S3: SEM micrograph of the surface of PVA-FLG film (a branch of network).

Author Contributions: Synthesis of FLG and composites, characterization of composites (TGA, mechanical tests, SEM), B.V.d.S.; DSC, conductivity measurements, H.E.M.; characterization of FLG (XPS, TGA, SEM), A.M.; conductivity measurements, P.L.; DSC characterization, C.S.; SEM microscopy, T.R.; principal investigator/project coordinator, supervision, writing, I.J. All authors have read and agreed to the published version of the manuscript.

Funding: This work was supported by Centre National de la Recherche Scientifique (CNRS), pre-maturation program.

Acknowledgments: The PCA accredited laboratory No. AB 1450, located at the Centre from Bioimmobilisation and Innovative Packaging Materials in Szczecin (Poland) is acknowledged for O_2 transmission measurements. Thierry Dintzer (ICPEES) are acknowledged for high electron intensity SEM microscopy measurements. Dris Ihiawakrim (IPCMS) is acknowledged for TEM microscopy, Damien Favier (ICS) is acknowledged for help with traction tests.

Conflicts of Interest: The authors declare no conflict of interest.

References

1. Dreyer, D.R.; Todd, A.D.; Bielawski, C.W. Harnessing the chemistry of graphene oxide. *Chem. Soc. Rev.* **2014**, *43*, 5288–5301. [CrossRef] [PubMed]
2. Rozada, R.; Paredes, J.I.; Villar-Rodil, S.; Martínez-Alonso, A.; Tascón, J.M. Towards full repair of defects in reduced graphene oxide films by two-step graphitization. *Nano Res.* **2013**, *6*, 216–233. [CrossRef]
3. Mo, S.; Peng, L.; Yuan, C.; Zhao, C.; Tang, W.; Ma, C.; Shen, J.; Yang, W.; Yu, Y.; Min, Y. Enhanced properties of poly (vinyl alcohol) composite films with functionalized graphene. *RSC Adv.* **2015**, *5*, 97738–97745. [CrossRef]
4. Guo, J.; Ren, L.; Wang, R.; Zhang, C.; Yang, Y.; Liu, T. Water dispersible graphene noncovalently functionalized with tryptophan and its poly (vinyl alcohol) nanocomposite. *Compos. Part B Eng.* **2011**, *42*, 2130–2135. [CrossRef]
5. Tang, Z.; Lei, Y.; Guo, B.; Zhang, L.; Jia, D. The use of rhodamine B-decorated graphene as a reinforcement in polyvinyl alcohol composites. *Polymer* **2012**, *53*, 673–680. [CrossRef]
6. Layek, R.K.; Samanta, S.; Nandi, A.K. The physical properties of sulfonated graphene/poly (vinyl alcohol) composites. *Carbon* **2012**, *50*, 815–827. [CrossRef]
7. Bao, C.; Guo, Y.; Song, L.; Hu, Y. Poly (vinyl alcohol) nanocomposites based on graphene and graphite oxide: A comparative investigation of property and mechanism. *J. Mater. Chem.* **2011**, *21*, 13942–13950. [CrossRef]
8. Zhao, X.; Zhang, Q.; Chen, D.; Lu, P. Enhanced mechanical properties of graphene-based poly (vinyl alcohol) composites. *Macromolecules* **2010**, *43*, 2357–2363. [CrossRef]
9. Kashyap, S.; Pratihar, S.K.; Behera, S.K. Strong and ductile graphene oxide reinforced PVA nanocomposites. *J. Alloys Compd.* **2016**, *684*, 254–260. [CrossRef]

10. Zhou, T.; Qi, X.; Fu, Q. The preparation of the poly (vinyl alcohol)/graphene nanocomposites with low percolation threshold and high electrical conductivity by using the large-area reduced graphene oxide sheets. *Express Polym. Lett.* **2013**, *7*, 747–755. [CrossRef]

11. Wang, J.; Wang, X.; Xu, C.; Zhang, M.; Shang, X. Preparation of graphene/poly (vinyl alcohol) nanocomposites with enhanced mechanical properties and water resistance. *Polym. Int.* **2011**, *60*, 816–822. [CrossRef]

12. Jan, R.; Habib, A.; Akram, M.A.; Khan, A.N. Uniaxial drawing of graphene-PVA nanocomposites: Improvement in mechanical characteristics via strain-induced exfoliation of graphene. *Nanoscale Res. Lett.* **2016**, *11*, 377. [CrossRef] [PubMed]

13. Liang, J.; Huang, Y.; Zhang, L.; Wang, Y.; Ma, Y.; Guo, T.; Chen, Y. Molecular-level dispersion of graphene into poly (vinyl alcohol) and effective reinforcement of their nanocomposites. *Adv. Funct. Mater.* **2009**, *19*, 2297–2302. [CrossRef]

14. Salavagione, H.J.; Martínez, G.; Gómez, M.A. Synthesis of poly (vinyl alcohol)/reduced graphite oxide nanocomposites with improved thermal and electrical properties. *J. Mater. Chem.* **2009**, *19*, 5027–5032. [CrossRef]

15. Ismail, Z.; Abdullah, A.H.; Abidin, A.S.Z.; Yusoh, K. Application of graphene from exfoliation in kitchen mixer allows mechanical reinforcement of PVA/graphene film. *Appl. Nanosci.* **2017**, *7*, 317–324. [CrossRef]

16. Marka, S.K.; Sindam, B.; Raju, K.J.; Srikanth, V.V. Flexible few-layered graphene/poly vinyl alcohol composite sheets: Synthesis, characterization and EMI shielding in X-band through the absorption mechanism. *RSC Adv.* **2015**, *5*, 36498–36506. [CrossRef]

17. Khan, M.; Khan, A.N.; Saboor, A.; Gul, I.H. Investigating mechanical, dielectric, and electromagnetic interference shielding properties of polymer blends and three component hybrid composites based on polyvinyl alcohol, polyaniline, and few layer graphene. *Polym. Compos.* **2018**, *39*, 3686–3695. [CrossRef]

18. Pattammattel, A.; Kumar, C.V. Kitchen chemistry 101: Multigram production of high quality biographene in a blender with edible proteins. *Adv. Funct. Mater.* **2015**, *25*, 7088–7098. [CrossRef]

19. Xiang, L.; Ma, S.Y.; Wang, F.; Zhang, K. Nanoindentation models and Young's modulus of few-layer graphene: A molecular dynamics simulation study. *J. Phys. D Appl. Phys.* **2015**, *48*, 395305. [CrossRef]

20. Celzard, A.; Mareche, J.; Furdin, G.; Puricelli, S. Electrical conductivity of anisotropic expanded graphite-based monoliths. *J. Phys. D Appl. Phys.* **2000**, *33*, 3094. [CrossRef]

21. Ba, H.; Truong-Phuoc, L.; Pham-Huu, C.; Luo, W.; Baaziz, W.; Romero, T.; Janowska, I. Colloid Approach to the Sustainable Top-Down Synthesis of Layered Materials. *ACS Omega* **2017**, *2*, 8610–8617. [CrossRef] [PubMed]

22. Janowska, I.; Truong-Phuoc, L.; Ba, H.; Pham-Huu, C. Nanocomposites Nanomatériau/Système Polymoléculaire Colloïdaux, et Méthodes de Préparation. WO Patent 2018087484A1; PCT/FR2017/053064, 9 November 2017.

23. Blyth, R.; Buqa, H.; Netzer, F.; Ramsey, M.; Besenhard, J.; Golob, P.; Winter, M. XPS studies of graphite electrode materials for lithium ion batteries. *Appl. Surf. Sci.* **2000**, *167*, 99–106. [CrossRef]

24. Majorek, K.A.; Porebski, P.J.; Dayal, A.; Zimmerman, M.D.; Jablonska, K.; Stewart, A.J.; Chruszcz, M.; Minor, W. Structural and immunologic characterization of bovine, horse, and rabbit serum albumins. *Mol. Immunol.* **2012**, *52*, 174–182. [CrossRef] [PubMed]

25. Mohanty, A.; Janowska, I. Tuning the structure of in-situ synthesized few layer graphene/carbon composites into nanoporous vertically aligned graphene electrodes with high volumetric capacitance. *Electrochim. Acta* **2019**, *308*, 206–216. [CrossRef]

26. Guan, G.; Zhang, S.; Liu, S.; Cai, Y.; Low, M.; Teng, C.P.; Phang, I.Y.; Cheng, Y.; Duei, K.L.; Srinivasan, B.M. Protein induces layer-by-layer exfoliation of transition metal dichalcogenides. *J. Am. Chem. Soc.* **2015**, *137*, 6152–6155. [CrossRef] [PubMed]

27. Putz, K.W.; Compton, O.C.; Palmeri, M.J.; Nguyen, S.T.; Brinson, L.C. High-nanofiller-content graphene oxide–polymer nanocomposites via vacuum-assisted self-assembly. *Adv. Funct. Mater.* **2010**, *20*, 3322–3329. [CrossRef]

28. Janowska, I. Evaporation-induced self-assembling of few-layer graphene into a fractal-like conductive macro-network with a reduction of percolation threshold. *Phys. Chem. Chem. Phys.* **2015**, *17*, 7634–7638. [CrossRef]

29. Jiang, L.; Shen, X.P.; Wu, J.L.; Shen, K.C. Preparation and characterization of graphene/poly (vinyl alcohol) nanocomposites. *J. Appl. Polym. Sci.* **2010**, *118*, 275–279. [CrossRef]

30. Patole, A.S.; Patole, S.P.; Jung, S.-Y.; Yoo, J.-B.; An, J.-H.; Kim, T.-H. Self assembled graphene/carbon nanotube/polystyrene hybrid nanocomposite by in situ microemulsion polymerization. *Eur. Polym. J.* **2012**, *48*, 252–259. [CrossRef]

31. Yang, X.; Li, L.; Shang, S.; Tao, X.-M. Synthesis and characterization of layer-aligned poly (vinyl alcohol)/graphene nanocomposites. *Polymer* **2010**, *51*, 3431–3435. [CrossRef]

Article

Supported Ultra-Thin Alumina Membranes with Graphene as Efficient Interference Enhanced Raman Scattering Platforms for Sensing

Montserrat Aguilar-Pujol [1], Rafael Ramírez-Jiménez [1,2], Elisabet Xifre-Perez [3], Sandra Cortijo-Campos [1], Javier Bartolomé [1,4], Lluis F. Marsal [3] and Alicia de Andrés [1,*]

[1]   Instituto de Ciencia de Materiales de Madrid, Consejo Superior de Investigaciones Científicas Cantoblanco, 28049 Madrid, Spain; xochitl.aguilarpujol@gmail.com (M.A.-P.); ramirez@fis.uc3m.es (R.R.-J.); s.cortijo@csic.es (S.C.-C.); jbvilchez@gmail.com (J.B.)
[2]   Departamento de Física, Escuela Politécnica Superior, Universidad Carlos III de Madrid, Avenida Universidad 30, Leganés, 28911 Madrid, Spain
[3]   Department of Electronic Engineering, Universitat Rovira i Virgili, Avda. Països Catalans 26, 43007 Tarragona, Spain; elisabet.xifre@urv.cat (E.X.-P.); lluis.marsal@urv.cat (L.F.M.)
[4]   Departamento de Física de Materiales, Universidad Complutense de Madrid, Plaza Ciencias 1, 28040 Madrid, Spain
*   Correspondence: ada@icmm.csic.es; Tel.: +34-913-349016

Received: 6 April 2020; Accepted: 23 April 2020; Published: 27 April 2020

Abstract: The detection of Raman signals from diluted molecules or biomaterials in complex media is still a challenge. Besides the widely studied Raman enhancement by nanoparticle plasmons, interference mechanisms provide an interesting option. A novel approach for amplification platforms based on supported thin alumina membranes was designed and fabricated to optimize the interference processes. The dielectric layer is the extremely thin alumina membrane itself and, its metallic aluminum support, the reflecting medium. A CVD (chemical vapor deposition) single-layer graphene is transferred on the membrane to serve as substrate to deposit the analyte. Experimental results and simulations of the interference processes were employed to determine the relevant parameters of the structure to optimize the Raman enhancement factor (E.F.). Highly homogeneous E.F. over the platform surface are obtained, typically 370 ± (5%), for membranes with ~100 nm pore depth, ~18 nm pore diameter and the complete elimination of the Al_2O_3 bottom barrier layer. The combined surface enhanced Raman scattering (SERS) and interference amplification is also demonstrated by depositing ultra-small silver nanoparticles. This new approach to amplify the Raman signal of analytes is easily obtained, low-cost and robust with useful enhancement factors (~400) and allows only interference or combined enhancement mechanisms, depending on the analyte requirements.

Keywords: interference; enhanced Raman scattering; alumina membrane; graphene; nanoparticles; optical simulations; AFM; SEM

---

1. Introduction

The search of novel platforms for the amplification of Raman signal is still an objective since Raman spectroscopy is one of the most powerful techniques to identify analytes through the characteristic vibration modes of molecules and crystals. The detection and imaging of extremely diluted and/or complex materials still require further research and development to get cheap, reliable, reproducible and stable over time systems that can be easily reused several times. The enhancement achieved using localized plasmons, the so-called surface enhanced Raman scattering (SERS) [1–4] is definitely the most efficient process allowing to reach single molecule detection through complex structures and strategies [3,5,6]. Nevertheless, reproducibility, stability, complexity and reusability are still issues [7]. Another crucial challenge is obtaining homogeneous enhancement across the platforms to allow reliable quantitative detection [8]. Recently, the fabrication of hybrid platforms that combine nano- or micro-structured semiconductors (nanopillars, nanorods, etc.) with metallic nanoparticles (NPs) has been proposed to further enhance amplification [9,10]; however, homogeneity is achieved by employing expensive lithographic methods. Another method proposed to increase the specific area is by the formation of nanoporous metals [11].The combination of metallic NPs or nanostructures with graphene is an interesting approach since graphene may also provide an extra enhancement of chemical nature, the chemical mechanism (CM) [9,12]. A related but different approach consists in the deposition of nanostructured metallic arrays coupled with metallic films separated by a very thin dielectric spacer layer [13] as the case of gold nanopyramid arrays coupled to a gold film separated by a silica layer, which led to strong light absorption confined in the space layer with E.F. up to 233 [14]. The high enhancement of plasmon intensity at the gap is of interest because of its applications in metamaterials, energy transfer, sensors and solar energy harvesting [15].

In this context, the amplification of the electromagnetic signal based on interference processes has been scarcely explored. The interference of light occurring at the interfaces of multilayered heterostructures that combine materials with very different refractive indices (n) can be tuned depending the application. The interference enhanced Raman scattering (IERS) was applied in the 1980s to detect the phonons of ultrathin films [16,17] obtaining enhancement factors (E.F.) of around 20. It is now commonly used to increase the Raman signal of graphene and other 2D materials typically by using a SiO_2 dielectric layer on silicon single crystals with gains up to around 40 [18–21]. E.F. of up to 70 was reported for graphene bubbles on copper [22]. Some examples can be found on the combination of interference and SERS mechanisms [23,24] with low amplification factors, and, recently the design of optimized amplification platforms using aluminum as reflecting surface and Al_2O_3 as dielectric layer demonstrated efficient IERS and combined IERS and SERS amplifications [25].

Here we propose a new concept for an interference amplification platform which is robust and versatile. It is based on supported porous alumina membranes where the dielectric layer is the alumina membrane and the air of its pores and the reflecting layer is the metallic aluminum foil at the base of the membrane. Graphene is transferred on top of the porous alumina membrane and serves as the support where the analyte is deposited. Graphene is an excellent bio-compatible platform [26–28] with interesting characteristics since it can by-pass metal-biomaterial interactions occurring in SERS and quenches molecular fluorescence, highly inopportune for Raman spectroscopy. We have studied the different parameters that control the amplification factor such as the depth and density of the pores or the presence of an alumina barrier at the bottom of them. Interference is extremely sensitive to the thickness of the dielectric layer, which in this case is the depth of the pores but still we obtained highly efficient IERS with factors up to 400 and homogeneous amplification over the sample surface.

2. Materials and Methods

Porous alumina was obtained by a two-step electrochemical anodization process of aluminum to obtain a porous alumina layers with highly ordered pore distribution [29]. Before any anodization, pure aluminum substrates (99.999% purity) were cleaned with acetone, water and ethanol and electropolished in a mixture of ethanol (EtOH) and perchloric acid ($HClO_4$) 4:1 (*v/v*) at 20 V for 5 min in order to eliminate the surface roughness of the commercial aluminum substrates [30]. Subsequently, the first electrochemical anodization step took place in an aqueous solution of sulphuric acid (H_2SO_4) 0.3 M at 10 V and 3 °C. After 20 h, an alumina layer with disordered pores was obtained. It was dissolved by wet chemical etching in a mixture of 0.4 M phosphoric acid (H_3PO_4) and 0.2 M chromic acid (H_2CrO_7) at 70 °C for 3 h [31]. The second anodization step was performed under the same conditions as the previous one. The anodization time of this step determined the pore depth of the final ordered pore alumina layer [32]. The above mentioned combination of sulphuric acid, anodization voltage and electrolyte temperature has been specially selected for obtaining ultrathin alumina layers. With this selection we manage to extraordinarily slow down the speed of the anodization process and fabricate extremely thin layers with controlled thickness. Finally, the barrier layer formed at the bottom of the pores, inherent to the electrochemically obtained alumina, needed to be eliminated to increase the E.F. of the designed platforms. A simple and rapid method for its elimination was developed for these extremely small pores that consisted of the steady decrease of the anodization current followed by a pore-widening treatment performed by wet chemical etching in H_3PO_4 5 wt % at 35 °C for 4.5 min.

PMMA (polymethilmetacrylate)/Graphene on copper foil from *Graphenea* company (San Sebastián, Spain) was transferred onto the porous alumina. Cu was first eliminated in a 2.1 M $FeCl_3$ and 1.3 M HCl solution for 15 min. The PMMA/Graphene stack was rinsed in a deionized water bath twice, immerged in a 10% HCl solution and then again into deionized water three times with the last bath for 20 h to complete the removal of Cu. Subsequently, the Gr/PMMA film was fished onto the porous alumina samples and baked on a hot plate at 90 °C. PMMA was then removed by immersing in warm acetone at 50 °C followed by further vacuum thermal treatment at 250 °C.

Silver nanoparticles were deposited at room temperature with the gas aggregation technique [4,33] using a magnetron sputtering source (Nanogen50, from Mantis Ltd., Manchester, United Kingdom) and an Ag target 99.95% purity. The ejection of atoms and nucleation of clusters is assisted by a mixture of Ar/He gas, carried along the aggregation chamber through an orifice and reaching the substrate. The base pressure was 5×10^{-9} mbar and the work pressures were 2.5×10^{-1} mbar inside the aggregation chamber and 2×10^{-3} mbar in the deposition chamber, with an Ar/He ratio of 1:2.4. These parameters give rise to spherical single crystalline nanoparticles with average diameter ~4 nm [34]. The Ag NPs were simultaneously deposited on the membrane/graphene and on the fused silica/graphene reference sample. Rhodamine 6G (R6G) films were then deposited by spin coating from a methanol solution (10^{-3} M) on the two previous samples (membrane/graphene/AgNPs and fused silica/graphene/AgNPs).

Micro-Raman experiments were performed at room temperature with the 488 nm line of an Ar^+ laser, incident power in the 0.1–8 mW range, an Olympus microscope (×100 and ×20 objectives) and a "super-notch-plus" filter from Kaiser. The scattered light was analyzed with a Horiba monochromator coupled to a Peltier cooled Synapse CCD. The estimated Raman spatial resolution is around 0.7 μm at 488 nm for the high NA (0.95) ×100 objective. The Raman signal of graphene transferred on fused silica is used as the reference to calculate the enhancement factors. Raman signal of single layer graphene is very regular over any standard substrate so the average of three measurements was used.

The morphology and roughness of the samples were examined using atomic force microscopy (AFM) (equipment and software from Nanotec^{TM}, Madrid, Spain) [35] Topographic characterization was carried out in the tapping mode, using commercial Si tips (Nanosensors PPP-NCH-w) with a cantilever resonance frequency $f_0 \approx 270$ kHz and $k \sim 30$ Nm^{-1}. Several regions were probed to confirm homogeneity of the surfaces to the micrometer scale. An estimation of the roughness is obtained from the full width at half-maximum (FWHM) of the height distribution of the analyzed topographic images.

The simulations of the Raman signal amplification in the membranes were performed calculating the propagation of light through multilayered media using the matrix transfer method. In this method, the amplitude of the electromagnetic waves at two different depths inside the structure are related by a complex matrix—the transfer matrix is constructed taking into account the geometry and the refractive indexes of each layer, the effect on the electromagnetic (EM) field amplitude when traveling through different layers is calculated by matrix multiplication. The method provides a solution to Maxwell equations that considers the interference of the infinite number of multiple reflections occurring for light propagating through multilayered media. In the particular case of Raman scattering, multiple interference impacts in two ways, firstly the amplitude of the light arriving at a particular location in the structure, where the scattering process takes place should be calculated, secondly the scattered light should travel to the detector outside the structure, the calculation of the intensity involves again the transfer matrix method. Both aspects are taken into account in the calculations carried out previously for different metals and dielectrics [25].

3. Results and Discussion

The minimum requirement to obtain light multireflections able to produce interference is the presence of two parallel interfaces, one at the analyte to be detected and the other at a highly reflecting surface located at a certain distance, which can be tuned to obtain constructive or destructive interference.

The maximum interference for visible excitation lasers is obtained for aluminum as the reflecting layer, because of its high imaginary part n_i ($\underline{n} = n_r + in_i$), combined with a material with no absorption ($n_i \sim 0$) and the smallest n_r possible [25]. According to the simulations, air is the optimum dielectric medium, thus we have evaluated the amplification related to alumina membranes which are not detached from their aluminum base. Figure 1a sketches the section of a supported membrane with a single layer graphene transferred on top of it. The relevant parameters are indicated—the depth of the pore h, the diameter of the pore d, the distance between pores D and the thickness of the bottom Al_2O_3 barrier layer b. Figure 1b shows the calculated Raman amplification (for calculation details see Ref. [26]) for graphene phonons considering the simplest possible system where the dielectric layer is either Al_2O_3 (red triangles) or air (green circles). The ideal platform would thus include a dielectric layer formed by air with 120 nm thickness for 488 nm laser excitation. The supported membranes are an intermediate situation where air is present in a fraction of the dielectric layer, at the pores. The most Interference is strongly dependent on its thickness, in this case the depth of the pores, h in Figure 1a, so it is challenging to get enough control of this parameter in the fabrication process. Several sets of supported membranes were obtained with pore diameters, d, in the range 10–20 nm and different depths (h)—around 60, 100 and 200 nm (SEM images of the membrane with $h \sim 200$ nm is shown in Figure 1c). The presence of Al_2O_3 at the bottom of the pores is detrimental for the interference process so this barrier layer is reduced by a steady decrease of the anodization current followed by a pore-widening treatment, this later treatment attempts to increase the air fraction of the membrane to be closer to the ideal situation determined by the simulations.

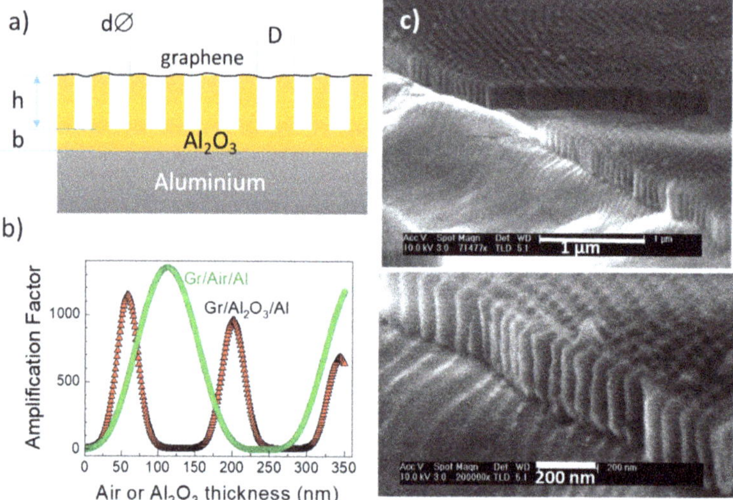

Figure 1. (**a**) Schema of the section of a supported membrane with a single layer graphene transferred on top of it, (**b**) calculated Raman amplification with Al_2O_3 (red triangles) or air (green circles) as the dielectric layer for 488 nm laser excitation, (**c**) scanning electron microscopy (SEM) images of an alumina supported membrane with $h \approx 200$ nm.

Figure 2a,b show in-plane and tilted SEM images of a supported alumina membrane with pore depth around 100 nm (sample named $h = 100$ nm) and ~18 nm pore diameter. The transference of a single layer graphene on top of the membrane strongly modifies its AFM topographic image

(Figure 2c,d). The analysis of the SEM and AFM images and the AFM profiles allow to estimate the pore to pore distance to be around 35 nm. An accurate determination of the pore diameter from the AFM images is challenging, since the tip size (20–30 nm) limits the estimation of pore lateral dimensions, however, obtaining pore to pore distance is quite precise. For the samples with transferred graphene, the profiles (and the height statistics) evidence how graphene mimics the membrane surface but strongly limits the oscillations to around ± 3 nm, much smaller than the pore depth (100 nm). The chosen small pore diameter, $d < 20$ nm, is therefore adequate to get small fluctuations of the overall dielectric thickness defined by the graphene top layer (the molecules to be sensed are deposited on top of the graphene).

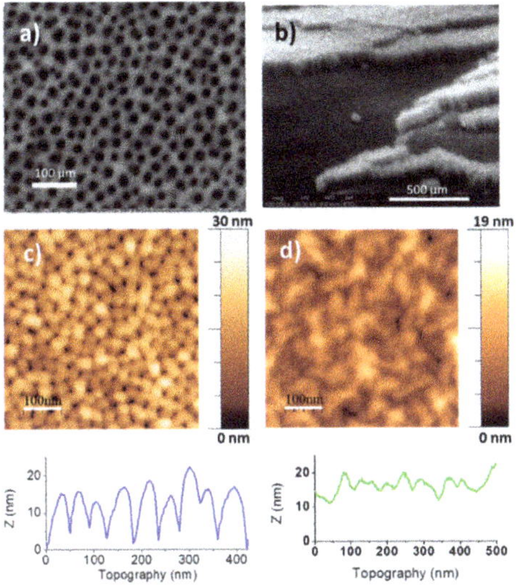

Figure 2. (**a,b**) SEM images of a membrane with pore diameter $d \approx 18$ nm and $h \approx 100$ nm, (**c,d**) atomic force microscopy (AFM) topographic images and height profiles of the pristine membrane and after graphene transfer, respectively.

The graphene transfer process and final graphene quality are first checked by optical microscopy. Figure 3a corresponds to the alumina membrane after fishing the graphene/PMMA film. Once the process to eliminate PMMA is concluded, typical micron sized regions of bi/tri-layer graphene can be easily seen (darker micron-sized spots in the optical images) as well as graphene wrinkles (Figure 3b,c). The interference process can also increase the contrast of optical images, as it occurs here and it is the first indication that the system will provide Raman amplification. In Figure 3c the edge of the transferred single-layer graphene is easily seen showing its high quality up to the very edge.

The quality of the transferred graphene is checked analyzing the Raman spectra. In Figure 3d the spectra of the $h = 100$ nm membrane at distant points are presented showing the characteristic D ($\approx$1580 cm^{-1}) and 2D ($\approx$2700 cm^{-1}) peaks with an intensity ratio $I_{2D}/I_G > 2$ indicating the single layer character of the transferred graphene. A defect peak (D at $\approx$1350 cm^{-1}) is detected with very small intensities. The black line in Figure 3d corresponds to a dark point in Figure 3b and clearly signals to a graphene bi-layer with the characteristic almost identical intensities of G and 2D peaks ($I_G \approx I_{2D}$), being I_G two times that of the single layer spectra and a higher intensity of the defect peak D.

Membranes with $h = 60$ nm were fabricated to optimize the E.F., however, for such small pore depth the quality of the membranes is compromised in terms of the order of the pores as well as in the height uniformity. The AFM topographic images show an increased disorder of the pores for the $h = 60$ nm membrane compared to the $h = 100$ nm one (Figure 4a,b), also, the height distribution of the AFM image in the case of the 60 nm sample is increased significantly, especially in comparison with the total nominal hole depth (~25%) (Figure 4c). Attempts to increase the pore fraction by extending the pore-widening treatment time to 9 min, leads to further disorder (Figure 4d).

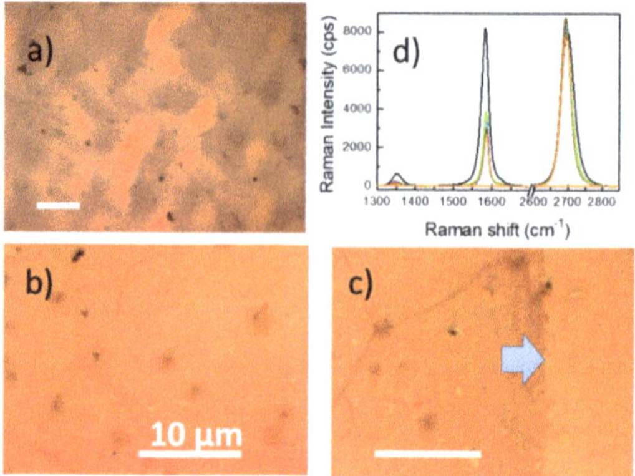

Figure 3. Optical images of an alumina supported membrane ($h \approx 100$ nm) (**a**) with graphene/PMMA, (**b**) with graphene once PMMA is eliminated showing the wrinkles and bi/tri-layer graphene spots and (**c**) graphene edge (indicated by a blue arrow). (**d**) Raman spectra of graphene at different positions. Black curve corresponds to a graphene bilayer (darker spot in the images). The scale bar corresponds to 10 µm in all cases.

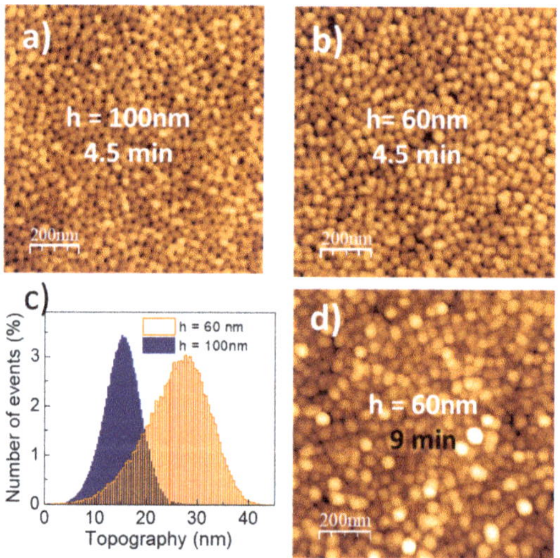

Figure 4. AFM images of (**a**) $h = 100$ nm (**b**) $h = 60$ nm $t = 4.5$ min and (**d**) $h = 60$ nm $t = 9$ min membranes. (**c**) Height distributions of the membranes in (**a,b**).

The efficiency of the interference process of the platforms is evaluated by measuring graphene Raman spectra compared to those of graphene transferred onto fused silica. The enhancement factor is defined as E.F. $= I_{2D}$ (membrane)/I_{2D} (fused silica). The obtained amplification for the $h = 60$ nm $t = 4.5$ min is E.F. ≈ 265 and for $h = 100$ nm membrane, higher values are obtained with average value E.F. ≈ 370.

One important point is to evaluate the uniformity of the amplification over the sample at different scales. Therefore, besides recording a set of spectra at distant points, we obtained, for the $h = 100$ nm sample (optical image in Figure 5a), 121 spectra from 10 µm × 10 µm squares, 1 µm steps, that we can transform into Raman images of the 2D and D peak intensities (Figure 5b,c) and of the Raman enhancement factor E.F (Figure 5d). The 2D peak image (Figure 5b) shows a very uniform intensity with slightly lower intensity regions that correspond to the darker points of the optical image related to micron-sized areas of bi/tri-layer graphene (Figure 5a) as we commented before. Just the opposite occurs to the D peak whose intensity is increased at the bilayer graphene regions. Finally, the amplification, except at the small bilayer regions, is very uniform with an average value of 370 ± 5%. Such Raman amplification factors are remarkable and, interestingly, are obtained without the use of nanoparticles.

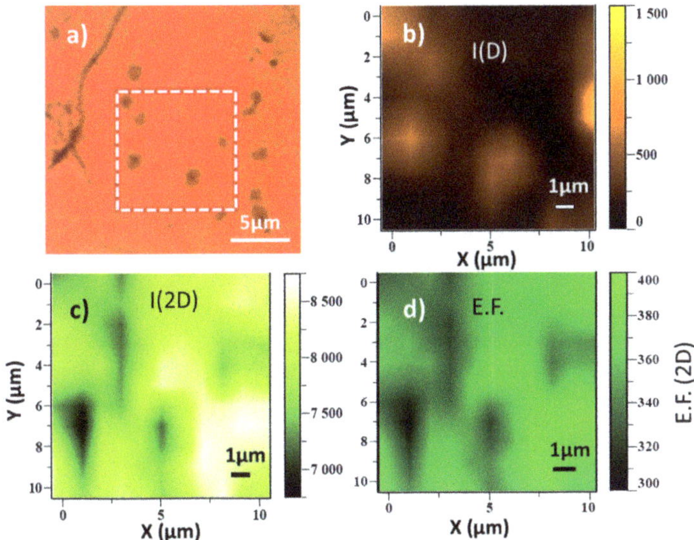

Figure 5. (**a**) Optical image of graphene on a 100 nm membrane, the 10 µm × 10 µm white dashed square is the tested area in the Raman images of (**b**) D peak intensity, (**c**) 2D peak intensity and (**d**) enhancement factor E.F. = I_{2D} (membrane)/I_{2D} (fused silica).

Several characteristics of the membranes reduce the amplification for a perfect "air layer" obtained in the simulations. On one hand, the calculations indicate that an alumina barrier layer present at the bottom of the pores drastically depletes amplification (Figure 6a). In this figure the amplification is calculated as a function of the dielectric layer thickness, the dielectric being formed by air and different fixed alumina barrier layers (thickness: 2, 5, 15 and 30 nm). Even a thin 5 nm of Al_2O_3 barrier already reduces the maximum amplification to values below 400, similar to the experimental E.F. On the other hand, unfortunately, the pores (the air) occupy only a fraction of the membrane, so that the dielectric layer is an in-plane combination of alumina and air, which also modifies the interference response (Figure 6b). The estimated fraction of pores in the $h = 100$ nm sample obtained from the SEM images is around 20%. In Figure 6c both issues are combined and E.F. has been calculated for 20% pore fraction and different values of the alumina bottom barrier (0, 2 and 5 nm). The horizontal green line indicates the experimental E.F. obtained for the $h = 100$ nm membrane with good coincidence with 20% pore fraction and 0 nm alumina barrier. According to calculations the actual depth of the pores is around 90 nm rather than the 100 nm estimated from the SEM images and the treatment to reduce the bottom barrier is effective.

Finally, 4 nm silver nanoparticles (NPs) were deposited [see Refs. 4 and 34 for the NPs characterization] onto the graphene layer on top of the membrane to check the possibility to further enhance the Raman signal of graphene and of an analyte. A schema of the resulting membrane based system is plotted in Figure 7a). In this case we used Rhodamine 6G (R6G) to test the amplification capability. Ag NPs were simultaneously deposited on the membrane platform and on the reference sample (fused silica/graphene) previously used and R6G was then deposited by spin coating. Ultra-small Ag NPs were used to preserve the transparency required for the interference process and because we recently demonstrated that, for the same plasmon absorption, the SERS effect is more efficient for these 4 nm NPs than for larger ones due to higher hot-spot density [34].

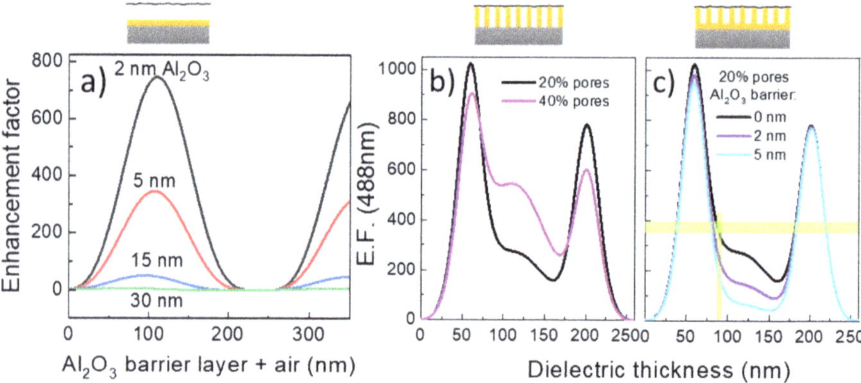

Figure 6. E.F. calculated at 488 nm laser excitation for (**a**) a dielectric formed by air and an Al$_2$O$_3$ barrier of the indicated thickness, (**b**) different pore fractions without barrier and (**c**) pore fraction of 20% and alumina barrier layer of 0, 2 and 5 nm. The horizontal green line corresponds to the experimental E.F. value.

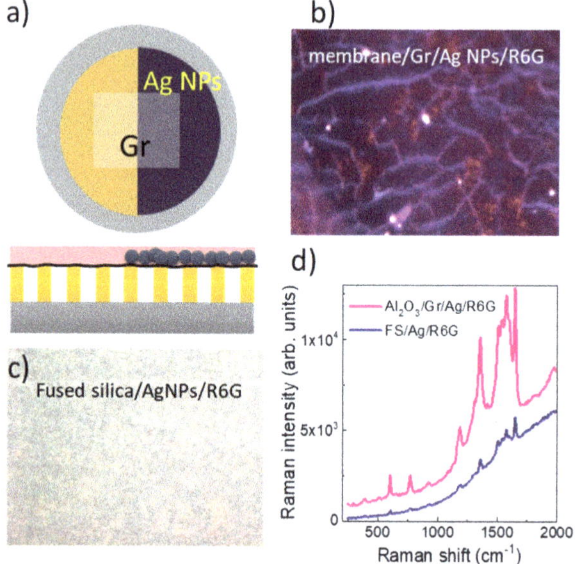

Figure 7. (**a**) Schema of the membrane based amplification platform: supported membrane/graphene/Ag NPs/R6G, top view and cross section, (**b**) optical image of the Ag NPs region and (**c**) of the fused silica/graphene/Ag NPs/R6G reference sample (images of 36 μm × 27 μm regions). (**d**) Typical R6G Raman spectra from both samples.

Optical images of the membrane and the fused silica surfaces are shown in Figure 7b,c. The R6G fluorescence is almost quenched allowing to easily detect the R6G Raman modes. In Figure 7d, a representative Raman spectrum obtained for the membrane/graphene/Ag NPs/R6G (pink curve) is compared to that for the fused silica/graphene/Ag Nps/R6G sample (dark blue curve). A 10^{-3}

M solution of R6G was deposited by spin coating on both substrate which correspond finally to the equivalent concentration of one R6G monolayer on the substrates [4]. The spectra present the characteristic Raman peaks of R6G. In both cases the same SERS and CM amplifications are occurring so that the intensity increase in the membrane spectra is due to the interference process demonstrating the cooperative amplification of SERS and interference effects.

4. Conclusions

Supported alumina membranes where specifically designed and fabricated to be used as amplification platforms to enhance the Raman signal of analytes by interference processes of the incoming and scattered light beams. The metallic aluminum support is used as the reflecting medium and the dielectric layer is the combination of the air of the pores and the alumina of the walls. Pore diameters in the 10–20 nm range are adequate to transfer a CVD single-layer graphene to serve as the substrate to deposit the analyte to be detected. Graphene mimics the membrane surface but presents a flat surface with small height fluctuations ~3 nm, which is found to be adequate for interference efficiency. The theoretical optimum pore depth depends on the pore fraction of the membrane, it is around 60 nm for the first interference order and around 200 nm for the second order for 20% pore fraction. Platforms based on membranes with pores height around 60, 100 and 200 nm were prepared, however, the control of the membrane quality in the 60 nm range is not enough. The elimination of the alumina barrier layer (at the bottom of the pores) is crucial according to calculations and the employed process for its elimination is found to be totally efficient. The Raman signals of Rhodamine 6G, spin-coated on graphene and graphene itself were used to test the platforms. E.F. up to 400 is obtained for membranes with ~100 nm pore depth, ~18 nm pore diameter and the complete elimination of the Al_2O_3 bottom barrier layer. The most limiting parameter is the pore fraction in the membrane, which reaches around 20% for 18 nm pore diameter. Further pore widening, which is favorable to increase E.F. in principle, produces larger in-plane disorder and surface roughness (height distribution).

We demonstrated the possibility to further enhance the Raman signal of R6G by depositing ultra-small (4 nm diameter) silver nanoparticles on the graphene layer prior to spin-coating the analyte. Combined SERS and IERS processes is observed.

This new approach to amplify the Raman signal of analytes by means of interference is cheap and robust with useful enhancement factors (~400). It allows the combination of plasmonic and interference amplifications, however, these platforms are also appropriate to amplify Raman signals in the cases where the use of nanoparticles is to be avoided.

Author Contributions: Conceptualization, A.d.A.; investigation, M.A.-P., R.R.-J., E.X.-P., L.F.M., S.C.-C., J.B.; writing—original draft preparation, A.d.A., L.F.M.; writing—review and editing, A.d.A., L.F.M., J.B., S.C.-C.; funding acquisition, A.d.A. and L.F.M. All authors have read and agreed to the published version of the manuscript.

Funding: The research leading to these results has received funding from Ministerio de Ciencia, Innovación y Universidades (RTI2018-096918-B-C41) and RTI2018-094040-B-I00) and by the Agency for Management of University and Research Grants (AGAUR) 2017-SGR-1527. S.C. acknowledges the grant BES-2016-076440 from MINECO.

References

1. Stiles, P.L.; Dieringer, J.A.; Shah, N.C.; Van Duyne, R.P. Surface-Enhanced Raman Spectroscopy. *Annu. Rev. Anal. Chem. (Palo Alto. Calif.)* **2008**, *1*, 601–626. [CrossRef]
2. Sharma, B.; Frontiera, R.R.; Henry, A.-I.; Ringe, E.; Van Duyne, R.P. SERS: Materials, Applications, and the Future. *Mater. Today* **2012**, *15*, 16–25. [CrossRef]
3. Xie, W.; Schlücker, S. Rationally Designed Multifunctional Plasmonic Nanostructures for Surface-Enhanced Raman Spectroscopy: A Review. *Rep. Prog. Phys.* **2014**, *77*, 116502. [CrossRef]

4. Jimenez-Villacorta, F.; Climent-Pascual, E.; Ramirez-Jimenez, R.; Sanchez-Marcos, J.; Prieto, C.; de Andrés, A. Graphene–ultrasmall Silver Nanoparticle Interactions and Their Effect on Electronic Transport and Raman Enhancement. *Carbon* **2016**, *101*, 305–314. [CrossRef]

5. Qian, X.-M.; Nie, S.M. Single-Molecule and Single-Nanoparticle SERS: From Fundamental Mechanisms to Biomedical Applications. *Chem. Soc. Rev.* **2008**, *37*, 912–920. [CrossRef] [PubMed]

6. Wang, P.; Liang, O.; Zhang, W.; Schroeder, T.; Xie, Y.-H. Ultra-Sensitive Graphene-Plasmonic Hybrid Platform for Label-Free Detection. *Adv. Mater.* **2013**, *25*, 4918–4924. [CrossRef] [PubMed]

7. Kouba, K.; Proška, J.; Procházka, M. Gold film over SiO_2 nanospheres—New thermally resistant substrates for surface-enhanced Raman scattering (SERS) spectroscopy. *Nanomaterials* **2019**, *9*, 1426. [CrossRef] [PubMed]

8. Yang, L.; Lee, J.H.; Rathnam, C.; Hou, Y.; Choi, J.W.; Lee, K.B. Dual-Enhanced Raman Scattering-Based Characterization of Stem Cell Differentiation Using Graphene-Plasmonic Hybrid Nanoarray. *Nano Lett.* **2019**, *19*, 8138–8148. [CrossRef] [PubMed]

9. Schmidt, M.S.; Hübner, J.; Boisen, A. Large area fabrication of leaning silicon nanopillars for Surface Enhanced Raman Spectroscopy. *Adv. Mater.* **2012**, *24*, 11–18. [CrossRef]

10. Barbillon, G. Fabrication and SERS performances of metal/Si and metal/ZnO nanosensors: A review. *Coatings* **2019**, *9*, 86. [CrossRef]

11. Garoli, D.; Schirato, A.; Giovannini, G.; Cattarin, S.; Ponzellini, P.; Calandrini, E.; Zaccaria, R.P.; D'Amico, F.; Pachetti, M.; Yang, W.; et al. Galvanic replacement reaction as a route to prepare nanoporous aluminum for UV plasmonics. *Nanomaterials* **2020**, *10*, 102. [CrossRef]

12. Tatarkin, D.E.; Yakubovsky, D.I.; Ermolaev, G.A.; Stebunov, Y.V. Surface-Enhanced Raman Spectroscopy on Hybrid Graphene/Gold Substrates near the Percolation Threshold. *Nanomaterials* **2020**, *10*, 164. [CrossRef] [PubMed]

13. Cesario, J.; Quidant, R.; Badenes, G.; Enoch, S. Electromagnetic coupling between a metal nanoparticle grating and a metallic surface. *Opt. Lett.* **2015**, *30*, 3404. [CrossRef] [PubMed]

14. Zheng, P.; Kasani, S.; Wu, N. Converting plasmonic light scatterinf to confined light absorption and creating plexcitons by coupling a gold nano-pyramid onto a silica-gold film. *Nanoscale Horiz.* **2019**, *4*, 516–525. [CrossRef]

15. Kasani, S.; Curtin, K.; Wu, N. 2D and 3D plasmonic nanostructure array patterns. *Nanophotonics* **2019**, *8*, 2065–2089. [CrossRef]

16. Nemanich, R.J.; Tsai, C.C.; Conell, A.N. Interference-Enhanced Raman Scattering of Very Thin Titanium and Titanium Oxide Films. *Phys. Rev. Lett.* **1980**, *44*, 273–276. [CrossRef]

17. Connell, G.A.N.; Nemanich, R.J.; Tsai, C.C. Interference Enhanced Raman Scattering from Very Thin Absorbing Films. *Appl. Phys. Lett.* **1980**, *36*, 31–33. [CrossRef]

18. Wang, Y.Y.; Ni, Z.H.; Shen, Z.X.; Wang, H.M.; Wu, Y.H.; Wang, Y.Y.; Ni, Z.H.; Shen, Z.X. Interference Enhancement of Raman Signal of Graphene. *Appl. Phys. Lett.* **2008**, *92*, 043121. [CrossRef]

19. Yoon, D.; Moon, H.; Son, Y.-W.; Choi, J.S.; Park, B.H.; Cha, Y.H.; Kim, Y.D.; Cheong, H. Interference Effect on Raman Spectrum of Graphene on SiO_2/Si. *Phys. Rev. B* **2009**, *80*, 125422. [CrossRef]

20. Ni, Z.; Wang, Y.; Yu, T.; Shen, Z. Raman Spectroscopy and Imaging of Graphene. *Nano Res.* **2008**, *1*, 273–291. [CrossRef]

21. Vančo, L.; Kotlár, M.; Kadlečíková, M.; Vretenár, V.; Vojs, M.; Kováč, J. Interference-enhanced Raman scattering in SiO_2/Si structures related to reflectance. *J. Raman Spectrosc.* **2019**, *50*, 1502–1509. [CrossRef]

22. Ramírez-Jiménez, R.; Álvarez-Fraga, L.; Jimenez-Villacorta, F.; Climent-Pascual, E.; Prieto, C.; De Andrés, A. Interference enhanced Raman effect in graphene bubbles. *Carbon N. Y.* **2016**, *105*, 556–565. [CrossRef]

23. Abidi, I.H.; Cagang, A.A.; Tyagi, A.; Riaz, M.A.; Wu, R.; Sun, Q.; Luo, Z. RSC Advances Oxidized Nitinol Substrate for Interference Enhanced Raman Scattering of Monolayer. *RSC Adv.* **2016**, *6*, 7093–7100. [CrossRef]

24. Gao, L.; Ren, W.; Liu, B.; Saito, R.; Wu, Z.; Li, S. Surface and Interference Coenhanced Raman Scattering of Graphene. *ACS Nano* **2009**, *3*, 933–939. [CrossRef] [PubMed]

25. Alvarez-Fraga, L.; Climent-Pascual, E.; Aguilar-Pujol, M.; Ramirez, R.; Jiménez-Villacorta, F.; Prieto, C.; de Andres, A. Efficient heterostructures for combined interference and plasmon resonance Raman amplification. *ACS Appl. Mater. Interfaces* **2017**, *9*, 4119–4125. [CrossRef] [PubMed]

26. Xu, W.; Mao, N.; Zhang, J. Graphene: A Platform for Surface-Enhanced Raman Spectroscopy. *Small* **2013**, *9*, 1206–1224. [CrossRef]

27. Jesion, I.; Skibniewski, M.; Skibniewska, E.; Strupiński, W.; Szulc-Dąbrowska, L.; Krajewska, A.; Pasternak, I.; Kowalczyk, P.; Pińkowski, R. Graphene and Carbon Nanocompounds: Biofunctionalization and Applications in Tissue Engineering. *Biotechnol. Biotechnol. Equip.* **2015**, *29*, 415–422. [CrossRef]

28. Thompson, B.C.; Murray, E.; Wallace, G.G. Graphite Oxide to Graphene. Biomaterials to Bionics. *Adv. Mater.* **2015**, *27*, 7563–7582. [CrossRef]

29. Santos, A.; Ferré-Borrull, J.; Pallarès, J.; Marsal, L.F. Hierarchical nanoporous anodic alumina templates by asymmetric two-step anodization. *Phys. Status Solidi (a)* **2011**, *208*, 668–674. [CrossRef]

30. Kumeria, T.; Santos, A.; Rahman, M.M.; Ferreé-Borrull, J.; Marsal, L.F.; Losic, D. Advanced structural engineering of nanoporous photonic structures: Tailoring nanopore architecture to enhance sensing properties. *ACS Photonics* **2014**, *1*, 1298–1306. [CrossRef]

31. Santos, A.; Formentín, P.; Ferré-Borrull, J.; Pallarès, J.; Marsal, L.F. Nanoporous anodic alumina obtained without protective oxide layer by hard anodization. *Mater. Lett.* **2012**, *67*, 296–299. [CrossRef]

32. Santos, A.; Formentín, P.; Pallarès, J.; Ferré-Borrull, J.; Marsal, L.F. Structural engineering of nanoporous anodic alumina funnels with high aspect ratio. *J. Electroanal. Chem.* **2011**, *655*, 73–78. [CrossRef]

33. Huttel, Y. *Gas-Phase Synthesis of Nanoparticles*; Wiley-VCH Verlag GmbH & Co.: Weinheim, Germany, 2017.

34. Cortijo-Campos, S.; Ramírez-Jiménez, R.; Climent-Pascual, E.; Aguilar-Pujol, M.; Jiménez-Villacorta, F.; Martínez, L.; Jiménez-Riobóo, R.; Prieto, C.; De Andrés, A. Raman amplification in the ultra-small limit of Ag nanoparticles on SiO_2 and graphene: Size and inter-particle distance effects. *Mater. Des.* **2020**, *192*, 108702. [CrossRef]

35. Horcas, I.; Fernández, R.; Gómez-Rodríguez, J.M.; Colchero, J.; Gómez-Herrero, J.; Baró, A.M. WSXM: A software for scanning probe microscopy and a tool for nanotechnology. *Rev. Sci. Instrum.* **2007**, *78*, 013705. [CrossRef]

nanomaterials

MDPI

Article

Noble Metal-Free TiO$_2$-Coated Carbon Nitride Layers for Enhanced Visible Light-Driven Photocatalysis

Bo Zhang, Xiangfeng Peng and Zhao Wang *

National Engineering Research Center of Industry Crystallization Technology,
School of Chemical Engineering and Technology, Tianjin University, Tianjin 300072, China;
yinyan@tjufe.edu.cn (B.Z.); tassadar528@eyou.com (X.P.)
* Correspondence: wangzhao@tju.edu.cn; Tel.: +86-138-2052-5018

Received: 2 April 2020; Accepted: 21 April 2020; Published: 23 April 2020

Abstract: Composites of g-C$_3$N$_4$/TiO$_2$ were one-step prepared using electron impact with dielectric barrier discharge (DBD) plasma as the electron source. Due to the low operation temperature, TiO$_2$ by the plasma method shows higher specific surface area and smaller particle size than that prepared via conventional calcination. Most interestingly, electron impact produces more oxygen vacancy on TiO$_2$, which facilitates the recombination and formation of heterostructure of g-C$_3$N$_4$/TiO$_2$. The composites have higher light absorption capacity and lower charge recombination efficiency. g-C$_3$N$_4$/TiO$_2$ by plasma can produce hydrogen at a rate of 219.9 μmol·g^{-1}·h^{-1} and completely degrade Rhodamine B (20mg·L^{-1}) in two hours. Its hydrogen production rates were 3 and 1.5 times higher than that by calcination and pure g-C$_3$N$_4$, respectively. Electron impact, ozone and oxygen radical also play key roles in plasma preparation. Plasma has unique advantages in metal oxides defect engineering and the preparation of heterostructured composites with prospective applications as photocatalysts for pollutant degradation and water splitting.

Keywords: dielectric barrier discharge plasma; oxygen vacancy; g-C$_3$N$_4$/TiO$_2$; photodegradation; H$_2$ evolution

---

1. Introduction

Photocatalytic technology, a promising strategy for addressing energy shortages and environmental pollution, is important for the production of hydrogen via water splitting and the degradation of organic pollutants [1–4]. TiO$_2$, discovered by Fujishima in 1972 [5], is the most widely studied and applied semiconductor photocatalyst and is non-toxic, stable, and cheap [6–8]. Many preparation methods of TiO$_2$ and TiO$_2$ composite have been used, such as the sol-gel method [9], solvothermal method [10], and chemical vapor deposition method [11] and so on. However, a green, simple, cheap and energy-efficient way for catalyst preparation is still necessary.

Non-thermal plasma, which has relatively low bulk temperature and extremely high electron temperature, has excellent advantages in preparing catalysts [12–14]. The catalyst can be prepared quickly without serious agglomeration due to the low temperature and high energy of plasma. On the other hand, the nucleation and crystallization process of the catalyst is very unique in plasma [13]. Furthermore, the catalysts prepared using the plasma method have small particles, strong interaction and specific structures [15,16]. However, TiO$_2$ alone can only absorb ultraviolet light (only 4% of solar energy), even if the light absorption properties of TiO$_2$ are improved. In addition, higher photo-generated charge recombination efficiency also affects its photocatalytic activity [17].

Strategies have been proposed to increase the photocatalytic activity of TiO$_2$ under visible light. For example, compounding TiO$_2$ with a narrow bandgap semiconductor catalyst can enhance its absorption of visible light and construct a special heterostructure [18,19]. Doping elements can reduce the TiO$_2$ bandgap and increase its light absorption range [20,21]. Loading noble metals to TiO$_2$ as the co-catalyst can act as its active site to enhance photocatalytic activity [22,23]. TiO_2-coated carbon nitride layers as composites for enhanced photocatalytic activity is a more attractive approach. The g-C_3N_4 is widely used in the degradation of organic pollutants and water splitting by visible light irradiation due to its strong visible light response, high thermal resistance and chemical stability [24–27]. g-C_3N_4/TiO_2 composites can not only transfer the photo-generated charge of g-C_3N_4 to TiO_2 to increase its charge separation efficiency, but also reduce bandgap to increase its visible light absorption region [28–30]. Ma et al. got highly photocatalytic water splitting performance with g-C_3N_4/TiO_2 composites by solvothermal method under visible light [10]. Papailias et al. utilized high temperature calcination to synthesize g-C_3N_4/TiO_2 nanocomposites for NO_x removal [31].

In this work, g-C_3N_4/TiO_2 composites were prepared by using dielectric barrier discharge plasma, and their photocatalytic activities were evaluated by degrading RhB and hydrogen evolution under visible light irradiation. Due to the characteristics of the plasma preparation method, g-C_3N_4/TiO_2 composites have many different properties compared to that by the traditional calcination method. Meanwhile, the mechanism of photocatalytic process and dielectric barrier discharge plasma preparation was proposed, respectively. It is predictable that the plasma method for catalysts preparation will be a very promising field.

2. Materials and Methods

2.1. Materials

Tetrabutyl titanate (TBT), melamine, absolute ethanol, rhodamine B (RhB), and triethanolamine were purchased from Shanghai Aladdin Biochemical Technology Co. (Shanghai, China). All chemicals were used directly.

Synthesis of TiO_2 and g-C_3N_4/TiO_2

g-C_3N_4 was dispersed by ultrasound in 20 mL of ethanol. We added 3ml TBT slowly into the as-prepared g-C_3N_4 suspension under adequate stirring. Then, the as-prepared hybrid suspensions were stand still for 24 h. The sample was filtered after 24 h. and drying at 80 °C. The samples were divided into two portions.

One portion was treated by DBD plasma to obtain g-C_3N_4/TiO_2 composites according to the procedure reported in previous work [32,33]. Figure S1a shows the DBD device. The plasma was generated by the high voltage generator that can provide a sinusoidal waveform at 22 kHz with a voltage range of 0 to 30 kV. There are two quartz plates and a quartz ring between the two electrodes, and the sample was placed in the quartz ring. The gas atmosphere of the DBD reactor was air. The average power and average voltage of DBD during catalyst preparation were 200 W and 100 V, respectively. One-time plasma operation proceeded for 3 min to restrict the heat effect, followed by manually stirring to expose the untreated samples outside. The operation was repeated 20 times, until total plasma treatment time was 1 h. As shown in Figure S1b, the infrared (IR) image taken by the IR camera (Ircon, 100PHT, Everett, WA, USA) shown that the temperature of DBD plasma was below 106 °C. Finally, the obtained samples were denoted as TCNX-D (X = 10, 30, 50, 70, 90). X is equal to the weight ratio of g-C_3N_4 in the composites. D represents the samples prepared by DBD plasma. Figure S2 shows the schematic illustration of preparation of g-C_3N_4/TiO_2 composite.

For comparison, another portion of the sample was calcined 450 °C at a rate of 5 °C/min for 2 h and designated as TCN50-C (50 wt% g-C_3N_4/TiO_2).

The preparation method of pure TiO_2 is the same as the above method. The samples obtained were denoted as TiO_2-D and TiO_2-C according to the different preparation methods.

2.2. Characterization

A Rigaku D/Max-2500 V/PC diffractometer with Cu Kα1 radiation (Cu Kα1 α = 0.154 nm, 40 kV, 40 mA, 8°·min^{-1}, Rigaku, Tokyo, Japan) was used to analyze the crystal phase of samples. Fourier transform infrared (FTIR) spectra was analyzed by a Bruker Alpha FTIR-attenuated total reflection (ATR) instrument (Bruker, Karlsruhe, Germany). A Biaode SSA-7000 analyzer (Biaode Electronic Technology Ltd., Beijing, China) was applied to determine specific surface area and pore size. Thermogravimetry analysis was conducted on Perkin-Elmer TGA/DTA thermo-gravimetric analyzer (Waltham, MA, USA) in O_2 atmosphere. The temperature range was 20–800 °C. An ULVAC-PHI-5000versaprobe instrument (Tokyo, Japan) was used to obtain X-ray photoelectron spectra. The adventitious carbon C1s used for element correction is located at 284.8 eV. Electron paramagnetic resonance measurement was obtained on a Bruker A300 spectrometer (Karlsruhe, Germany) at room temperature. The ultraviolet–visible (UV–vis) diffuse-reflectance spectroscopy was recorded by a UV-2600 spectrophotometer (Shimadzu, Kyoto, Japan). Photoluminescence (PL) spectroscopy was analyzed on Hitachi F-4600 spectrometer (Tokyo, Japan) with excitation at 350 nm. Scanning electron microscopy (SEM) and energy dispersive X-ray spectroscopy were conducted on Hitachi S4800 instrument (Tokyo, Japan). High-resolution transmission electron microscopy (HRTEM) was obtained using a JEM-2100F instrument (JEOL, Tokyo, Japan).

2.3. Measurement of Photocatalytic Activity

2.3.1. RhB degradation

We mixed 50 mg samples and 100 mL RhB (20 mg·L^{-1}) solution uniformly in continuous stirring. First the mixture was stirred without light for 0.5 h to obtain the balance of adsorption and desorption. Visible light was provided by Xenon lamp (HSX-F300, 300 W) with a 420 nm cut filter. The illumination power was 100 mW/cm^2. The light source was placed about 10 cm above the RhB solution. In the experiment, 4 mL solution was taken every 20 min for RhB degradation rate test. The degradation rate of the obtained RhB solution was determined using a UV-2600 instrument.

2.3.2. Hydrogen Generation

The H_2 evolution experiment was conducted on the PerfectLight 3AG instrument (Beijing, China). The amount of H_2 was determined by online gas chromatography. 100 mg sample was added to 70 mL of solution containing 10% triethanolamine and 133 μL of H_2PtCl_6 (0.5 wt%). The mixed solution was placed in 100 mL closed glass container. The 300 W xenon lamp was equipped with total reflection and 420 nm filter to provide full-spectrum light and visible light, respectively. The light source was placed 10 cm above the suspension. After full-spectrum irradiation for 3 h in vacuum, the Pt was loaded on the samples. Then the hydrogen production reaction started, and the content of H_2 in the system was measured by online gas chromatography per hour.

2.3.3. Photoelectrochemical Investigation

The photocurrent density was conducted at the LK2010 electrochemical system. It is a traditional three-electrode system, tin oxide mixed with fluorine (FTO) conductive glass loaded with sample as working electrode. The working electrode was made as follows: 50 mg of the sample and 10 μL of nafion (5%) were added to 500 μL absolute ethanol, 200 μL of the suspension was loaded on FTO glass after mixing evenly. Then the FTO working electrode was obtained by drying at room temperature for 12 h. The reference electrode and counter electrode were the saturated Hg/HgO electrode and Pt wire, respectively. The 0.1M Na_2SO_4 solution was used as electrolyte. Visible light was provided by 300 W Xe lamp equipped with 420 nm filter. At the beginning of the test, we fully introduced N_2 into the electrolyte to remove dissolved oxygen. The experimental potential is 0.4 eV, which is the optimal value by testing (see Figure S3).

3. Results and Discussion

3.1. Physico-Chemical Properties

The crystal phases of g-C$_3$N$_4$, TiO$_2$, and g-C$_3$N$_4$/TiO$_2$ composites were analyzed by using wide-angle X-ray diffraction (XRD). As shown in Figure 1a, g-C$_3$N$_4$ exhibits (100) and (002) planes at 13.2° and 27.6°, which correspond to tri-s-triazine structure and stacking of the conjugated aromatic system, respectively [34]. The as-prepared TiO$_2$ present the diffraction peaks at 25.1°, 37.7°, 47.8°, 54.0°, 55.1° and 62.5°, which corresponded to the (101), (004), (200), (105), (211) and (204) crystal planes of anatase TiO$_2$ (JCPDS 71-1166), respectively [35]. It indicated TiO$_2$ was successfully prepared by DBD plasma. As the proportion of g-C$_3$N$_4$ increased in g-C$_3$N$_4$/TiO$_2$ composites, the intensities of g-C$_3$N$_4$ peaks increased gradually. However, the no peak position shift of TiO$_2$ indicates that g-C$_3$N$_4$ has no influence on the crystal structure of TiO$_2$. Moreover, compared with samples (TiO$_2$ and g-C$_3$N$_4$/TiO$_2$ composites) prepared by high-temperature calcination, the samples by DBD plasma have lower peak intensity and wider peak width. On the basis of the full width at half maximum (FWHM) of (101) crystal plane and Debye–Scherrer equation, the average crystalline sizes of TiO$_2$ in the g-C$_3$N$_4$/TiO$_2$ composites were calculated and are listed in Table 1 [36]. The crystalline sizes of TiO$_2$ in the composites decreased gradually as g-C$_3$N$_4$ increased. It is worth noting that TiO$_2$ has smaller crystallite size than that by the calcination method. The relative low temperature of plasma is the main reason restricting the agglomeration of particles.

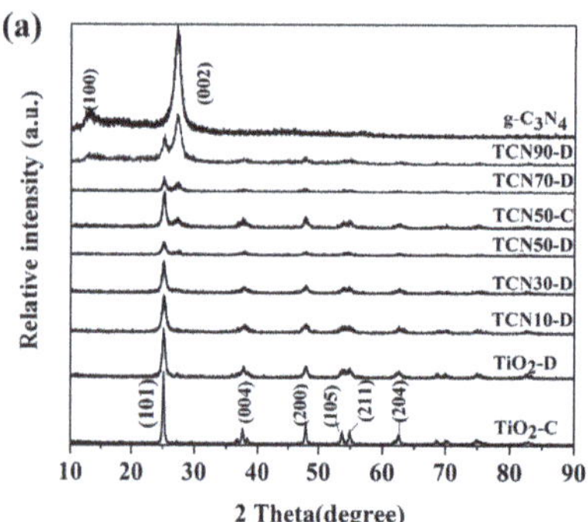
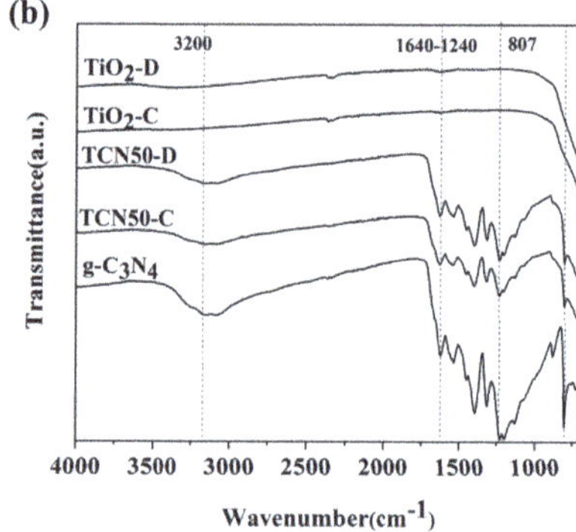

Figure 1. (**a**) XRD patterns of TiO$_2$, g-C$_3$N$_4$ and TCNX samples, (**b**) FTIR spectra of TiO$_2$, g-C$_3$N$_4$, TCN50 samples.

Table 1. Surface and structural characterization of TiO$_2$, g-C$_3$N$_4$ and TCNX composites.

Sample	S_{BET} (m^2/g)	Pore Volume (cm^3/g)	Average Pore Radius (nm)	Crystallite Size (nm)
TiO$_2$-D	64.5649	0.1566	2.99	14.3
TiO$_2$-C	28.1876	0.1023	6.79	17.8
TCN50-D	72.8473	0.1451	2.86	12.3
TCN50-C	29.2735	0.1117	6.02	14.4
g-C$_3$N$_4$	7.1313	0.0604	11.34	-

Fourier transform infrared spectra (FTIR) of g-C$_3$N$_4$, TiO$_2$, and g-C$_3$N$_4$/TiO$_2$ composites are shown in Figure 1b. It reveals the composition and chemical bonding of samples. For pure g-C$_3$N$_4$, the peak of 807 cm^{-1} is due to the tri-s-triazine unit structure, and four intense bands in the region 1240–1640 cm^{-1} are attributed to the stretching of the C–N heterocycle in g-C$_3$N$_4$ [37]. For TiO$_2$-C and TiO$_2$-D, they showed similar peaks, and the shoulder bands between 3000–3500 cm^{-1} can be attributed to –OH stretching vibration [38]. The shoulder bands near 3200 cm^{-1} in TCN50 composites are contributed by the N-H stretching vibration modes [37,39]. TCN50 composite exhibited both the g-C_3N_4 and TiO_2 characteristic peaks. This shows that g-C_3N_4/TiO_2 composites were directly prepared by DBD plasma and is agreement with XRD analysis results. During plasma process, high energy electron bombard samples. It can be deduced that the surface hydroxyl and amino groups enhance interaction between TiO_2 and g-C_3N_4 [40]. The strong interfacial connection can be used as the channel for charge conduction to improve charge separation efficiency.

Nitrogen adsorption–desorption isotherms and pore-size distribution curves of g-C_3N_4, TiO_2 and TCN50 samples are shown in Figure S4. It is clearly observed that samples prepared by plasma have larger N_2 adsorption capacity. The specific data are listed in Table 1. The specific surface area of TiO_2-D (64.56 m^2/g) is approximately 2.3 times that of TiO_2-C (28.18 m^2/g), which can be ascribed to the low agglomeration of the sample's particles prepared by low-temperature plasma. The specific surface area of g-C_3N_4 merely is 7.13 m^2/g. As shown in Table S1, the specific surface area and pore volume of g-C_3N_4/TiO_2 composites decreased and the average pore radius increased gradually as the proportion of g-C_3N_4 increased. Compared with TCN50-C, TCN50-D has higher specific surface area, pore volume, and smaller average pore radius, which indicates that the surface of the plasma-treated sample can expose more active sites. In addition to the low temperature, the repulsion between the electrons attached to the particles prevents the agglomerating of particles [12], resulting in the excellent dispersion of TiO_2 on g-C_3N_4 to enhance photocatalytic activity.

Thermogravimetry (TG) was used to analyze thermal stability of the samples. As shown in Figure S5, the weight of TiO_2 prepared by DBD plasma only reduced by 2%, which indicated that DBD plasma can complete decompose TBT to TiO_2 at moderate temperature. The g-C_3N_4 gradually lost weight from 550 °C to 730 °C. The composites TCNX gradually lost weight from 520 °C to 650 °C, mainly due to the burning of g-C_3N_4. The thermal stability of g-C_3N_4 was reduced after coating with TiO_2. The reason is catalytic action of TiO_2 and the cross-linking ring of g-C_3N_4 after compounding [41]. Ignoring the slight weight loss due to water, the actual proportions of g-C_3N_4 in the composites TCNX-D ($X = 10, 30, 50, 50, 70, 90$) were 13.0, 30.3, 52.9, 71.3, and 90.0 wt%, respectively. The proportion of g-C_3N_4 in TCN50-C was 53.6%, which was approximately equal to the ideal ratio of TCN50-D.

3.2. Characterizations of Oxygen Vacancies

Chemical states of elements in g-C_3N_4, TiO_2, and g-C_3N_4/TiO_2 composites were investigated by X-ray photoelectron spectroscopy (XPS). Figure S6a shows the high-resolution C 1s spectrum of the prepared samples. All samples have two C 1s peaks at 284.8 and 288.3 eV, which can be ascribed to the inherent adventitious carbon and N-C-N coordination, respectively [42]. Compared with the N-C-N peak of pure g-C_3N_4, TCN50 exhibit weaker peak intensity and positive shifts of 0.4 eV in binding energies, which indicates there is a chemical interaction between TiO_2 and g-C_3N_4, hence, g-C_3N_4 has close surface contact with TiO_2 [43]. As shown in Figure S5b, for the N 1 s high-resolution spectrum of pure g-C_3N_4 and g-C_3N_4/TiO_2 composites, three peaks were observed at about 398.8, 399.6, and 401.2 eV. The first peak is attributed to sp2 hybridized nitrogen (C=N–C), the second peak is due to the tertiary N in N–(C)^3 groups, and the last peak corresponds to the existence of amino groups (C–N–H) [44,45].

Figure 2a,b shows O 1s and Ti 2p spectrum region of pure TiO_2. Figure 2a shows the O 1s spectrum region of TiO_2-D have three peaks at about 529.7, 532.0 and 533.4 eV, which corresponds to the lattice oxygen, the oxygen vacancy and the adsorbed oxygen, respectively [46,47]. By calculating the ratio of the area occupied by oxygen vacancies, it was 23.4% and 18.6% in TiO_2-D and TiO_2-C, respectively, indicating that plasma method can produce more oxygen vacancies than that of the calcination method in TiO_2. Figure 2b shows the Ti 2p spectrum of pure TiO_2, There are four peaks at about 458.2, 463.7, 458.9, and 464.6 eV, corresponding to Ti^{3+} $2p_{3/2}$, Ti^{3+} $2p_{1/2}$, Ti^{4+} $2p_{3/2}$ and Ti^{4+} $2p_{1/2}$, respectively [46,47]. It was found that the area of Ti^{3+} in TiO_2-D and TiO_2-C were 15.6% and 11.1%, respectively. Compared with TiO_2 by calcination, the valence of more Ti in TiO_2 by plasma is reduced from Ti^{4+} to Ti^{3+}. This is consistent with the observation for the oxygen vacancy. During the plasma preparation of TiO_2, the oxygen atoms escaped to form oxygen vacancies and trivalent Ti [46].

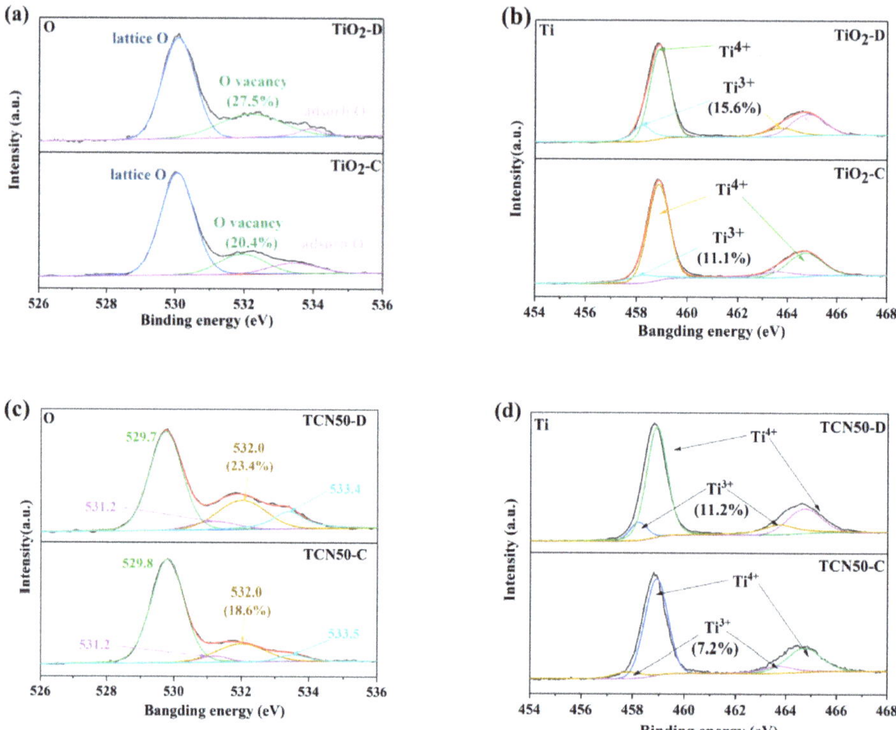

Figure 2. The high-resolution XPS spectra: (**a**) O 1s, (**b**) Ti 2p, of TiO_2 samples. (**c**) O 1s, (**d**) Ti 2p of TCN50 samples.

Figure 2c,d shows O 1s and Ti 2p spectrum of TCN50 composites. TCN50 and TiO_2 show similar peaks at O 1s and Ti 2p. However, the intensity of the peaks representing oxygen vacancies and Ti^{3+} decreased. The ratio of oxygen vacancies and Ti^{3+} were 23.4% and 11.2% in TCN50-D, 18.6% and 7.2% in TCN50-C, respectively. It is indicated that g-C_3N_4 occupies the oxygen vacancies of TiO_2 after TiO_2 coated on g-C_3N_4. This will undoubtedly strengthen the interaction of TiO_2 and g-C_3N_4. It is in agreement with previous results.

To further verify the existence of oxygen vacancies, electron paramagnetic resonance (EPR) analysis of the TCN50 composites was performed. As shown in Figure 3a, both TCN50-C and TCN50-D have the EPR signal at g = 2.003, which means the appearance of oxygen vacancies [48]. The peak intensity of TCN50-D is stronger than TCN50-C, indicating the plasma method produces more oxygen vacancies. As can be seen from Figure S7, the EPR spectra of TiO_2-C and D exhibit the same characteristics as that of TCN50. This is consistent with XPS results. In addition, Figure 3b shows different colors between TiO_2-D and TiO_2-C. The color of TiO_2-C is white, while TiO_2-D is gray. This indicates oxygen vacancies can narrow band gap to promote light harvesting, leading significant color change [49].

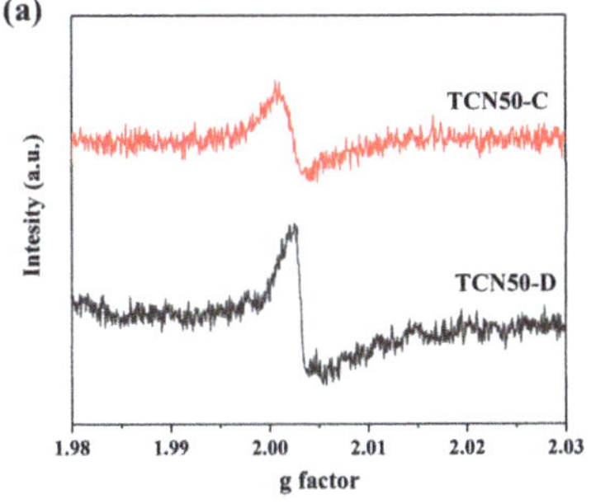

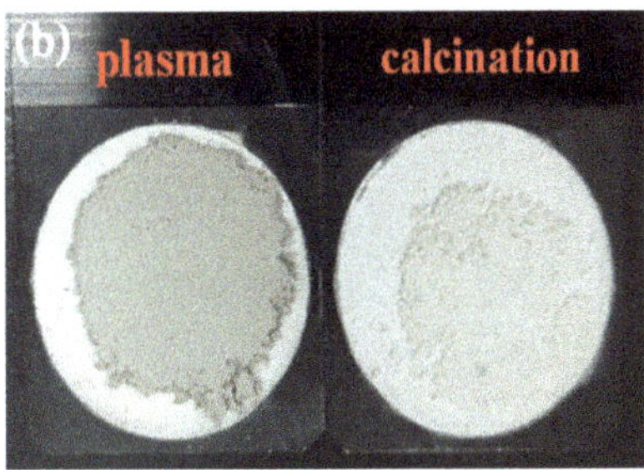

Figure 3. (**a**) EPR spectra of TCN50-C and TCN50-D samples, (**b**) images of TiO_2 prepared by plasma and calcination.

3.3. Optical Properties

Ultraviolet–visible diffuse reflection spectrum (UV-vis DRS) is shown in Figure 4a and Figure S8 reveal the light absorption capacity and bandgaps of the as-prepared g-C_3N_4, TiO_2, and TCNX. The critical values of the light response of TiO_2-D and TiO_2-C were found at 398 nm and 388 nm, indicating that bandgaps were 3.12 and 3.19 eV, respectively [50]. The absorption starting point of g-C_3N_4 was located at 458 nm. The absorption range of visible light of TCN50 composites was significantly expanded due to introducing g-C_3N_4. As shown in Figure 4b, on the basis of the Kubelka-Munk formula, the bandgaps of g-C_3N_4, TCN50-C, and TCN50-D were calculated to be 2.71, 2.70, and 2.63 eV, respectively. Based on the above results, the samples prepared by plasma had a narrower bandgap. This can be attributed to oxygen vacancies [51]. Oxygen vacancies can introduce a defect status below the conduction band, and thus narrows the bandgap to improve the light absorption range [46,52]. Hence, TCN50-D has a relatively strong visible light response capacity.

Photoluminescence (PL) spectra was shown in Figure 4c. The PL peak intensity is proportional to the recombination rates of the photogenerated electron-hole pairs [53]. g-C_3N_4 has strong peak intensity, indicating that its charge separation efficiency is low. TiO_2-D has weaker PL peak strength than that of TiO_2-C, indicating that TiO_2-D is more conducive to charge transport. After TiO_2 coated on g-C_3N_4, PL peak intensity was greatly reduced, because electrons can be transmitted through the interface of g-C_3N_4 and TiO_2 [54]. TCN50-D shows lower peak intensity than TCN50-C, resulting in that TCN50-D has lower charge recombination rate and stronger photocatalytic activity. Meanwhile, the presence of oxygen vacancies can promote the separation of electrons and holes [55]. The oxygen vacancies not only restrain the recombination of charges, but also narrow the band gap to increase light absorption [52,56]. As shown in Figure 4d, TCN50-D exhibits the highest photocurrent density than TCN50-C, indicating that TCN50-D has higher photoelectric conversion efficiency [57].

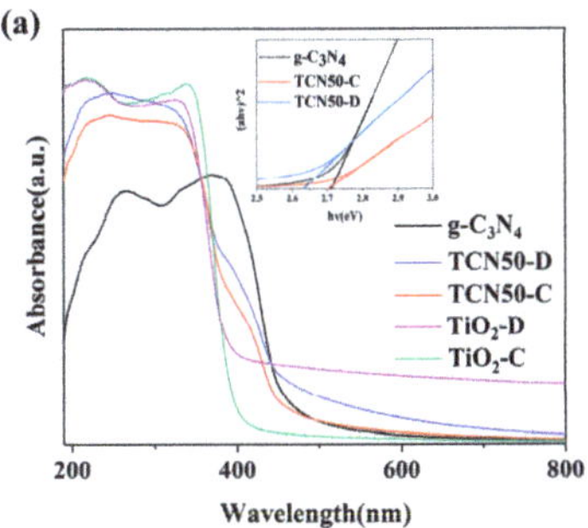

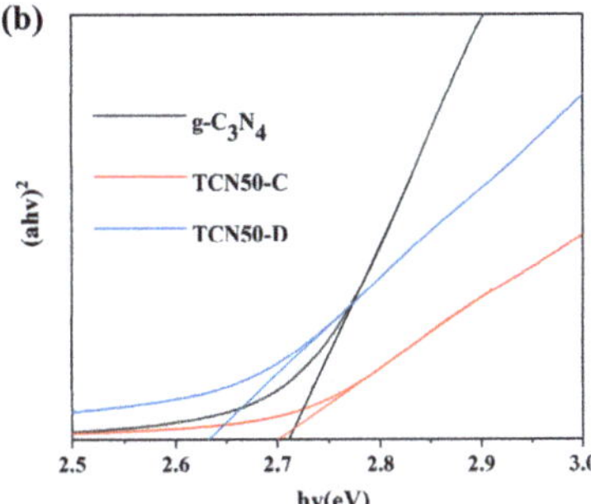

Figure 4. *Cont.*

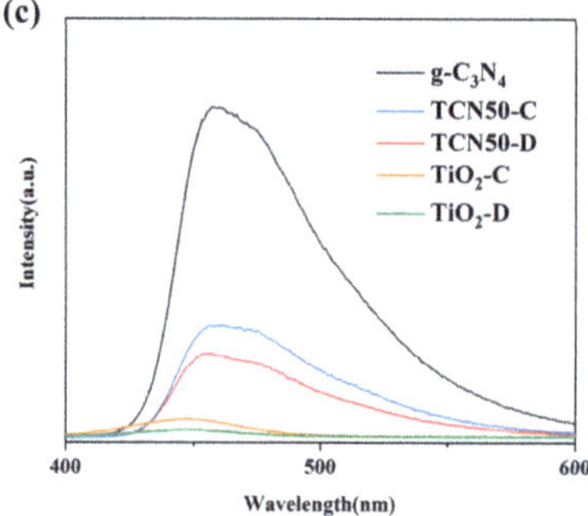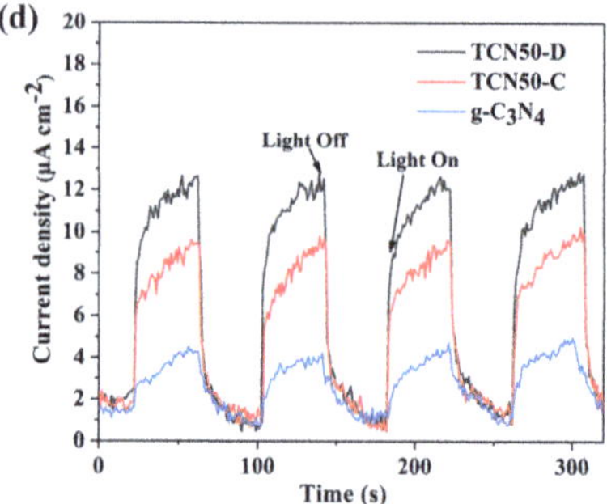

Figure 4. (**a**) UV–vis DRS of TiO$_2$, g-C$_3$N$_4$, TCN50-C and TCN50-D samples. (**b**) The relationship between (ahv)^2 and photo energy. (**c**) Photoluminescence spectra of g-C$_3$N$_4$, TiO$_2$ and TCN50 samples. (**d**) Photocurrent density vs. time for g-C$_3$N$_4$/FTO, TCN50-C/FTO and TCN50-D/FTO.

3.4. Morphologies

Figure 5 shows the morphology and microstructure of TiO$_2$, g-C$_3$N$_4$ and g-C$_3$N$_4$/TiO$_2$ composites. As shown in Figure 5a,b, TiO$_2$-D has less particle aggregation than TiO$_2$-C, which agreed well with previous result. As shown in Figure 5c,d, g-C$_3$N$_4$ has an anomalous layered structure and smooth flat surface. After TiO$_2$ coating process by plasma, TiO$_2$ was uniformly dispersed on the surface of g-C$_3$N$_4$, and the surface of g-C$_3$N$_4$ changes from smooth to rough. Figure 5e and Figure S9 show the energy dispersive X-ray spectroscopy (EDS) mapping of TCN50-D, the four elements (C, N, O, Ti) are well-dispersed in the TCN50-D composite. It can form a heterostructure for electron transport and reduce charge recombination efficiency.

To further investigate the microstructure of g-C$_3$N$_4$/TiO$_2$ composites, HRTEM was conducted. As can be seen from Figure 5f,g, the particle size of TiO$_2$ in TCN50-C composite is around 18 nm, that in the TCN50-D composite is around 12 nm. This proves that plasma treated sample has smaller particle size. The lattice spacing of TiO$_2$ is 0.35 nm representing the (101) plane of anatase titanium oxide. Figure 5h also shows the clear lattice fringe of g-C$_3$N$_4$ is 0.32 nm, corresponding to the (002) lattice plane of g-C$_3$N$_4$ [54]. Compared with TCN50-C composite, TCN50-D have more and tighter heterojunctions, which also means extensive interfacial contacts to enhance photocatalytic activity.

3.5. Photocatalytic Activity

RhB degradation and H$_2$ production under visible light were conducted to evaluate the photocatalytic performance of the samples. As shown in Figure 6a, the concentration of RhB hardly changed within 0.5 h of dark adsorption. For TiO$_2$-C and TiO$_2$-D, RhB was degraded by 14.3% and 33.7% under two hours of the irradiation of visible light, respectively, which can be ascribed to the sensitization of dye [58]. Plasma-treated TiO$_2$ show better photoactivity. Pure g-C$_3$N$_4$ degraded 63.7% of RhB under two hours of visible light irradiation. With the increase of proportion of g-C$_3$N$_4$ in g-C$_3$N$_4$/TiO$_2$, the degradation efficiency increased first and then decreased; 50 wt% g-C$_3$N$_4$ in g-C$_3$N$_4$/TiO$_2$ composite exhibited the highest RhB degradation efficiency. When the amount of g-C$_3$N$_4$ is low, TiO$_2$ covers its surface and prevents absorption of visible light. However, when the amount of TiO$_2$ is low, it is insufficient to promote charge separation. Therefore, the appropriate ratio of g-C$_3$N$_4$/TiO$_2$ has higher photocatalytic activity. Compared with 79.7% RhB degradation efficiency of TCN50-C, TCN50-D can completely degrade RhB in two hours, due to its stronger visible light absorption and charge separation efficiency.

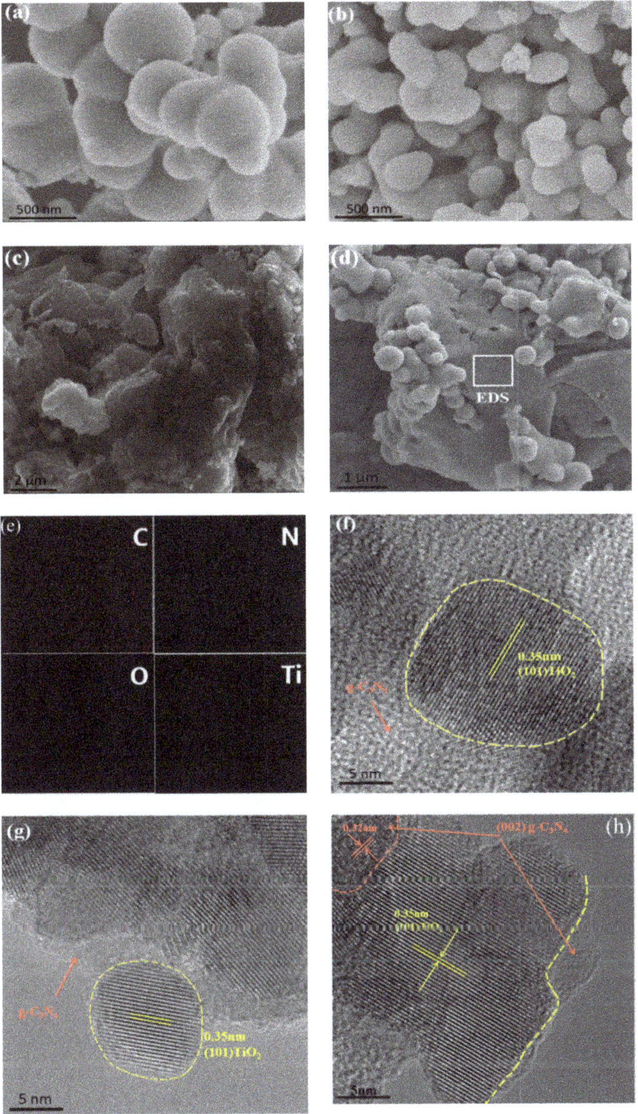

Figure 5. SEM images of (**a**) TiO$_2$-C, (**b**) TiO$_2$-D, (**c**) g-C$_3$N$_4$, (**d**) TCN50-D and (**e**) EDS elemental mappings of TCN50-D samples; high-resolution transmission electron microscopy (HRTEM) images of (**f**) TCN50-C sample, (**g,h**) TCN50-D sample.

Furthermore, Figure 6b shows the reusability experiment of the TCN50-D sample. The photocatalytic activity for degradation of RhB decreased from 95.5% to 90.2% within 6 h under visible light after the third cycle. This indicated that the g-C$_3$N$_4$/TiO$_2$ composite by plasma has high stability.

As shown in Figure 6c, the hydrogen production of pure TiO$_2$ was not detected in the system. Pure g-C$_3$N$_4$ only produced 72.1 μmol·g^{-1} H$_2$ per hour. The hydrogen production of TCN50-D and TCN50-C composites was 219.9 and 174.3 μmol·g^{-1}·h^{-1}, respectively. TCN50-D exhibited higher photocatalytic activity and approximately 3 times that of pure g-C$_3$N$_4$ and 1.26 times that of TCN50-C.

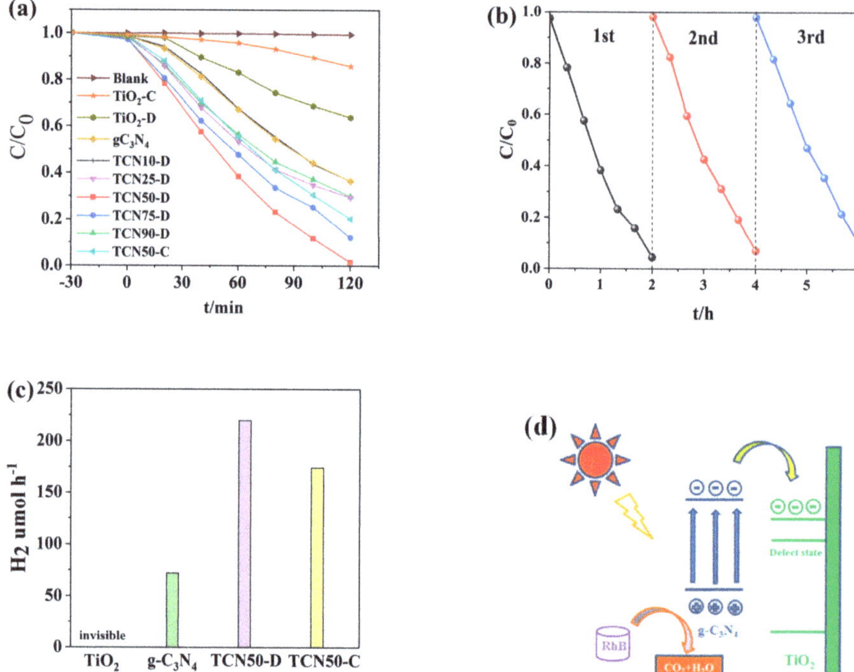

Figure 6. (**a**) Photocatalytic degradation of RhB under visible light irradiation. (**b**) The photodegradation stability of RhB over TCN50-D sample. (**c**) Photocatalytic H$_2$ evolution rates of TiO$_2$, g-C$_3$N$_4$, TCN50-C and TCN50-D under visible light irradiation. (**d**) Photocatalytic mechanism for the charge transfer between g-C$_3$N$_4$ and TiO$_2$ under visible light irradiation.

3.6. Mechanism

In order to investigate the phtotcatalytic degradation mechanism of the TCN50-D composite, different scavengers were added before RhB degraded by TCN50-D. As shown in the Figure S10, the degradation activity of TCN50-D changed slightly after the addition of the hydroxyl radical scavenger isopropyl alcohol (IPA) and the superoxide radical scavenger p-benzoquinone (BQ). However, after the addition of the hole scavenger Ethylene Diamine Tetraacetic Acid-2Na (EDTA-2Na), a significant decrease in photocatalytic activity for RhB degradation was observed, indicating that the hole is the main active species for degrading RhB. By measuring the intersection of the slope of the XPS valence band curve and the X axis, the valence band (VB) of the samples can be determined. As shown in Figure S11, the VB of TCN50-C and TCN50-D is 2.32 eV and 2.38 eV, respectively, indicating that TCN50-D possesses stronger oxidizing ability [40]. The VB of TiO$_2$-D and TiO$_2$-C is 2.91 eV and 2.84 eV, respectively. The conduction band (CB) of TiO$_2$-D can be calculated that is −0.20 eV, according to its band gap of 3.11 eV. The VB and CB of g-C$_3$N$_4$ are 2.24 eV and −0.47 eV, respectively [40]. Therefore, the energy band structure model of TCN50-D can be constructed and shown in Figure 6d. Under visible light irradiation, g-C$_3$N$_4$ captures photons to generate electron-hole pairs, then electrons can be transported from the CB of g-C$_3$N$_4$ to the CB of TiO$_2$-D to reduce H$^+$ to produce H$_2$. Meanwhile, the holes in the VB of g-C$_3$N$_4$ can oxidize RhB to exert photodegradation [59].

The mechanism of TiO$_2$ and g-C$_3$N$_4$/TiO$_2$ composite prepared by dielectric barrier discharge plasma is proposed as follows. The high-energy electron bombardment plays a key role in the process.

During the plasma process, there are a large number of electrons that can decompose titanium hydroxide through breaking bonds by electron bombardment (Equation (1)). This also can produce oxygen vacancies.

$$Ti(OH)_4 + e \rightarrow TiO_2 + H_2O \tag{1}$$

Under the influence of plasma, the ozone and oxygen atoms can be generated by using Equations (2)–(4) [16,60].

$$e + O_2 \rightarrow e + O(^3P) + O(^3P) \tag{2}$$

$$e + O_2 \rightarrow e + O(^1D) + O(^3P) \tag{3}$$

$$O + O_2 + M \rightarrow O^{3*} + M \rightarrow O^3 + M \tag{4}$$

As shown in Equations (5) and (6), the oxygen atom and ozone can also lead to $Ti(OH)_4$ decomposition.

$$Ti(OH)_4 + O \rightarrow TiO_2 + 2H_2O + 1/2O_2 \tag{5}$$

$$Ti(OH)_4 + O_3 \rightarrow TiO_2 + 2H_2O + 3/2O_2 \tag{6}$$

4. Conclusions

TiO_2 and g-C_3N_4/TiO_2 composites were prepared by DBD plasma, which is a green, easy, and efficient method for catalyst preparation. Here, TiO_2 by plasma has enriched oxygen vacancies and larger specific surface area due to the electron impact. The electron impact, ozone and oxygen radical play important role in plasma preparation, which facilitate the interaction of TiO_2 and g-C_3N_4 and forms heterojunctions. g-C_3N_4/TiO_2 composites by plasma have stronger light absorption capacity and higher charge separation efficiency. TCN50-D exhibited the highest photocatalysis activity on RhB degradation and hydrogen production, which was ascribed to enriched oxygen vacancies and special heterostructures. Plasma, green and convenient technology, provides a promising strategy for oxide defect engineering and composite preparation.

Supplementary Materials: The following are available online at http://www.mdpi.com/2079-4991/10/4/805/s1: Figure S1: The dielectric barrier discharge (DBD) plasma reactor and infrared (IR) image of DBD plasma, Figure S2: The schematic illustration for the preparation of TCN composites photocatalysts, Figure S3: Photocurrent tests of TCN50-D at different potentials, Figure S4: Nitrogen adsorption–desorption isotherms (a) and pore-size distribution curves (b) of g-C_3N_4, TiO_2 and TCN50 samples, Figure S5: Thermogravimetric analysis curves, Figure S6: The high-resolution XPS spectra C 1s and N 1s, Figure S7: The EPR spectra of TiO_2-C and D samples, Figure S8: FTIR spectra, UV–vis DRS and XPS valence band (VB) spectra of TCNX, Figure S9: EDS patterns and element mapping of TCN50-D sample, Figure S10: Trapping experimental of photogenerated radicals and holes in TCN50-D sample for the RhB degradation, Figure S11: VB XPS of TiO_2-C, TiO_2-D, TCN50-C and TCN50-D, Table S1: Surface and structural characterization of TiO_2, g-C_3N_4 and TCNX composites.

Author Contributions: Z.W. conceived the experiments. B.Z. conducted the preparation and photocatalysis experiment. X.P. set up the experimental device. B.Z. and Z.W. wrote the original draft of the manuscript. All authors contributed to the discussion of the results. All authors have read and agreed to the published version of the manuscript.

Funding: This work was supported by National Key Research and Development Program of China (No.2016YFF0102503), National Natural Science Foundation of China (No.21878214) and State Key Laboratory of Efficient Utilization for Low Grade Phosphate Rock and Its Associated Resources (No.WFKF2019-03).

Conflicts of Interest: The authors declare no conflict of interest.

References

1. Chava, R.K.; Son, N.; Kim, Y.S.; Kang, M. Controlled Growth and Bandstructure Properties of One Dimensional Cadmium Sulfide Nanorods for Visible Photocatalytic Hydrogen Evolution Reaction. *Nanomaterials* **2020**, *10*, 619. [CrossRef] [PubMed]
2. Pan, Y.; You, Y.; Xin, S.; Li, Y.; Fu, G.; Cui, Z.; Men, Y.; Cao, F.; Yu, S.; Goodenough, J.B. Photocatalytic CO_2 reduction by carbon-coated indium-oxide nanobelts. *J. Am. Chem. Soc.* **2017**, *139*, 4123–4129. [CrossRef] [PubMed]

3. Wang, Z.; Li, C.; Domen, K. Recent developments in heterogeneous photocatalysts for solar-driven overall water splitting. *Chem. Soc. Rev.* **2019**, *48*, 2109–2125. [CrossRef] [PubMed]

4. Lin, L.; Yu, Z.; Wang, X. Crystalline carbon nitride semiconductors for photocatalytic water splitting. *Angew. Chem. Int. Ed.* **2019**, *131*, 6225–6236. [CrossRef]

5. Fujishima, A.; Honda, K. Electrochemical photolysis of water at a semiconductor electrode. *Nature* **1972**, *238*, 37. [CrossRef]

6. Nawaz, R.; Kait, C.F.; Chia, H.Y.; Isa, M.H.; Huei, L.W. Glycerol-mediated facile synthesis of colored titania nanoparticles for visible light photodegradation of phenolic compounds. *Nanomaterials* **2019**, *9*, 1586. [CrossRef]

7. Yu, X.; Fan, X.; An, L.; Liu, G.; Li, Z.; Liu, J.; Hu, P. Mesocrystalline Ti^{3+}-TiO_2 hybridized g-C_3N_4 for efficient visible-light photocatalysis. *Carbon* **2018**, *128*, 21–30. [CrossRef]

8. Xing, Z.; Zhang, J.; Cui, J.; Yin, J.; Zhao, T.; Kuang, J.; Xiu, Z.; Wan, N.; Zhou, W. Recent advances in floating TiO_2-based photocatalysts for environmental application. *Appl. Catal. B-Environ.* **2018**, *225*, 452–467. [CrossRef]

9. Moongraksathum, B.; Chen, Y. Anatase TiO_2 co-doped with silver and ceria for antibacterial application. *Catal. Today* **2018**, *310*, 68–74. [CrossRef]

10. Ma, J.; Tan, X.; Jiang, F.; Yu, T. Graphitic C_3N_4 nanosheet-sensitized brookite TiO_2 to achieve photocatalytic hydrogen evolution under visible light. *Catal. Sci. Technol.* **2017**, *7*, 3275–3282. [CrossRef]

11. Tan, Y.; Shu, Z.; Zhou, J.; Li, T.; Wang, W.; Zhao, Z. One-step synthesis of nanostructured g-C_3N_4/TiO_2 composite for highly enhanced visible-light photocatalytic H_2 evolution. *Appl. Catal. B-Environ.* **2018**, *230*, 260–268. [CrossRef]

12. Li, J.; Ma, C.; Zhu, S.; Yu, F.; Dai, B.; Yang, D. A review of recent advances of dielectric barrier discharge plasma in catalysis. *Nanomaterials* **2019**, *9*, 1428. [CrossRef] [PubMed]

13. Wang, Z.; Zhang, Y.; Neyts, E.C.; Cao, X.; Zhang, X.; Jang, B.W.L.; Liu, C. Catalyst preparation with plasmas: How does it work? *ACS Catal.* **2018**, *8*, 2093–2110. [CrossRef]

14. Dou, S.; Tao, L.; Wang, R.; El Hankari, S.; Chen, R.; Wang, S. Plasma-assisted synthesis and surface modification of electrode materials for renewable energy. *Adv. Mater.* **2018**, *30*, 1705850. [CrossRef]

15. Yan, X.; Zhao, B.; Liu, Y.; Li, Y. Dielectric barrier discharge plasma for preparation of Ni-based catalysts with enhanced coke resistance: Current status and perspective. *Catal. Today* **2015**, *256*, 29–40. [CrossRef]

16. Sun, Q.; Yu, B.; Liu, C. Characterization of ZnO nanotube fabricated by the plasma decomposition of $Zn(OH)_2$ via dielectric barrier discharge. *Plasma Chem. Plasma Process.* **2012**, *32*, 201–209. [CrossRef]

17. Zada, A.; Qu, Y.; Ali, S.; Sun, N.; Lu, H.; Yan, R.; Zhang, X.; Jing, L. Improved visible-light activities for degrading pollutants on TiO_2/g-C_3N_4 nanocomposites by decorating SPR Au nanoparticles and 2,4-dichlorophenol decomposition path. *J. Hazard. Mater.* **2018**, *342*, 715–723. [CrossRef]

18. Tahir, M.B.; Sagir, M.; Shahzad, K. Removal of acetylsalicylate and methyl-theobromine from aqueous environment using nano-photocatalyst WO_3-TiO_2 @g-C_3N_4 composite. *J. Hazard. Mater.* **2019**, *363*, 205–213. [CrossRef]

19. Zhao, W.; Yang, X.; Liu, C.; Qian, X.; Wen, Y.; Yang, Q.; Sun, T.; Chang, W.; Liu, X.; Chen, Z. Facile construction of all-solid-state Z-scheme g-C_3N_4/TiO_2 thin film for the efficient visible-light degradation of organic pollutant. *Nanomaterials* **2020**, *10*, 600. [CrossRef]

20. Marques, J.; Gomes, T.D.; Forte, M.A.; Silva, R.F.; Tavares, C.J. A new route for the synthesis of highly-active N-doped TiO_2 nanoparticles for visible light photocatalysis using urea as nitrogen precursor. *Catal. Today* **2019**, *326*, 36–45. [CrossRef]

21. Lu, Z.; Zeng, L.; Song, W.; Qin, Z.; Zeng, D.; Xie, C. In situ synthesis of C-TiO_2/g-C_3N_4 heterojunction nanocomposite as highly visible light active photocatalyst originated from effective interfacial charge transfer. *Appl. Catal. B-Environ.* **2017**, *202*, 489–499. [CrossRef]

22. Sanabria-Arenas, B.E.; Mazare, A.; Yoo, J.; Nhat, T.N.; Hejazi, S.; Bian, H.; Diamanti, M.V.; Pedeferri, M.P.; Schmuki, P. Intrinsic AuPt-alloy particles decorated on TiO_2 nanotubes provide enhanced photocatalytic degradation. *Electrochim. Acta* **2018**, *292*, 865–870. [CrossRef]

23. Hampel, B.; Kovacs, G.; Czekes, Z.; Hernadi, K.; Danciu, V.; Ersen, O.; Girleanu, M.; Focsan, M.; Baia, L.; Pap, Z. Mapping the photocatalytic activity and ecotoxicology of Au, Pt/TiO_2 composite photocatalysts. *ACS Sustain. Chem. Eng.* **2018**, *6*, 12993–13006. [CrossRef]

24. Ong, W.J.; Tan, L.L.; Ng, Y.H.; Yong, S.T.; Chai, S.P. Graphitic carbon nitride (g-C$_3$N$_4$)-based photocatalysts for artificial photosynthesis and environmental remediation: Are we a step closer to achieving sustainability? *Chem. Rev.* **2016**, *116*, 7159–7329. [CrossRef] [PubMed]

25. Yu, S.; Wang, Y.; Sun, F.; Wang, R.; Zhou, Y. Novel mpg-C$_3$N$_4$/TiO$_2$ nanocomposite photocatalytic membrane reactor for sulfamethoxazole photodegradation. *Chem. Eng. J.* **2018**, *337*, 183–192. [CrossRef]

26. Lee, G.; Lee, X.; Lyu, C.; Liu, N.; Andandan, S.; Wu, J.J. Sonochemical synthesis of copper-doped BiVO$_4$/g-C$_3$N$_4$ nanocomposite materials for photocatalytic degradation of bisphenol a under simulated sunlight irradiation. *Nanomaterials* **2020**, *10*, 498. [CrossRef]

27. Wang, Z.; Peng, X.; Tian, S.; Wang, Z. Enhanced hydrogen production from water on Pt/g-C$_3$N$_4$ by room temperature electron reduction. *Mater. Res. Bull.* **2018**, *104*, 1–5. [CrossRef]

28. Yan, J.; Wu, H.; Chen, H.; Zhang, Y.; Zhang, F.; Liu, S.F. Fabrication of TiO$_2$/C$_3$N$_4$ heterostructure for enhanced photocatalytic Z-scheme overall water splitting. *Appl. Catal. B-Environ.* **2016**, *191*, 130–137. [CrossRef]

29. Liu, H.; Zhang, Z.; He, H.; Wang, X.; Zhang, J.; Zhang, Q.; Tong, Y.; Liu, H.; Ramakrishna, S.; Yan, S.; et al. One-step synthesis heterostructured g-C$_3$N$_4$/TiO$_2$ composite for rapid degradation of pollutants in utilizing visible light. *Nanomaterials* **2018**, *8*, 842. [CrossRef]

30. Smýkalová, A.; Sokolová, B.; Foniok, K.; Matějka, V.; Praus, P. Photocatalytic degradation of selected pharmaceuticals using g-C$_3$N$_4$ and TiO$_2$ nanomaterials. *Nanomaterials* **2019**, *9*, 1194. [CrossRef]

31. Papailias, I.; Todorova, N.; Giannakopoulou, T.; Yu, J.; Dimotikali, D.; Trapalis, C. Photocatalytic activity of modified g-C$_3$N$_4$/TiO$_2$ nanocomposites for NOx removal. *Catal. Today* **2017**, *280*, 37–44. [CrossRef]

32. Zhang, B.; Wang, Z.; Peng, X.; Wang, Z.; Zhou, L.; Yin, Q. A novel route to manufacture 2D layer MoS$_2$ and g-C$_3$N$_4$ by atmospheric plasma with enhanced visible-light-driven photocatalysis. *Nanomaterials* **2019**, *9*, 1139. [CrossRef] [PubMed]

33. Peng, X.; Wang, Z.; Wang, Z.; Gong, J.; Hao, H. Electron reduction for the preparation of rGO with high electrochemical activity. *Catal. Today* **2019**, *337*, 63–68. [CrossRef]

34. Yang, H.; Jin, Z.; Hu, H.; Bi, Y.; Lu, G. Ni-Mo-S nanoparticles modified graphitic C$_3$N$_4$ for efficient hydrogen evolution. *Appl. Surf. Sci.* **2018**, *427*, 587–597. [CrossRef]

35. Du, X.; Bai, X.; Xu, L.; Yang, L.; Jin, P. Visible-light activation of persulfate by TiO$_2$/g-C$_3$N$_4$ photocatalyst toward efficient degradation of micropollutants. *Chem. Eng. J.* **2020**, *384*, 123245. [CrossRef]

36. Hao, R.; Wang, G.; Jiang, C.; Tang, H.; Xu, Q. In situ hydrothermal synthesis of g-C$_3$N$_4$/TiO$_2$ heterojunction photocatalysts with high specific surface area for Rhodamine B degradation. *Appl. Surf. Sci.* **2017**, *411*, 400–410. [CrossRef]

37. Di, T.; Zhu, B.; Cheng, B.; Yu, J.; Xu, J. A direct Z-scheme g-C$_3$N$_4$/SnS$_2$ photocatalyst with superior visible-light CO$_2$ reduction performance. *J. Catal.* **2017**, *352*, 532–541. [CrossRef]

38. Lei, J.; Chen, Y.; Shen, F.; Wang, L.; Liu, Y.; Zhang, J. Surface modification of TiO$_2$ with g-C$_3$N$_4$ for enhanced UV and visible photocatalytic activity. *J. Alloys Compd.* **2015**, *631*, 328–334. [CrossRef]

39. Wang, H.; Liang, Y.; Liu, L.; Hu, J.; Cui, W. Highly ordered TiO$_2$ nanotube arrays wrapped with g-C$_3$N$_4$ nanoparticles for efficient charge separation and increased photoelectrocatalytic degradation of phenol. *J. Hazard. Mater.* **2018**, *344*, 369–380. [CrossRef]

40. Ma, J.; Tan, X.; Yu, T.; Li, X. Fabrication of g-C$_3$N$_4$/TiO$_2$ hierarchical spheres with reactive {001} TiO$_2$ crystal facets and its visible-light photocatalytic activity. *Int. J. Hydrogen Energy* **2016**, *41*, 3877–3887. [CrossRef]

41. Tong, Z.; Yang, D.; Xiao, T.; Tian, Y.; Jiang, Z. Biomimetic fabrication of g-C$_3$N$_4$/TiO$_2$ nanosheets with enhanced photocatalytic activity toward organic pollutant degradation. *Chem. Eng. J.* **2015**, *260*, 117–125. [CrossRef]

42. Ma, L.; Wang, G.; Jiang, C.; Bao, H.; Xu, Q. Synthesis of core-shell TiO$_2$@g-C$_3$N$_4$ hollow microspheres for efficient photocatalytic degradation of rhodamine B under visible light. *Appl. Surf. Sci.* **2018**, *430*, 263–272. [CrossRef]

43. Li, F.T.; Liu, S.J.; Xue, Y.B.; Wang, X.J.; Hao, Y.J.; Zhao, J.; Liu, R.H.; Zhao, D.S. Structure modification function of g-C$_3$N$_4$ for Al$_2$O$_3$ in the in situ hydrothermal process for enhanced photocatalytic activity. *Chem. A Eur. J.* **2015**, *21*, 10149–10159. [CrossRef]

44. Wu, F.; Li, X.; Liu, W.; Zhang, S. Highly enhanced photocatalytic degradation of methylene blue over the indirect all-solid-state Z-scheme g-C$_3$N$_4$-RGO-TiO$_2$ nanoheterojunctions. *Appl. Surf. Sci.* **2017**, *405*, 60–70. [CrossRef]

45. Shi, L.; Yang, L.; Zhou, W.; Liu, Y.; Yin, L.; Hai, X.; Song, H.; Ye, J. Photoassisted construction of holey defective g-C$_3$N$_4$ photocatalysts for efficient visible-light-driven H$_2$O$_2$ production. *Small* **2018**, *14*, 1703142. [CrossRef] [PubMed]

46. Zhang, F.; Feng, G.; Hu, M.; Huang, Y.; Zeng, H. Liquid-plasma hydrogenated synthesis of gray titania with engineered surface defects and superior photocatalytic activity. *Nanomaterials* **2020**, *10*, 342. [CrossRef]

47. Shi, H.; Long, S.; Hu, S.; Hou, J.; Ni, W.; Song, C.; Li, K.; Gurzadyan, G.G.; Guo, X. Interfacial charge transfer in 0D/2D defect-rich heterostructures for efficient solar-driven CO$_2$ reduction. *Appl. Catal. B-Environ.* **2019**, *245*, 760–769. [CrossRef]

48. Feng, X.; Wang, P.; Hou, J.; Qian, J.; Ao, Y.; Wang, C. Significantly enhanced visible light photocatalytic efficiency of phosphorus doped TiO$_2$ with surface oxygen vacancies for ciprofloxacin degradation: Synergistic effect and intermediates analysis. *J. Hazard. Mater.* **2018**, *351*, 196–205. [CrossRef]

49. Zhang, Y.C.; Afzal, N.; Pan, L.; Zhang, X.; Zou, J.J. Structure-activity relationship of defective metal-based photocatalysts for water splitting: Experimental and theoretical perspectives. *Adv. Sci.* **2019**, *6*, 1900053. [CrossRef]

50. Chen, X.; Wei, J.; Hou, R.; Liang, Y.; Xie, Z.; Zhu, Y.; Zhang, X.; Wang, H. Growth of g-C$_3$N$_4$ on mesoporous TiO$_2$ spheres with high photocatalytic activity under visible light irradiation. *Appl. Catal. B-Environ.* **2016**, *188*, 342–350. [CrossRef]

51. Li, Y.; Wang, W.; Wang, F.; Di, L.; Yang, S.; Zhu, S.; Yao, Y.; Ma, C.; Dai, B.; Yu, F. Enhanced photocatalytic degradation of organic dyes via defect-rich TiO$_2$ prepared by dielectric barrier discharge plasma. *Nanomaterials* **2019**, *9*, 720. [CrossRef]

52. Lv, Y.; Pan, C.; Ma, X.; Zong, R.; Bai, X.; Zhu, Y. Production of visible activity and UV performance enhancement of ZnO photocatalyst via vacuum deoxidation. *Appl. Catal. B-Environ.* **2013**, *138–139*, 26–32. [CrossRef]

53. Xue, K.; He, R.; Yang, T.; Wang, J.; Sun, R.; Wang, L.; Yu, X.; Omeoga, U.; Pi, S.; Yang, T.; et al. MOF-based In2S3-X2S3 (X=Bi; Sb)@TFPT-COFs hybrid materials for enhanced photocatalytic performance under visible light. *Appl. Surf. Sci.* **2019**, *493*, 41–54. [CrossRef]

54. Jiang, X.; Xing, Q.; Luo, X.; Li, F.; Zou, J.; Liu, S.; Li, X.; Wang, X. Simultaneous photoreduction of Uranium(VI) and photooxidation of Arsenic(III) in aqueous solution over g-C$_3$N$_4$/TiO$_2$ heterostructured catalysts under simulated sunlight irradiation. *Appl. Catal. B-Environ.* **2018**, *228*, 29–38. [CrossRef]

55. Wang, Y.; Cai, J.; Wu, M.; Chen, J.; Zhao, W.; Tian, Y.; Ding, T.; Zhang, J.; Jiang, Z.; Li, X. Rational construction of oxygen vacancies onto tungsten trioxide to improve visible light photocatalytic water oxidation reaction. *Appl. Catal. B-Environ.* **2018**, *239*, 398–407. [CrossRef]

56. Li, J.; Weng, B.; Cai, S.; Chen, J.; Jia, H.; Xu, Y. Efficient promotion of charge transfer and separation in hydrogenated TiO$_2$/WO$_3$ with rich surface-oxygen-vacancies for photodecomposition of gaseous toluene. *J. Hazard. Mater.* **2018**, *342*, 661–669. [CrossRef]

57. Deng, Y.; Tang, L.; Zeng, G.; Wang, J.; Zhou, Y.; Wang, J.; Tang, J.; Wang, L.; Feng, C. Facile fabrication of mediator-free Z-scheme photocatalyst of phosphorous-doped ultrathin graphitic carbon nitride nanosheets and bismuth vanadate composites with enhanced tetracycline degradation under visible light. *J. Colloid Interface Sci.* **2018**, *509*, 219–234. [CrossRef]

58. Chang, F.; Zhang, J.; Xie, Y.; Chen, J.; Li, C.; Wang, J.; Luo, J.; Deng, B.; Hu, X. Fabrication, characterization, and photocatalytic performance of exfoliated g-C$_3$N$_4$-TiO$_2$ hybrids. *Appl. Surf. Sci.* **2014**, *311*, 574–581. [CrossRef]

59. Yu, J.; Wang, S.; Low, J.; Xiao, W. Enhanced photocatalytic performance of direct Z-scheme g-C$_3$N$_4$-TiO$_2$ photocatalysts for the decomposition of formaldehyde in air. *Phys. Chem. Chem. Phys.* **2013**, *15*, 16883. [CrossRef]

60. Kogelschatz, U. Dielectric-barrier discharges: Their history, discharge physics, and industrial applications. *Plasma Chem. Plasma Process.* **2003**, *23*, 1–46. [CrossRef]

 *nanomaterials*

Article

Piezoelectric BiFeO$_3$ Thin Films: Optimization of MOCVD Process on Si

Quentin Micard, Guglielmo Guido Condorelli and Graziella Malandrino *

Dipartimento di Scienze Chimiche, Università degli Studi di Catania, INSTM UdR Catania,
Viale Andrea Doria 6, I-95125 Catania, Italy; qmicard@gmail.com (Q.M.); guido.condorelli@unict.it (G.G.C.)
* Correspondence: gmalandrino@unict.it; Tel.: +39-095-7385055

Received: 29 February 2020; Accepted: 25 March 2020; Published: 28 March 2020

Abstract: This paper presents a simple and optimized metal organic chemical vapor deposition (MOCVD) protocol for the deposition of perovskite BiFeO$_3$ films on silicon-based substrates, in order to move toward the next generation of lead-free hybrid energy harvesters. A bi-metal mixture that is composed of Bi(phenyl)$_3$, and Fe(tmhd)$_3$ has been used as a precursor source. BiFeO$_3$ films have been grown by MOCVD on IrO$_2$/Si substrates, in which the conductive IrO$_2$ functions as a bottom electrode and a buffer layer. BiFeO$_3$ films have been analyzed by X-ray diffraction (XRD) for structural characterization and by field-emission scanning electron microscopy (FE-SEM) coupled with energy dispersive X-ray (EDX) analysis for the morphological and chemical characterizations, respectively. These studies have shown that the deposited films are polycrystalline, pure BiFeO$_3$ phase highly homogenous in morphology and composition all over the entire substrate surface. Piezoelectric force microscopy (PFM) and Piezoelectric Force Spectroscopy (PFS) checked the piezoelectric and ferroelectric properties of the film.

Keywords: BiFeO$_3$; MOCVD; Si substrate; thin film; perovskite; lead-free piezoelectric; energy harvesting

1. Introduction

Multiferroics are materials in which at least two of the three ferroic orders, ferroelectricity, ferromagnetism or antiferromagnetism, and ferroelasticity coexist. Among multiferroics, perovskite bismuth ferrite (BiFeO$_3$) and its derived systems are of special interest for keeping their properties in extreme temperature environments because of their Curie (T_C = 1103 K) and Neel temperatures (T_N = 643 K) well above room temperature [1,2]. In addition to ferroelectric and antiferromagnetic properties, BiFeO$_3$ nanostructures, combined with graphene, show appealing photocatalytic activity [3,4]. Photoelectric, pyroelectric, and piezoelectric properties are the most studied and appealing characteristics, which make BiFeO$_3$ an important material for energy harvesting applications [5]. The possibility of combining the above-mentioned properties in a single device, a hybrid energy harvester, makes BiFeO$_3$ one of the most promising materials for the next generation of lead-free harvesters. For piezoelectric harvester, Pb(Zr$_x$Ti$_{1-x}$)O$_3$ (PZT) has been commonly used [6,7], but the rising of environmental issues and questions on process sustainability has brought the light on lead-free perovskites, such as BiFeO$_3$, LiNbO$_3$ [8,9], and (K,Na)NbO$_3$ [10], because of their encouraging capabilities for hybrid energy harvesting [11–13].

Different deposition techniques have been used to obtain thin films of BiFeO$_3$, and their derived systems [14–16]. Most of the synthetic routes continue to rely on expensive single crystal substrates, such as SrTiO$_3$, SrTiO$_3$:Nb, and LaAlO$_3$ [17,18], but recently flexible or even bendable substrates are investigated [19].

The main deposition techniques that have been applied to the production of BiFeO$_3$ films are: chemical solution deposition [20,21], pulsed laser deposition (PLD) [22,23], sputtering [24], sol-gel [15,16], and metal organic chemical vapor deposition (MOCVD) [25–28]. So far, only sol-gel and sputtering have been applied to deposit BiFeO$_3$ film used in a piezoelectric harvester [28,29], due to their simplicity and low operating temperature, which enable the use of a wide variety of substrates. The latest results for photovoltaic oriented devices have been obtained by PLD deposited films [23]. However, these techniques may have some drawbacks due to the substrate dimensions and difficulty in process scalability. MOCVD is a very appealing technique in terms of homogeneous and conformal deposition on large area substrates and easy possibility of scaling up, thus appointing itself as one of the best industrially applicable process.

This paper aims to optimize a method for the fabrication of highly homogeneous BiFeO$_3$ (from now on BFO) thin films on Si (001) that were buffered with an IrO$_2$ layer, compatible with conventional industrial processes in order to create a functional hybrid energy harvester. Moreover, special attention has been given to the impact of the different deposition parameters on the quality of the films that were grown on Si. Si-based substrates present many advantages, the first one is the important cost reduction when compared to the widely spread single crystals. Subsequently, when considering that micro electro-mechanical systems (MEMs) microfabrication on Si implies well known protocols and techniques, keeping Si as base material for future functional structures is a key factor in lowering the cost and avoid the time-consuming process development for single crystals.

X-ray diffraction (XRD) has been used for structural characterization, and field-emission scanning electron microscopy (FE-SEM) and energy dispersive X-ray analysis (EDX) have been used for the morphological and chemical analysis, respectively. Finally, local piezoresponse force spectroscopy (PFS) [30] and the domain mapping with piezoresponse force microscopy (PFM) have confirmed the piezoelectric/ferroelectric properties of the samples.

2. Materials and Methods

2.1. Thin Film Deposition

Bi(phenyl)$_3$ and Fe(tmhd)$_3$ (phenyl = –C$_6$H$_5$, H-tmhd = 2,2,6,6-tetramethyl-3,5-heptandione), were mixed and used as a multicomponent precursor mixture. Bi(phenyl)$_3$ and Fe(tmhd)$_3$ were purchased from Strem Chemicals Inc. (Bischheim, France) and used without any further purification. Thin film depositions were performed in a customized, horizontal, hot wall MOCVD reactor with a 20° sample holder inclination. The bi-metallic mixture was placed in an alumina boat and heated at 120 °C. Argon and oxygen were used, respectively, as carrier and reactant gasses, by varying their flow from 150 sccm to 900 sccm (standard cubic centimeter per minute) for both species. The depositions were carried out in the temperature range 600–800 °C for 60 min. BFO films were deposited on a 10 mm × 10 mm Si (001) substrate coated with a 200 nm film of IrO$_2$ acting, at the same time, as bottom electrode for piezoelectric characterization and as buffer layer between BFO and Si.

2.2. Thin Film Characterisation

Analyses of crystalline phases were done through X-ray diffraction (XRD) measurements. θ–2θ XRD patterns were recorded in grazing incidence mode (0.8°) while using a Smartlab diffractometer (Rigaku, Tokyo, Japan), which was equipped with a rotating anode of Cu Kα radiation operating at 45 Kv and 200 Ma. The morphologies were examined through field emission gun scanning electron microscopy (FE-SEM), using a SUPRA VP 55 microscope (ZEISS, Jena, Germany). The films were analyzed by energy dispersive X-ray (EDX) analysis using an INCA-Oxford windowless detector (Oxford Instruments, Abingdon, UK) with an electron beam energy of 15 keV.

Film roughness and ferroelectric properties were measured on an atomic force microscopy NT-MDT solver PRO with a conductive gold coated silicon cantilever, CSG10/Au purchased from NT-MDT (NT-MDT, Moscow, Russia). The BFO samples were grounded to the IrO$_2$ bottom electrode and a good contact was obtained with silver paint.

3. Results and Discussion

3.1. Metal Organic Chemical Vapor Deposition (MOCVD) Grown BiFeO₃ Films on IrO₂/Si Substrate

Starting deposition experiments, as presently reported, are based on earlier works performed on single crystals $SrTiO_3$ [26]. Good results were obtained in the temperature range 750–800 °C, with the best film quality and piezoelectric performance being obtained for the films that were deposited at 800 °C on $SrTiO_3$. Thus, preliminary depositions have been carried out at 750 °C and 800 °C with Ar and O_2 gas flows of 150 sccm for both species. Nevertheless, films that are deposited at 800 °C are not homogeneous and show delamination due to the effect of the high temperature on the IrO_2 bottom layer, which causes its corrugation. Films deposited at 750 °C on a 10×10 mm² IrO_2/Si substrate, while using a susceptor with an inclination of 20°, show good properties in term of homogeneity and adhesion. The susceptor inclination has no effect on adherence, but it might have some effect on the homogeneity of the film. The sample structure has been checked using XRD and patterns have been recorded in the range 20–60° (Figure 1a) and are in agreement with the ICDD data (Card No 20-0169) based on the rhombohedal structure, space group R3c, of the BFO phase.

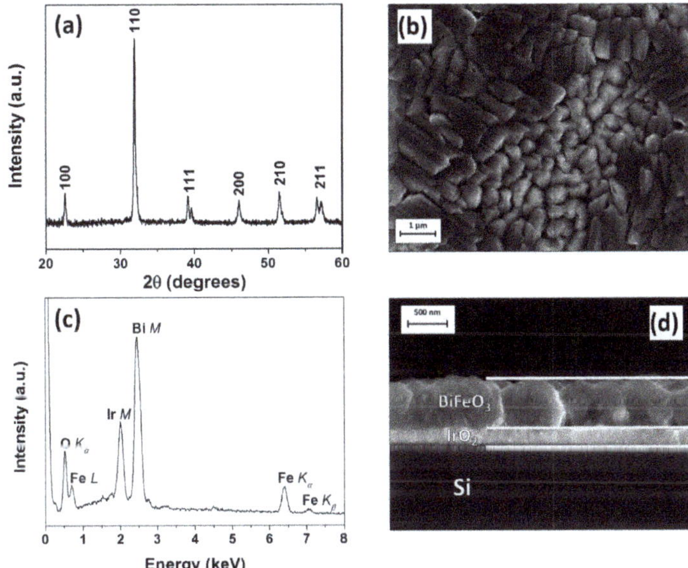

Figure 1. (**a**) X-ray diffraction (XRD) pattern, (**b**) field-emission scanning electron microscopy (FE-SEM) top view image, (**c**) energy dispersive X-ray analysis (EDX) analysis, and (**d**) FE-SEM cross section image of the BiFeO₃ thin film deposited on IrO₂/Si substrate.

The main diffraction peaks are observed at $2\theta = 22.60°$, $31.90°$, and $39.15°$, which are associated, respectively, with the 100, 110, and 111 reflections of the BFO phase, while considering a pseudocubic structure. $BiFeO_3$ might be considered pseudocubic, having the rhombohedral structure cell parameters of $a_{rh} = 3.965$ Å and $\alpha_{rh} = 89.4°$ [2]. Thus, the deposited thin films are polycrystalline and comparison of the peak intensities with the database values indicates a slight preferential orientation along the <110> direction.

FE-SEM has been used to monitor film morphology and, when coupled with energy dispersive X-ray analysis (EDX), to check chemical homogeneity and assess the elemental quantification in the films. The BFO films show a very uniform and dense morphology with massive and well coalesced grains. They adopt elongated shapes with two different preferential orientations (Figure 1b). EDX analysis shows chemical composition homogeneity on large sample areas. The sample average composition indicates a Bi:Fe ratio equal to 1 (Figure 1c). Film thickness has been also checked by several cross sections and shows an average value of 600 nm (Figure 1d). The IrO_2 layer of 200 nm is clearly identifiable between the MOCVD grown BFO film and Si substrate (Figure 1d).

3.2. Impact of Gas Flows on $BiFeO_3$ Film Growth

Basic deposition parameters produced good quality films in terms of structure and composition. Nevertheless, the impact of the gas flows (either carrier or reactant) on the film morphology and quality is one of the major steps to develop a fully optimized process. Starting from the initial parameters, deposition temperature = 750 °C, Flow(Ar) = 150 sccm, Flow(O_2) = 150 sccm, and duration 1h, the effect of the gas flow variation has been investigated by changing one gas flow (Ar or O_2) and maintaining constant all of the other parameters to understand the impact of each one of them on the final product. In Figure 2, the results of BFO films deposited with diverse Ar or O_2 flows are summarized based on the FE-SEM comparative images.

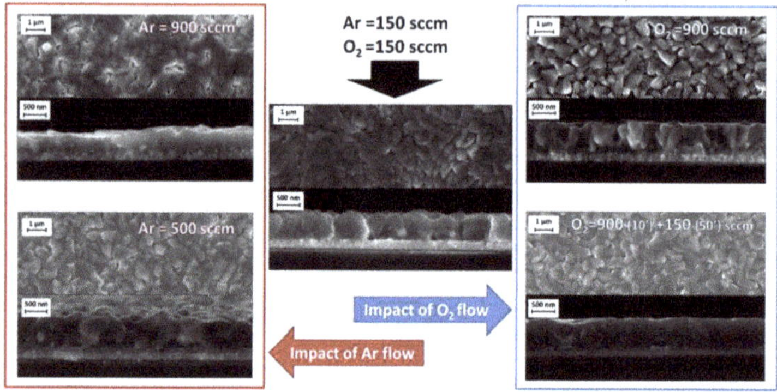

Figure 2. Comparison of $BiFeO_3$ (BFO) thin films deposited under different conditions. BFO reference sample is in the center, on the left the impact of the argon flow and on the right the impact of the oxygen flow on BFO are presented. For every condition, FE-SEM top view and cross sections are reported.

The morphologies of films deposited with 500 sccm or 900 sccm Ar flows, keeping all the other parameters constant, are reported on the left side of Figure 2. The FE-SEM plan-view images show homogeneous samples with grains of 700–800 nm. The cross-section images show that the deposition at 500 sccm does not present major changes in terms of density and thickness with respect to the 150 sccm Ar deposited film. On the other hand, at 900 sccm, the cross section of the film indicates an important increase of film density, but the main drawback is that the film growth rate is significantly decreased when compared to our reference experimental conditions. The average growth rates drop from 9 nm/min to 4 nm/min. This variation might be related to a dilution effect, since the carrier flow increase determines the precursor dilution. The coupled influence of the carrier flow increase and precursor dilution is responsible for the lower growth rate and, consequently, for the formation of thinner but denser films at the highest carrier flow (900 sccm).

When considering previous results with argon, a significant flow of oxygen, 900 sccm, has firstly been tested. BFO growth rate is not limited by the high O_2 flow and, at the same time, film top-view and cross section show a denser film composed of grains of about 800–850 nm (Figure 2, right side). According to these first observations, and to the promising impact that the high oxygen flow has on the film quality, a deposition experiment has been tailored while using a first step of 10 min. with Flow (O_2) = 900 sccm followed by a second one of 50 min. with Flow (O_2) = 150 sccm. The first step enables the creation of a seed layer with smaller nucleation sites, the second allows for these nuclei to grow, giving rise to coalesced grains forming a dense and homogeneous film. FE-SEM analysis (Figure 2, right side) confirms this observation, the grain size is smaller, and cross section shows an important diminution of the roughness and extremely dense BFO film. As expected, the growth rate is unchanged when compared to the reference sample. Thus, a visible improvement of the film quality is observed by simply varying the oxygen flow during the deposition of the film and, at the same time, the process remains straightforward without adding extra steps.

3.3. Impact of the Temperature on BiFeO₃ Film

Attempts to tune the sample orientation by changing deposition temperature have been done in the 600–800 °C temperature range with an Ar flow of 150 sccm and an O_2 flow of 900 sccm for 10 min. and 150 sccm for 50 min. Depositions that were performed at 600 °C lead to a deterioration of film quality and homogeneity. FE-SEM image shows flat iron rich islands on the BFO film surface. Moreover, material quality and density are also heavily impacted by the low temperature process, coalescence is incomplete, and cracks are visible. At 800 °C, the IrO_2 layer is no longer stable and it starts to interact with the film. The buffer layer corrugates in several points, forming bumps on the surface and leading to the delamination of the $Si/IrO_2/BFO$ structure. Defects that are caused by extreme conditions make the samples realized outside of the temperature range 650–750 °C, difficult to be analyzed, and compared to previous results. Within this ideal 100 °C deposition range, XRD patterns, as reported in Figure 3a–c, indicate that polycrystalline single phase BFO films are deposited.

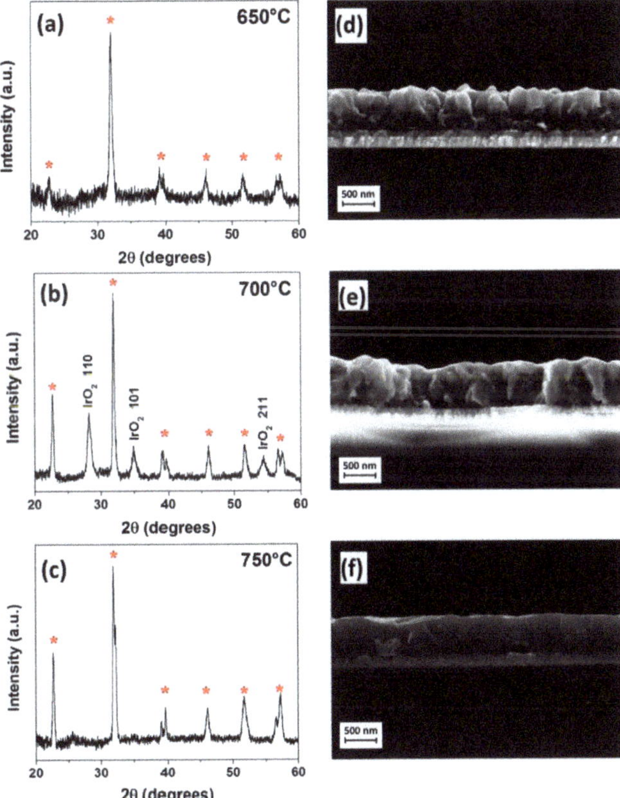

Figure 3. XRD diffraction patterns and FE-SEM cross section images of BFO thin films deposited at (**a,d**) 650 °C, (**b,e**) 700 °C, and (**c,f**) 750 °C.

When comparing the various sample patterns, it has been observed that a preferential orientation competition takes place between the (100) and (110) planes, but no complete orientation is observed, whatever the deposition parameters. The cross-section images (Figure 3d–f) of the films deposited in this temperature range show similar thicknesses, thus pointing to similar growth rates. Specifically, growth rates have been evaluated every 50 °C between 600 °C and 750 °C, yielding growth rates of 7, 10, 11, and 10 nm/min., respectively, for 600, 650, 700, and 750 °C.

This independence between film thickness and substrate temperature in the 600–750 °C range seems to point to a mass transport regime. Figure 4 represents the plot of the ln growth rate vs. 1000/T. The apparent activation energy of 22 kJ/mol, as derived from the Arrhenius plot, clearly indicates that, in the used deposition temperature range, BFO film growth occurs in the mass transport-limited regime.

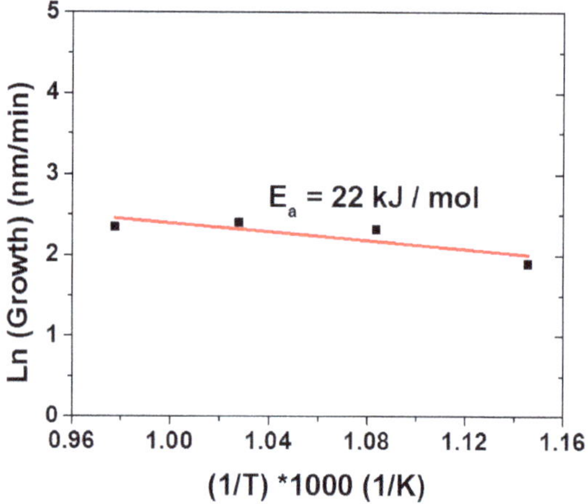

Figure 4. Arrhenius plot of the ln growth rate vs. 1000/T for the BiFeO$_3$ MOCVD process starting from the Bi(phenyl)$_3$ and Fe(tmhd)$_3$ precursors.

3.4. Functional Properties

BFO thin film topography on the IrO$_2$ buffer layer has been recorded by a classical contact mode vertical atomic force microscopy (AFM) scan (Figure 5a). The 5 μm × 5 μm investigated area exhibits the same morphology as the one that was observed by FE-SEM (Figure 1) with a root mean square (RMS) roughness of 28.6 nm. The piezoelectric property of the deposited film on the IrO$_2$/Si substrate has been investigated through piezoresponse force spectroscopy (PFS). Piezoresponse (in terms of the amplitude of the out of plane displacement, Mag), as a function of the alternating voltage, showed the typical butterfly loop [30,31] for an applied bias from −9 V to 9 V between microscope cantilever tip and BFO bottom electrode IrO$_2$ (Figure 5b), indicating the piezoelectric behavior of the film.

In Figure 5b the unit for the Mag is nA, because the instrument measures the vertical displacement from the photocurrent of the laser beam reflected by the displaced tip. In particular, the tip displacement is proportional to the difference between the photocurrent incident on different photodiode sections. Figure 5b shows a hysteresis loop, since, after applying a −9 V bias and then reducing the bias until 0 V (black curve), the piezoresponce is different from that observed when applying a +9 V bias and then reducing it to 0 V (red curve).

Following this first observation, a study of ferroelectric domain switching [32] has been carried out on a 2.5 μm × 2.5 μm area. Figure 6a reports the AFM topography of the scanned area. At first, a PFM image of the ferroelectric domain of the "as-deposited" BFO film was obtained in terms of phase difference between the vertical piezoresponse signal and an applied alternating voltage, before the application of any bias voltage (Figure 6b). Subsequently, to observe the switching of the domains, a bias voltage of −9 V was applied to the entire area through the scanning tip and map of the ferroelectric domains was recorded with a 0 V bias voltage (Figure 6c). Subsequently, a similar PFM image was obtained at 0V (Figure 6d), after the application of a +9 V bias. The polycrystalline nature of the film might limit phase scan interpretation and, indeed, the measured signal is the average of all the ferroelectric domains that were placed between the cantilever tip and the bottom electrode. Very few differences between domain phases of the "as prepared" BFO film and the film after application of the −9 V bias are visible (Figure 6c). On the other hand, the impact of the application of +9 V bias on the film polarization was much more important (Figure 6d), indicating the switching of several ferroelectric domains. As an example, the material has a visible response after bias voltage application because of domains switching when comparing the circled zones on Figure 6b–d.

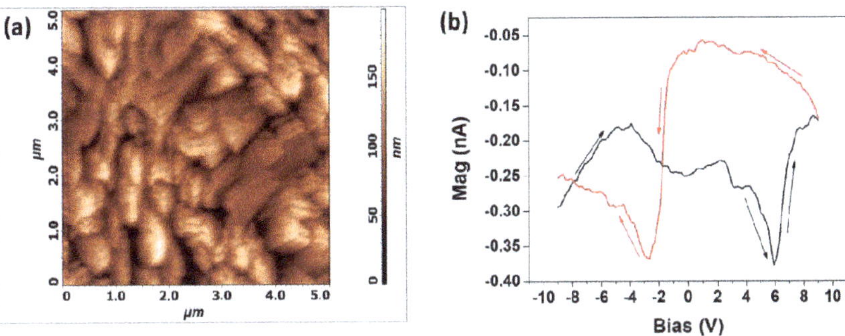

Figure 5. (**a**) Atomic force microscopy (AFM) scan of a 5 µm × 5 µm area and (**b**) butterfly loop (the "black" curve corresponds to the −9 V to +9 V scan and "red" curve corresponds to the +9 V to −9 V) of the "as deposited" BFO thin film on Si obtained by PFS.

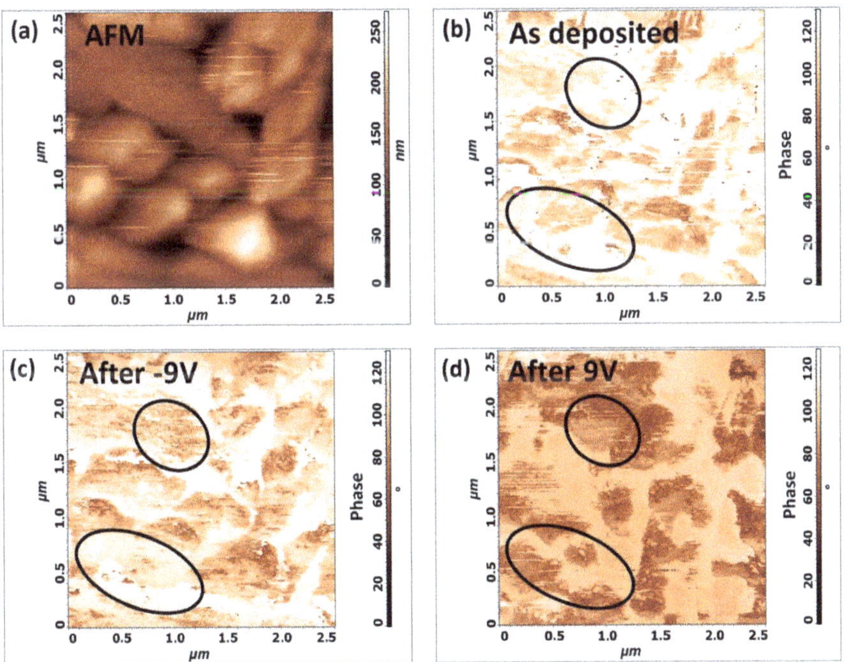

Figure 6. Ferroelectric domains switching: (**a**) AFM image of the scanned 2.5 µm × 2.5 µm area and (**b**) phase scan of the "as deposited" BFO film; (**c**) phase scan of the BFO film after the application of a −9 V bias; and, (**d**) phase scan of the BFO film after the application of a +9 V bias.

PFS and PFM confirmed the piezoelectric properties of the as deposited BFO film. Furthermore, even if sample polycrystallinity might limit ferroelectric mapping, domains switching can be observed by reducing the working area, thus confirming the functional properties of the BFO thin film on Si.

4. Conclusions

In conclusion, BFO films have been successfully deposited by MOCVD on silicon substrate, with an IrO_2 bottom electrode, acting as a buffer layer since it can stand high temperatures. The morphology, density, thickness, and Bi:Fe ratio in the films are homogenous on the whole sample surface of 10 mm × 10 mm. The BFO thin films show different growth orientations, but no specific relationship has been found between orientation and experimental conditions. Various experiments indicate that the optimal deposition temperature range is between 650 °C and 750 °C with a fixed argon flow of 150 sccm and the use of high oxygen flow, 900 sccm for 10 min., in order to induce the formation of numerous BFO nucleation sites, and 150 sccm for 50 min. to trigger the growth of a denser film with smaller grains when compared to the other investigated conditions.

Thus, the BFO films can be successfully deposited on Si at a lower temperature and in a more cost-effective process, with respect to the previously reported methodologies. Moreover, the present approach offers the major advantage to be easily scalable and the use of IrO_2, as a conductive oxide, gives the opportunity for future characterizations and device microfabrications. Finally, it can be pointed out that the material quality and production cost of lead-free perovskites, as BFO, are key points for the scaling-up development of a new generation of hybrid energy harvesting devices. The MOCVD approach, as presently reported, answers to both demands and it is compatible with the current technologies.

Author Contributions: Investigation, Q.M.; writing-original draft preparation, Q.M; supervision, G.G.C. and G.M.; writing-review and editing, G.G.C. and G.M.; funding acquisition, G.M. All authors have read and agreed to the published version of the manuscript.

Funding: This work is supported by the European Community under the Horizon 2020 Programme in the form of the MSCA-ITN-2016 ENHANCE Project, Grant Agreement N.722496.

Acknowledgments: The authors thank the Bio-nanotech Research and Innovation Tower (BRIT) laboratory of the University of Catania (Grant no. PONa3_00136 financed by the Italian Ministry for Education, University and Research, MIUR) for the Smartlab diffractometer facility.

Conflicts of Interest: The authors declare no conflict of interest.

References

1. Choi, T.; Lee, S.; Choi, Y.J.; Kiryukhin, V.; Cheong, S.-W. Switchable ferroelectric diode and photovoltaic effect in BiFeO3. *Science* **2009**, *324*, 63–66. [CrossRef] [PubMed]
2. Catalan, G.; Scott, J.F. Physics and Applications of Bismuth Ferrite. *Adv. Mater.* **2009**, *21*, 2463–2485. [CrossRef]
3. Irfan, S.; Liang, G.; Li, F.; Chen, Y.; Rizwan, S.; Jin, J.; Zheng, Z.; Ping, F. Effect of graphene oxide nano-sheets on structural, morphological and photocatalytic activity of $BiFeO_3$-based nanostructures. *Nanomaterials* **2019**, *9*, 1337. [CrossRef] [PubMed]
4. Li, J.; Wang, Y.; Ling, H.; Qiu, Y.; Lou, J.; Hou, X.; Bag, S.P.; Wang, J.; Wu, H.; Chai, G. Significant enhancement of the visible light photocatalytic properties in 3D $BiFeO_3$/graphene composites. *Nanomaterials* **2019**, *9*, 65. [CrossRef] [PubMed]
5. Queraltó, A.; Frohnhoven, R.; Mathur, S.; Gómez, A. Intrinsic piezoelectric characterization of $BiFeO_3$ nanofibers and its implications for energy harvesting. *Appl. Surf. Sci.* **2020**, *509*, 144760/1–144760/8. [CrossRef]
6. Hwang, G.T.; Annapureddy, V.; Han, J.H.; Joe, D.J.; Baek, C.; Park, D.Y.; Kim, D.H.; Park, J.H.; Jeong, C.K.; Park, K.-I.; et al. Self-Powered Wireless Sensor Node Enabled by an Aerosol-Deposited PZT Flexible Energy Harvester. *Adv. Energy Mater.* **2016**, *6*, 1600237/1–1600237/9. [CrossRef]
7. Muralt, P.; Polcawich, R.G.; Trolier-McKinstry, S. Piezoelectric thin films for sensors, actuators, and energy harvesting. *MRS Bulletin* **2009**, *34*, 658–664. [CrossRef]

8. Bartasyte, A.; Margueron, S.; Baron, T.; Oliveri, S.; Boulet, P. Toward High-Quality Epitaxial $LiNbO_3$ and $LiTaO_3$ Thin Films for Acoustic and Optical Applications. *Adv. Mater. Interfaces* **2017**, *4*, 1600998/1–1600998/36. [CrossRef]

9. Almirall, A.; Oliveri, S.; Daniau, W.; Margueron, S.; Baron, T.; Boulet, P.; Ballandras, S.; Chamaly, S.; Bartasyte, A. High-frequency surface acoustic wave devices based on epitaxial Z-$LiNbO_3$ layers on sapphire. *Appl. Phys. Lett.* **2019**, *114*, 162905/1–162905/5. [CrossRef]

10. Tkach, A.; Santos, A.; Zlotnik, S.; Serrazina, R.; Okhay, O.; Bdikin, I.; Costa, M.E.; Vilarinho, P.M. Effect of Solution Conditions on the Properties of Sol–Gel Derived Potassium Sodium Niobate Thin Films on Platinized Sapphire Substrates. *Nanomaterials* **2019**, *9*, 1600. [CrossRef]

11. Vats, G.; Chauhan, A.; Vaish, R. Thermal Energy Harvesting Using Bulk Lead-Free Ferroelectric Ceramics. *Int. J. Appl. Ceram. Technol.* **2015**, *12*, E49–E54. [CrossRef]

12. Kumar, A.; Sharma, A.; Kumar, R.; Vaish, R.; Chauhan, V.S. Finite element analysis of vibration energy harvesting using lead-free piezoelectric materials: A comparative study. *J. Asian Ceram. Soc.* **2014**, *2*, 139–143. [CrossRef]

13. Vidal, J.V.; Turutin, A.V.; Kubasov, I.V.; Kislyuk, A.M.; Malinkovich, M.D.; Parkhomenko, Y.N.; Kobeleva, S.P.; Pakhomov, O.V.; Sobolev, N.A.; Kholkin, A.L. Low-Frequency Vibration Energy Harvesting with Bidomain $LiNbO_3$ Single Crystals. *IEEE T Ultrason. Ferr.* **2019**, *66*, 1480–1487. [CrossRef] [PubMed]

14. Zhang, Y.; Wang, Y.; Qi, J.; Tian, Y.; Sun, M.; Zhang, J.; Hu, T.; Wei, M.; Liu, Y.; Yang, J. Enhanced Magnetic Properties of $BiFeO_3$ Thin Films by Doping: Analysis of Structure and Morphology. *Nanomaterials* **2018**, *8*, 711. [CrossRef]

15. Hou, P.; Liu, B.; Guo, Z.; Zhou, P.; Wang, B.; Zhao, L. Effect of Ho doping on the crystal structure, surface morphology and magnetic property of $BiFeO_3$ thin films prepared via the sol-gel technology. *J. Alloys Comp.* **2019**, *775*, 59–62. [CrossRef]

16. Sheoran, N.; Kumar, A.; Kumar, V.; Banerjee, A. Structural, Optical, and Multiferroic Properties of Yttrium (Y^{3+})-Substituted $BiFeO_3$ Nanostructures. *J. Supercond. Nov. Magn.* **2020**. [CrossRef]

17. Scillato, D.; Licciardello, N.; Catalano, M.R.; Condorelli, G.G.; Lo Nigro, R.; Malandrino, G. $BiFeO_3$ Films Doped in the A or B Sites: Effects on the Structural and Morphological Properties. *J. Nanosci. Nanotechnol.* **2011**, *11*, 8221–8225. [CrossRef]

18. Zhang, Q.; Huang, H.H.; Sando, D.; Summers, M.; Munroe, P.; Standard, O.; Valanoor, N. Mixed-phase bismuth ferrite thin films by chemical solution deposition. *J. Mater. Chem. C* **2018**, *6*, 2882–2888. [CrossRef]

19. Qian, J.; Wang, Y.; Liu, R.; Xie, X.; Yan, X.; Leng, J.; Yang, C. Bendable $Bi(Fe_{0.95}Mn_{0.05})O_3$ ferroelectric film directly on aluminum substrate. *J. Alloys Comp* **2020**, *827*, 154381/1–154381/6. [CrossRef]

20. Xu, H.-M.; Wang, H.; Shi, J.; Lin, Y.; Nan, C. Photoelectrochemical Performance Observed in Mn-Doped $BiFeO_3$ Heterostructured Thin Films. *Nanomaterials* **2016**, *6*, 215. [CrossRef]

21. Xuemei, C.; Guangda, H.; Jing, Y.; Xi, W.; Changhong, Y.; Weibing, W. Enhanced multiferroic properties of $(1\,1\,0)$-oriented $BiFeO_3$ film deposited on $Bi_{3.5}Nd_{0.5}Ti_3O_{12}$-buffered indium tin oxide/Si substrate. *J. Phys. D Appl. Phys.* **2008**, *41*, 225402/1–225402/5. [CrossRef]

22. Tian, G.; Ojha, S.; Ning, S.; Gao, X.; Ross, C.A. Structure, Ferroelectricity, and Magnetism in Self-Assembled $BiFeO_3$–$CoFe_2O_4$ Nanocomposites on (110)-$LaAlO_3$ Substrates. *Adv. Electron. Mater.* **2019**, *5*, 1900012/1–1900012/8. [CrossRef]

23. Zhou, Y.; Wang, C.; Tian, S.; Yao, X.; Ge, C.; Guo, E.J.; He, M.; Yang, G.; Jin, K. Switchable ferroelectric diode and photovoltaic effects in polycrystalline $BiFeO_3$ thin films grown on transparent substrates. *Thin Solid Films* **2020**, *698*, 137851/1–137851/6. [CrossRef]

24. Zhu, H.; Zhao, Y.; Wang, Y. Orientation dependent leakage current behaviors and ferroelectric polarizations of off-axis sputtered $BiFeO_3$ thin films. *J. Alloys Comp.* **2019**, *803*, 942–949. [CrossRef]

25. Catalano, M.R.; Spedalotto, G.; Condorelli, G.G.; Malandrino, G. MOCVD Growth of Perovskite Multiferroic $BiFeO_3$ Films: The Effect of Doping at the A and/or B Sites on the Structural, Morphological and Ferroelectric Properties. *Adv. Mater. Interfaces* **2017**, *1601025/1–1601025/7*. [CrossRef]

26. Condorelli, G.G.; Catalano, M.R.; Smecca, E.; Lo Nigro, R.; Malandrino, G. Piezoelectric domains in $BiFeO_3$ films grown via MOCVD: Structure/property relationship. *Surf. Coat. Technol.* **2013**, *230*, 168–173. [CrossRef]

27. Deepak, N.; Carolan, P.; Keeney, L.; Zhang, P.F.; Pemble, M.E.; Whatmore, R.W. Bismuth Self-Limiting Growth of Ultrathin $BiFeO_3$ Films. *Chem. Mater.* **2015**, *27*, 6508–6515. [CrossRef]

28. Yoshimura, T.; Murakami, S.; Wakazono, K.; Kariya, K.; Fujimura, N. Piezoelectric Vibrational Energy Harvester Using Lead-Free Ferroelectric BiFeO$_3$ Films. *Appl. Phys. Express* **2013**, *6*, 051501. [CrossRef]
29. Aramaki, M.; Yoshimura, T.; Murakami, S.; Satoh, K.; Fujimura, N. Demonstration of high-performance piezoelectric MEMS vibration energy harvester using BiFeO$_3$ film with improved electromechanical coupling factor. *Sens. Actuat. A-Phys.* **2019**, *291*, 167–173. [CrossRef]
30. Wu, S.; Zhang, J.; Liu, X.; Lv, S.; Gao, R.; Cai, W.; Wang, F.; Fu, C. Micro-Area Ferroelectric, Piezoelectric and Conductive Properties of Single BiFeO$_3$ Nanowire by Scanning Probe Microscopy. *Nanomaterials* **2019**, *9*, 190. [CrossRef]
31. Tudisco, C.; Pellegrino, A.L.; Malandrino, G.; Condorelli, G.G. Surface anchoring of bi-functional organic linkers on piezoelectric BiFeO$_3$ films and particles: Comparison between carboxylic and phosphonic tethering groups. *Surf. Coat. Technol.* **2018**, *343*, 75–82. [CrossRef]
32. Iwanowska, M.; Stolichnov, I.; Colla, E.; Tagantsev, A.; Muralt, P.; Setter, N. Polarization Reversal in BiFeO3 Capacitors: Complex Behavior Revealed by PFM. *Ferroelectrics* **2011**, *421*, 54. [CrossRef]

Review

The Recent Advances in the Mechanical Properties of Self-Standing Two-Dimensional MXene-Based Nanostructures: Deep Insights into the Supercapacitor

Yassmin Ibrahim [1,†], Ahmed Mohamed [2,†], Ahmed M. Abdelgawad [3], Kamel Eid [3,*], Aboubakr M. Abdullah [1,*] and Ahmed Elzatahry [4,*]

[1]   Center for Advanced Materials, Qatar University, Doha 2713, Qatar; yi1511403@student.qu.edu.qa
[2]   Department of Mechanical and Industrial Engineering, College of Engineering, Qatar University, Doha 2713, Qatar; a.mohamed@qu.edu.qa
[3]   Gas Processing Center, College of Engineering, Qatar University, Doha 2713, Qatar; aabdelga14@gmail.com
[4]   Materials Science and Technology Program, College of Arts and Sciences, Qatar University, Doha 2713, Qatar
*   Correspondence: kamel.eid@qu.edu.qa (K.E.); bakr@qu.edu.qa (A.M.A.); aelzatahry@qu.edu.qa (A.E.)
†   These authors contributed equally to this work.

Received: 11 August 2020; Accepted: 2 September 2020; Published: 25 September 2020

Abstract: MXenes have emerged as promising materials for various mechanical applications due to their outstanding physicochemical merits, multilayered structures, excellent strength, flexibility, and electrical conductivity. Despite the substantial progress achieved in the rational design of MXenes nanostructures, the tutorial reviews on the mechanical properties of self-standing MXenes were not yet reported to our knowledge. Thus, it is essential to provide timely updates of the mechanical properties of MXenes, due to the explosion of publications in this filed. In pursuit of this aim, this review is dedicated to highlighting the recent advances in the rational design of self-standing MXene with unique mechanical properties for various applications. This includes elastic properties, ideal strengths, bending rigidity, adhesion, and sliding resistance theoretically as well as experimentally supported with various representative paradigms. Meanwhile, the mechanical properties of self-standing MXenes were compared with hybrid MXenes and various 2D materials. Then, the utilization of MXenes as supercapacitors for energy storage is also discussed. This review can provide a roadmap for the scientists to tailor the mechanical properties of MXene-based materials for the new generations of energy and sensor devices.

Keywords: MXene; mechanical properties; 2D materials; metal carbide; young modules; supercapacitors

---

1. Introduction

Carbon-based nanostructures (C-Ns) such as graphene, carbon nanotubes, and carbon nitride are of great interest due to their unique physiochemical merits such as high surface area, thermal stability, and outstanding mechanical properties [1–4]. These properties promoted the utilization of C-Ns in structural composites, protective coatings, fibers, energy storage, catalysis, and durable wearable sensors; however, their complicated fabrication process remains a major challenge [5–7]. Y. Gogotsi and M.W. Barsoum groups discovered a novel family of 2D transition metal carbides or nitrides called MXene (pronounced "maxenes") [8]. The general formula of MXene is $M_{n+1}X_nT_x$ (n = 1–4), where M represents transition metals, A is an A-group element of group 13 to 15 in the periodic table, X is carbon or nitrogen, and T_x is surface functional groups (OH, O, Cl, F) (Scheme 1) [8]. There are around three main structures of MXenes, including M_2XT_x, $M_3X_2T_x$, and $M_4X_3T_x$, derived from the selective etching of MAX phases (M, A, and X elements are in Scheme 1) including M_2AX, M_3AX_2, and M_4AX_3. To this end, more than 30 MXenes compositions have prepared, such as Ti_2CT_x, Nb_2CT_x, V_2CT_x, $Ti_3C_2T_x$, $Mo_2TiC_2T_x$, $Mo_2Ti_2C_3$, $Ti_yNb_{2-y}CT_x$, and $Nb_yV_{2-y}CT_x$, along with additional dozens were explored by computational methods [9–11].

MXenes possess unique physical and chemical merits such as great miscibility, high surface area to volume ratio, accessible active sites, surface charge state, electron-rich density, and absorption of electromagnetic waves [12]. This is besides the impressive properties of 2D carbide transition metal carbides/nitrides, such as multilayered structures with excellent mechanical properties, strength, flexibility, and high electrical conductivity [12]. Additionally, the fabrication process of MXene is scalable, productive, controllable, facile, and feasible for large-scale applications [12]. MXenes with high negative zeta potential are miscible in various solvents, polymeric materials, and other C-Ns materials resulting in the formation of unlimited composites with various properties [13]. The impressive mechanical properties of MXenes are one of the unique features for MXene [2,14–16]. Despite the significant progress in the synthesis of MXene nanostructures, $Ti_3C_2T_x$ compound is the most widely studied material, for various applications, due to its impressive electrical conductivity, mechanical properties, and electrochemical properties electromagnetic shielding [2,14–16].

There are numerous published reviews in the fields of MXenes for energy, catalysis, and environmental remediation [12,17–23]. However, the reviews on the mechanical properties of self-standing MXenes are not yet reported [24]. Many studies have shown that MXenes exhibits excellent mechanical ion adsorption properties, which in turn will set the stage for exploring the possibility of their use in sensors and flexible devices [6,24–26]. For instance, the strain-tunable electrochemical properties of MXenes enable them to be a propitious solution for flexible and stretchable devices [6,24–26]. Regarding the electrochemical properties of MXenes, their large specific surface area makes them a promising candidate for various applications such as supercapacitor, Li-ion and Sodium-ion batteries, hydrogen storage, adsorption, and catalysts [6,24–26]. Due to the abundant research and ceaseless publications on the mechanical properties of MXene (more than 146 articles, according to SciFinder), it is crucial to provide a timely update of research efforts in this area.

Inspired by this, the presented review summarizes the recent progress of research work on the mechanical properties of self-standing MXenes, from both theoretical and experimental views. This includes: (1) elastic properties and superior strengths, (2) bending rigidity, (3) adhesion, and sliding resistance with their fundamental mechanism supported with numerous representative paradigms. Also, there are deep insights into the utilization of MXenes as supercapacitors. The future perspective of the mechanical applications of MXene is also discussed.

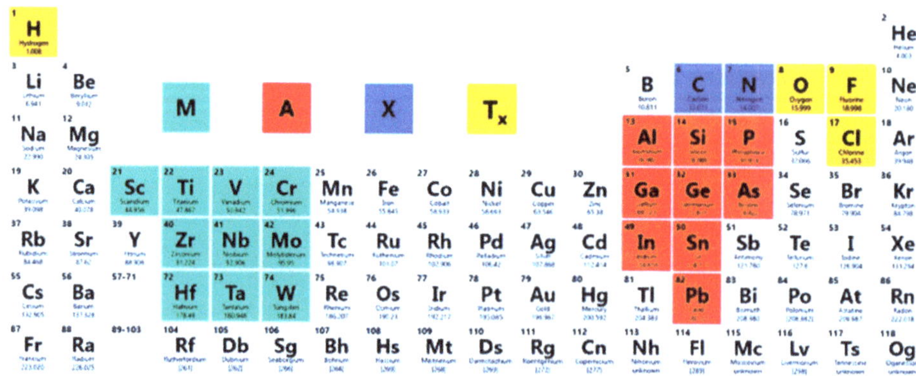

Scheme 1. The composition of MXenes and MAX phases from the periodic table.

2. Mechanical Properties of Self-Standing MXenes

In this section, the elastic properties and superior strengths of self-standing MXenes are briefly summarized, and we discuss the effect of other parameters such as layer thickness, functional groups, and presence of point defects, different transition metals, and substitutional doping. The mechanical properties of MXenes with different compositions are summarized in Table 1.

Table 1. The mechanical properties of MXenes with different compositions.

Materials	Morphology	Preparation Method (Experimentally/Theoretically)	Measurements	Elastic Constants c_{11} [GPa]	Young's Modulus E [GPa]	Strains along Uniaxial x (ε_x)	Strengths along Uniaxial x (σ_x) [GPa]	Ref.
$Ti_3C_2H_2$	2D unit cell	Theoretical calculations	VASP/PBE	419	392	-	-	[27]
$Zr_3C_2O_2$	2D hexagonal lattice	Etching Al layers in Zr_3AlC_5	DFT	392.9	-	-	-	[28]
Ti_2C	2D sheets	Theoretical calculations	VASP	609	-	-	-	[29]
Ti_2CO_2	2D sheets	Theoretical calculations	Nanoindentation process	-	983	-	-	[30]
Ta_2C	2D sheets	Theoretical calculations	CASTEP/Wu-Cohen	788	-	-	-	[31]
Ti_2C	2D sheets	Theoretical calculations	MD	-	597	-	-	[32]
Mo_2C	2D sheets	Chemical vapor deposition	VASP	-	312	-	-	[33]
Ti_2CO_2	2D sheets	Etching Al layers in Ti_2AlC_2	VASP/PBE	745	570	0.28	56	[34]
W_2C	2D sheets	Theoretical calculations	VASP/PBE	781.9	-	0.16	65.6	[35]
Materials	Morphology	Preparation method	Measurements	c_{11} (N/m)	E (N/m)	ε_x	σ_x (N/m)	Ref.
$Ti_3C_2O_2$	2D unit cell	Theoretical calculations	VASP/PBE	379	347	-	-	[27]
$W_2HfC_2O_2$	2D unit cell	Theoretical Calculations	VASP/PBE	-	-	-	47.3	[34]
Mo_2CO_2	hexagonal unit cell	Theoretical calculations	VASP/PBE	361	302	-	-	[35]
Ti_2CO_2	2D sheets	Theoretical Calculations	DFT	-	241	0.24	30.7	[36]

2.1. Elastic Properties and Ideal Strengths

2.1.1. Effect of Functional Terminations

Functional terminations (–O, –F, –OH) of carbides have a significant effect on the structural and mechanical properties of MXenes, as demonstrated extensively by DFT calculation. Figure 1a shows the variation of the calculated elastic constants c_{11} of M_2CT_2 MXenes as a function of the layer thickness and different functional terminations [27]. It can be seen that, except for Cr_2CO_2, the elastic constant for MXenes with oxygen functionalization showed higher elastic constants compared to those with hydroxyl and fluorine functional groups [27]. This is due to the stronger interaction between the oxygen and surface M atoms [27].

The stress-strain curves, as well as the deformation mechanisms, were investigated in response to tensile stress by DFT calculation for 2D $Ti_{n+1}C_n$ (n = 1–3) (Figure 1b) [37]. Three loading conditions were considered to measure the intrinsic mechanical responses to tensile strain in 2D Ti_2C, which are biaxial tension, uniaxial tension along the x-direction, and the y-direction [37]. The stress-strain relations for 2D Ti_2C under different loading conditions are shown in (Figure 1b) [37]. It was found that 2D Ti_2C is an elastically isotropic material, since the corresponding Young's modulus E_x and E_y were estimated to be 620 GPa and 600 GPa, respectively [37]. Moreover, 2D Ti_2CO_2 can sustain higher strains for the three loading conditions than 2D Ti_2C, which is even higher than that of graphene due to surface functionalizing oxygen [37]. Another large variation in mechanical properties was detected when different transition metal, along with surface functional groups, are used [28,38]. Furthermore, in comparison to other functional groups in Ti_3C_2, the oxygen group possesses the highest in-plane planar elastic modulus, as shown in (Figure 1c–e), leading to enhancement of strength, and adsorption energy, which indicates its good thermodynamic stabilization [39]. This can be attributed to the significant charge transfer from inner to outer surface bonds [39].

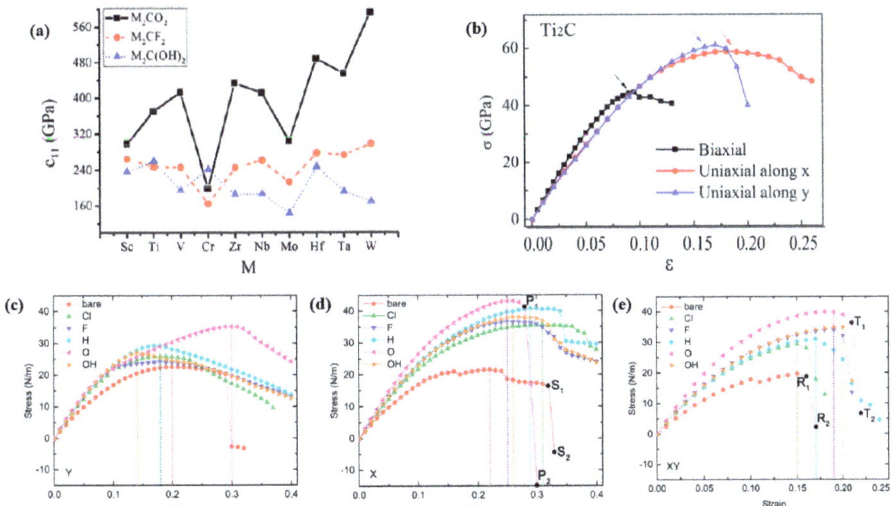

Figure 1. (**a**) The elastic constants of c11 for M_2CT_2 MXenes. Reproduced with permission from [27]. Copyright IOP Publishing, 2015. (**b**) Calculated stress-strain curves of 2D Ti_2C. Reproduced with permission from [37]. Copyright RSC, 2015 (**c**) The stress-strain curves in the uniaxial tension Y direction, (**d**) The stress-strain curves in the uniaxial tension X direction for Ti_3C_2 and P_1 and P_2 for $Ti_3C_2O_2$, (**e**) The stress-strain curves in biaxial tension for Ti_3C_2 and T_1 and T_2 for $Ti_3C_2O_2$. All the vertical lines mention the maximum stress values. Reproduced with permission from [39]. Copyright APS, 2016.

The effect of surface termination groups on the elastic constants of 2D Ti_2CT_2 and $Ti_3C_2T_2$ was investigated using first-principles calculation by Density-functional theory (DFT) simulation [40]. It was found that the stiffness is highly dependent on the termination group. The elastic stiffness of the MXenes is only maintained in the case of MXenes with O terminations while deteriorates in the case of F and OH terminations [40]. This can be explained by the in-plane lattice constant in the 2D MXenes with different termination groups. The in-plane lattice constant for both 2D Ti_2CT_2 and $Ti_3C_2T_2$ MXenes was shortest in the case of O termination. In contrast, F and OH termination had larger in-plane lattice constants indicating a strong interaction between Ti and terminating O atoms.

Another work in the literature [30] studied the effect of surface groups on the elastic properties of MXenes. The ionic mobility for MXenes with different termination groups was investigated under different strain conditions, using multiaxial loading schemes, biaxial and uniaxial tension along x-direction and y-direction. It was observed that Ti_2C (Ti_2CF_2) {Ti_2CO_2} can tolerate percentage of strains of 8 (20) {19}, 16 (29) {24}, and 18 (10) {29} under biaxial and uniaxial tensions along the x and y directions, respectively. Whereas Zr_2C (Zr_2CF_2) {Zr_2CO_2} can withstand strains of 12 (21) {21}%, 16 (29) {27}% and 17 (16) {28}%, respectively as shown in Figure 2 [30]. It can be seen that Ti_2CO_2 has higher critical strains than both Ti_2C and graphene [30]. Additionally, overall, the surface groups (O and F) increase the critical strain and provide more mechanical flexibility to the 2D MXenes by considerably slowing down the collapse of the transition metal layers. This makes MXenes with O and F termination groups potential candidates for high-performance lithium-ion batteries [30]. A recent study [32] investigated the effect of point defects on the elastic properties of MXenes using the atomistic simulation of nanoindentation of $Ti_{n+1}C_nO_2$ monolayer. The Young's modulus of $Ti_3C_2O_2$ was found to be 466 GPa, which is slightly lower than the obtained values by DFT and hybridized computational molecular dynamics (MD) simulations of 523 [41] and 502 [42] GPa, respectively. This can be attributed to the presence of surface terminations. Moreover, the breaking strength of $Ti_3C_2O_2$ was calculated as 25.2 N/m, lower than that of graphene (42 N/m) [43]. As seen in (Figure 3a,b), Ti_2CO_2 exhibited a more sudden fracture compared to $Ti_3C_2O_2$, at higher force and lower displacement [32]. This can be explained by the presence of two fewer atomic layers in Ti_2CO_2, resulting in a decreased resistance and more abrupt failure. The calculated elastic modulus of Ti_2CO_2 (983 GPa) is higher than that of $Ti_3C_2O_2$ and almost approaching the value of graphene [32]. However, this value is inconsistent with the previously reported values by DFT (636 GPa) [41] and hybridized MD (597 GPa) [42]. The calculated breaking force for Ti_2CO_2 of 33.6 N/m is approaching the levels of graphene [32].

Figure 3c–f shows simulation results of the nanoindentation of $Ti_3C_2O_2$ with titanium and carbon vacancies (V_{Ti} and V_C, respectively) [32]. It can be seen that the cracks failed to propagate to the edges of the samples contained 1% V_{Ti} and 10% V_C with the same extent of the pristine $Ti_3C_2O_2$, which explains the effect of defects on the fracture mechanism of the sheets [32]. Furthermore, the presence of defects results in a 17% reduction in elastic modulus (386 GPa), which is still higher than graphene oxide and in good agreement with the recent experimental studies [2].

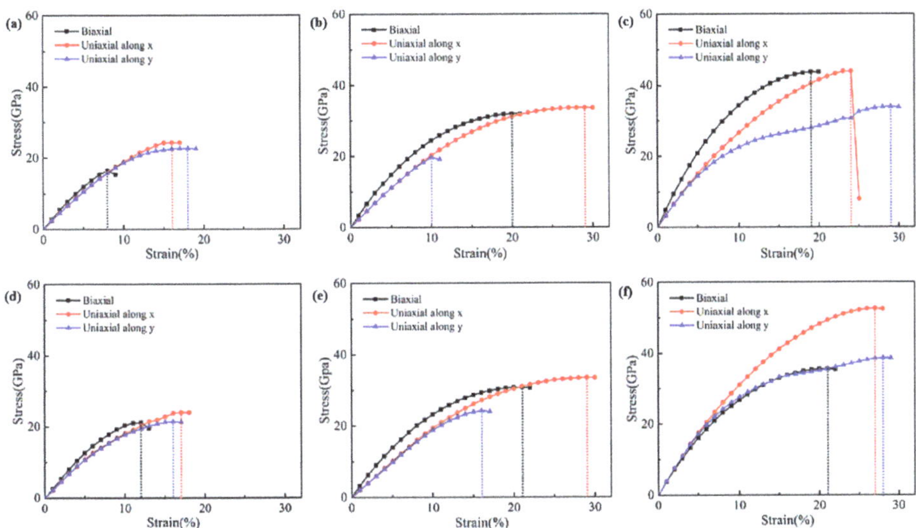

Figure 2. Strain-stress relationships for (**a**) Ti$_2$C, (**b**) Ti$_2$CF$_2$, (**c**) Ti$_2$CO$_2$, (**d**) Zr$_2$C, (**e**) Zr$_2$CF$_2$, and (**f**) Zr$_2$CO$_2$ under both biaxial and uniaxial load conditions. Reproduced with permission from [30]. Copyright PNAS, 2017.

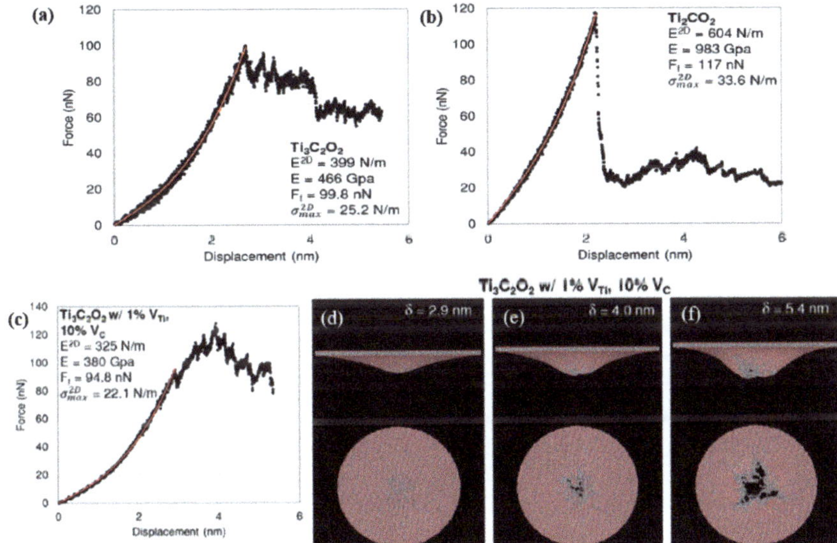

Figure 3. Force-Displacement curves for pristine monolayers of Ti$_3$C$_2$O$_2$ (**a**) and Ti$_2$CO$_2$ (**b**). (**c**) A representative force-displacement curve. (**d–f**) photographs are showing the progressive indentation and fracture of the same representative Ti$_3$C$_2$O$_2$ monolayer with 1% V$_{Ti}$ and 10% V$_C$. Reproduced with permission from [32] Copyright Elsevier, 2019.

2.1.2. Effect of the Mass of the Transition Metal

DFT calculation using the Vienna ab initio simulation package (VASP) code, the mechanical and dynamical properties were obtained for both pristine and terminated MXene (M$_2$XT$_2$) structures with M = Sc, Mo, Ti, Zr, Hf, X = C, N, and T = O, F [29]. It was found that for the pristine carbides, unlike nitride-based pristine, there is a positive correlation between the stiffness and the mass of the transition metal, as indicated by elastic constants [29]. Moreover, the Young Modulus for the nitrides was slightly higher than that of the carbides [29].

In a recent study [44], the effect of asymmetrical functionalization of F and OH groups on the mechanical properties of monolayer Janus MXenes M_2X (M = Sc, Ti, V, Mn, Nb, Mo, Hf; X = C, N) where the X atomic layer is sandwiched between 2 M layers was studied via DFT. It was found that mechanical properties depend on the mass of the transition metal and the surface functionalization. Results show that asymmetric functionalization has a consequential effect on the elastic properties of the MXenes. For all the pristine M_2X, the in-plane stiffness C of M_2C is slightly lower than that of M_2N due to the additional valence electron that the N atom provides than C atoms than in turn generate stiffer M-X bonds. However, due to the H structure of Mo_2X, the in-plane stiffness of Mo_2N is slightly lower than that of Mo_2C. Another finding was that by asymmetrical F/OH surface functionalization, the in-plane stiffness C of Sc_2C was increased from 92 Nm^{-1} to 192 Nm^{-1}, which agrees with what was found by [45]. Moreover, the in-plane stiffness C of monolayer M_2X is lower than that of both graphene [46] and single layer h-BN [47]. Upon asymmetrical surface functionalization, the mechanical stability and the in-plane stiffness C of monolayer M_2X can be enhanced [44]. Despite the fact that it is experimentally challenging to synthesize MXenes accompanied with mixed functional groups [48], eventually, the Janus MXenes could be synthesized experimentally, similar to the Janus graphene [33] and Janus graphene oxide [49].

2.1.3. New Types of MXenes

The enhanced mechanical properties of new types of MXenes, such as Mo_2C were predicted by DFT calculations [34]. The Mo_2C was fabricated via the chemical vapor deposition (CVD) method, where the carbon source was methane, and Cu-foil was selected to be the substrate for a molybdenum foil [50]. The lateral size of the fabricated Mo_2C was found to be >100 μm [50]. No significant structural changes were observed after immersing Mo_2C in several solvents such as isopropanol, ethanol, HCl, or after thermal annealing in air at 200 °C for 2 h, indicating its thermal and chemical stability [50]. Compared to the MoS_2, Mo_2C had a slightly higher biaxial elastic modulus of 312 ± 10 GPa [34]. The relatively large elastic modulus could be explained by the strong interactions between Mo and C atoms. The calculated stress-strain curve for Mo_2C (Figure 4a) shows mostly an elastic response until a critical strain of 0.086, then the Mo_2C exhibited creep deformation. Although this critical strain is less than that of MoS_2, the ideal strength of Mo_2C is predicted as 20.8 GPa, approaching the value of a monolayer of MoS_2 (23.8) GPa [34]. The impressive mechanical properties make Mo_2C a potential candidate for mechanical applications.

2.1.4. Effect of Doping

The effect of doping on the elastic properties of MXenes was investigated by DFT calculations [51]. Specifically, B and V atoms were substitutionally doped into Ti and C sites in Ti_2C, respectively, resulting in $Ti_2(C_{0.5}B_{0.5})$ and (Ti, V)C. While V-doping results only in marginal enhancement, B-doping yields improved the elastic properties by decreasing the in-plane Young's modulus and the yield strength. The reduction in the stiffness can be attributed to the weak-bond of Ti-B compared to the Ti-C bond (Figure 4b,c) [51]. Figure 5 shows the calculated stress-strain curves using the non-magnetic (NM) and the lowest energy antiferromagnetic (AFM) states [51]. It can be seen that a remarkable decrease of about 25–27% in Young's modulus and in-plane stiffness of $Ti_2(C_{0.5}B_{0.5})$ compared to Ti_2C. However, the doping of V at Ti sites results in the same stiffness of the undoped Ti_2C. Intriguingly, the stiffness of $Ti_2(C_{0.5}B_{0.5})$ was about 4.2, 1.5, 1.86, and 3.1 times higher than that of 2D MoS_2, graphene, *h*-BN, and SiC reported elsewhere, respectively, due to the B-doping effect [51,52]. In contrast, $Ti_2(C_{0.5}B_{0.5})$ and (Ti, V)C with O-termination groups exhibited improved elastic properties compared to undoped Ti_2C O-passivated or O-free, owing to the enhancement of the local strain, causing a consequent enlarging of the average thickness of O-passivated MXene by nearly 2% [51].

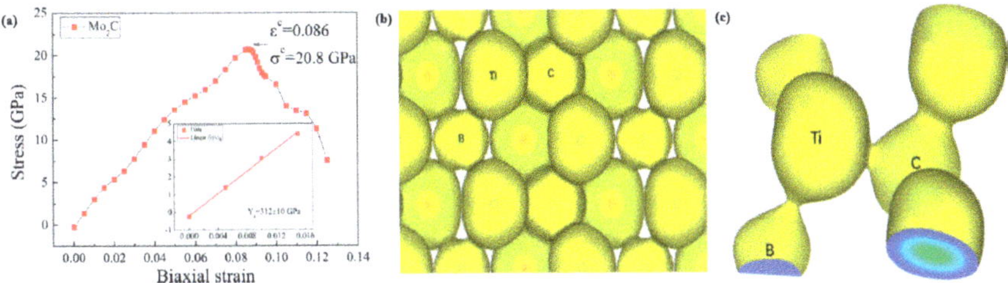

Figure 4. (**a**) Calculated stress versus biaxial strain for the Mo_2C. Reproduced with permission from [34]. Copyright ACS Publications, 2016. (**b**) Shows the plot in an extended region, (**c**) zoomed view focused on central Ti atom bonded with B and C. The stronger covalency of the Ti-C bond compared to the Ti-B bond is visible. Reproduced with permission from [51]. Copyright AIP Publishing, 2017.

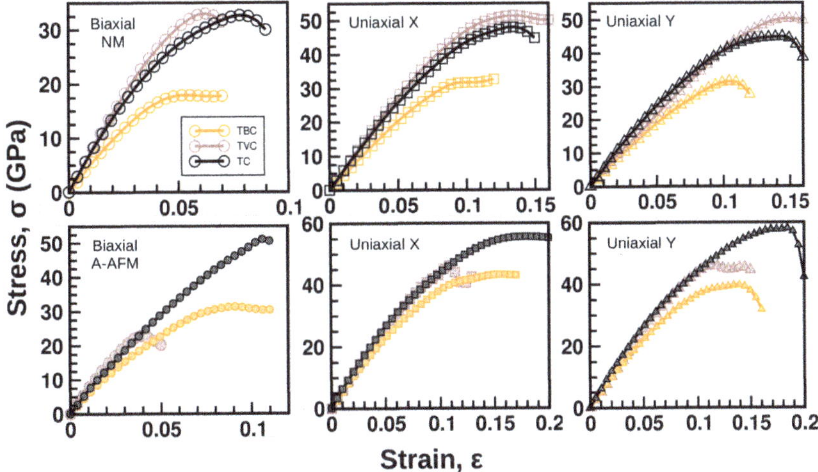

Figure 5. The stress-strain curve of Ti_2C, $Ti_2(C_{0.5}B_{0.5})$ and (Ti, V)C under biaxial and uniaxial tensile strains along with the X and Y directions. The top panels show the results for NM states, while the bottom panels show the results for the minimum energy AFM magnetic state. Reproduced with permission from [51]. Copyright AIP Publishing, 2017.

Another study [53] predicted enhanced elastic properties of a 2D Tungsten Carbide (W_2C) monolayer by DFT calculations. The calculated c_{11} of 781.9 GPa indicated that W_2C is mechanically stable as it satisfies the 2D materials criteria for mechanical stability [36]. The uniaxial tensile loading was applied along the armchair direction, where the correlation between the strain and stress was investigated. As the strain increases, the stress increases until approaching the ultimate tensile strength, as shown in (Figure 6a), then it decreases gradually. The same trend goes for the calculations along the zigzag direction. The calculated ultimate strength of W_2C is comparable to Ti_2C [37] but higher than that of MoS_2 [54]. The Young's modulus of W_2C along with the armchair and zigzag directions are 648 and 645 GPa, respectively, compared to graphene (1000 GPa) [43] and Ti_2C (600 GPa) [37]. Furthermore, W_2C was observed to have a high negative Poisson's ratio (NPR) as they exhibited a positive strain along the longitudinal direction while applying stretching force on the transverse direction (Figure 6b) [53]. This intrinsic NPR of W_2C could be explained by the robust coupling between C-p and W-d orbitals in the pyramid structural unit. Additionally, incorporating the surface functional groups to the calculations show that the NPR of W_2C was turned into PPR due to the weakening of M–C interactions [53].

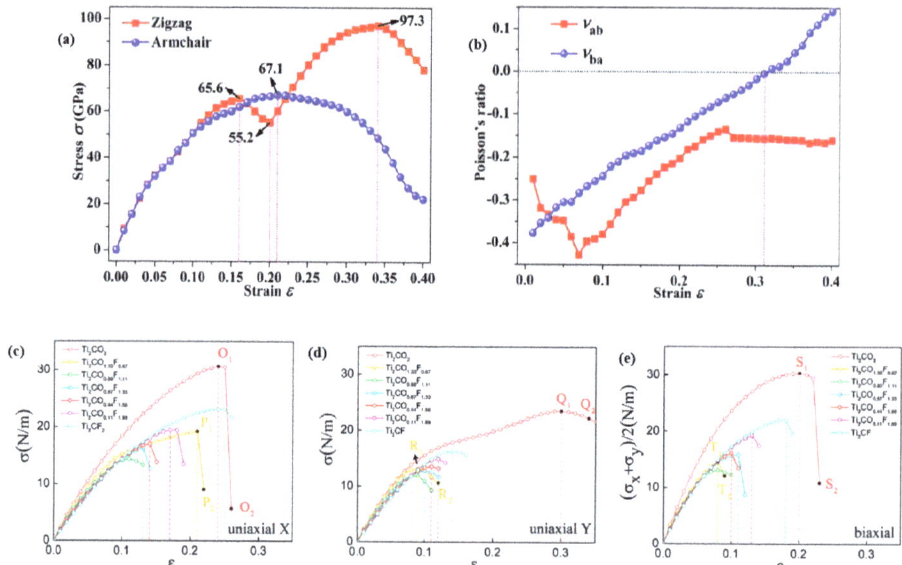

Figure 6. (**a**) Stress-strain curves of intrinsic W_2C under uniaxial stretching, (**b**) Poisson's ratios of W_2C stretched along with different directions. Reproduced with permission from [53]. Copyright RSC, 2018. The calculated stress σ versus strain ε curves for Ti_2CT_2 with pure and mixed terminations under (**c**) uniaxial x, (**d**) uniaxial y, and (**e**) biaxial loading conditions. Reproduced with permission from [55]. Copyright John Wiley & Sons, 2018.

2.1.5. Effect of Varying F/O Ratio

The mechanical properties of Ti_2CT_n terminated by O– and F– were manipulated by varying F/O ratio from 1:17 to 17:1 [55]. The mechanical properties of the five thermodynamically most stable structures with F/O ratios 1: 2, 5:4, 2:1, 7:2, and 17:1 were investigated. It was noticed that as the F/O ratio increases, the Young's, along the x- and y-directions, and shear moduli of Ti_2CT_2 gradually decrease. For instance, Young's modulus along the x-direction decreased from 222 to 159 (N m^{-1}), while the shear modulus decreased from 88 to 58 (N m^{-1}). Additionally, all the Ti_2CT_2, with pure and a mixture of surface terminations, exhibited Poisson ratios greater than that of graphene (0.224), ranging from 0.24 to 0.37. $Ti_2CO_{1.33}F_{0.67}$ exhibits the highest Young's modulus along the x-direction (222 N m^{-1}), among other Ti_2CT_2, which is comparable to Ti_2CO_2 (241 N m^{-1}). The calculated stress-strain curves shown in (Figure 6c–e) indicated that Ti_2CO_2 (24%, 30%, and 20% under uniaxial tensions along the x- and y-directions and under biaxial tensions) and Ti_2CF_2 (25%, 14%, and 18%) have higher critical strains than that of Ti_2CT_2 with mixed terminations. Thus, the mechanical flexibility could be decreased with mixed functional groups. Moreover, a lower critical strain of Ti_2CT_2 is observed with a higher degree of the mixture (Figure 6c–e), such as in $Ti_2CO_{0.11}F_{1.89}$, which could sustain only the lowest critical stains. A similar trend is seen for the ideal strength with the variation of the degree of mixture, indicating that the best mechanical strengths are assigned to $Ti_2CO_{1.33}F_{0.67}$ and $Ti_2CO_{0.11}F_{1.89}$. These findings pave the way to demonstrate more effective methods in tuning the electrochemical properties of MXenes by strains.

2.1.6. Effect of Number of Layers and Layer Thickness

The elastic modulus and breaking strength of monolayer and bilayer $Ti_3C_2T_x$ flakes were experimentally determined by AFM indentation [2]. It was shown that E^{2D} values determined for bilayer $Ti_3C_2T_x$ flakes are exactly twice that determined for monolayer MXene membranes, suggesting strong interaction between the layers due to hydrogen bonding. A single layer of $Ti_3C_2T_x$ has an effective Young's modulus of approximately 333 GPa, which is higher than that of graphene oxide (210 GPa) and some other MXenes. Meanwhile, the breaking strength of a single layer of $Ti_3C_2T_x$ was 17.3 ± 1.6 GPa. It was noted that Young's modulus obtained experimentally is lower than that from the MD simulation due to the presence of defects and surface functionalization.

The effect of layer thickness on the structural and elastic properties of 2D $Ti_{n+1}C_n$ was studied using MD calculations [42]. It was demonstrated that the Young's modulus of MXenes could be significantly increased by decreasing the layer thickness. The Young's modulus of Ti_2C, Ti_3C_2, and Ti_4C_3 was found to be 597, 502, and 534 GPa, respectively, with a strain ε less than 0.01 and within 10% interpolation error. As observed, the highest Young's modulus was reported to the thinnest Ti_2C carbide (3 atomic layers). These results are in agreement with other theoretical predictions from DFT [41].

Similar findings on the effect of monolayer thickness on the elastic properties of the carbide ($Ti_{n+1}C_n$) and nitride-based ($Ti_{n+1}N_n$) MXenes, by DFT calculations, were reported in another study [53]. It was shown that increasing the monolayer thickness decreases Young's moduli of MXenes. The Young's moduli of the $Ti_{n+1}C_n$ were found to be 601, 473, and 459 GPa for Ti_2C, Ti_3C_2, and Ti_4C_3, respectively, which is in good agreement with the values obtained previously by DFT calculations [37]. The bulk model was generated by increasing the layer of atoms to infinity, thus showing that bulk $Ti_{n+1}C_n$ has Young's modulus of 433 GPa, lower than that of rest of $Ti_{n+1}C_n$ MXenes. Although a similar trend was observed for the $Ti_{n+1}N_n$, higher Young's moduli of $Ti_{n+1}N_n$ over $Ti_{n+1}C_n$ was observed, which is consistent with previously reported experimental measurements for bulk TiN [56] and bulk TiC [57,58]. Furthermore, due to the 2D morphology of $Ti_{n+1}C_n$ and $Ti_{n+1}N_n$ with lower thickness, their calculated in-plane Poisson's ratios (ν) are 0.25 and 0.26, higher than that of the bulk TiC and TiN (~0.23) [56–58], which is indicative of increased elasticity.

2.1.7. Effect of Intercalated Ions and Electrolytes

In order to study the effect of the intercalated ions on the mechanical properties of MXenes, the mechanical properties were characterized at the nanoscale instead of at the macroscopic scale [59]. The elastic changes of a 2D $Ti_3C_2T_x$ based electrode in a direction normal to the basal plane were studied via in-situ contact resonance force microscopy (CRFM) imaging, combined with DFT during alkaline cation intercalation/extraction [59]. The DFT calculations agreed well with experiments since the presence of only 12.5% H_2O resulted in a drastic decrease of E from 126 GPa of the dry sample to 29 GPa. The out-of-plane elastic modulus significantly correlated with the cations content. The MXene electrode exhibited shrinkage of almost 10% (Figure 7a) in its lattice structure associated with a decrease in the interlayer distance after Li^+ intercalation [59].

The $Ti_3C_2T_x$ exhibited smaller volume changes when K^+ ions were intercalated, resulting in lower stiffness than in the case of Li^+ ions [59]. This is possible because the stiffness of the cation/water/MXene system is enhanced by the strong oxygen atoms bonds resulted from one hydrogen atom, from the surface hydroxyl group, being pushed out by the cations. Higher CR frequency values after Li^+ intercalation indicates a stiffer 2D structure, in the direction normal to the electrode surface, with elastic moduli ranging between 5 and 18 GPa (Figure 7b,c), twice that of water [59]. Additionally, it was found that the elastic modulus can be tuned using the right combination of the electrolyte and the electrode. The use of both the CRFM technique and DFT calculations revealed that the interface between the electrode/electrolyte could be controlled by probing the mechanical properties associated with the cation storage for applications such as supercapacitors and various types of batteries [59].

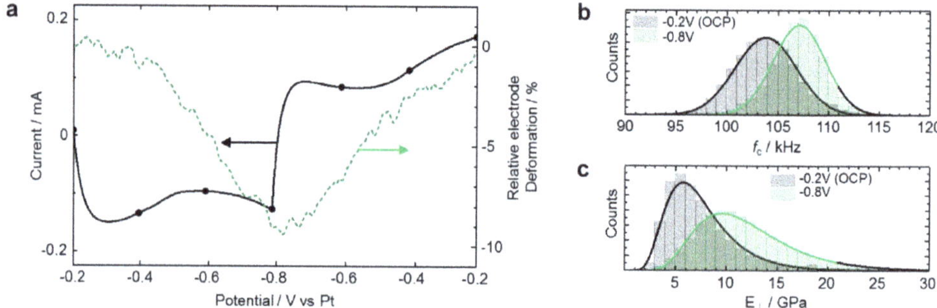

Figure 7. Elastic changes of $Ti_3C_2T_x$ in Li_2SO_4 electrolyte. (**a**) Electrochemical profile of Li^+ intercalation/ extraction showing the current and single point relative electrode deformation profiles as a function of potential. (**b**) Frequency distribution histograms at charged and discharged states. (**c**) Corresponding elastic modulus distribution histograms. Reproduced with permission from [59]. Copyright John Wiley & Sons, 2016.

2.2. Bending Rigidity

The mechanical response of 2D materials obtained under bending deformations is a critical quantity called the bending rigidity [24]. The bending rigidity of MXenes is poorly investigated. To the best of our knowledge, only two papers discussing this quantity have been published so far. The bending rigidity is also affected by some parameters, such as the layer thickness and the functionalization group. The bending rigidity of MXenes was first quantified in 2018 using classical MD simulation for three different 2D titanium carbides (Ti_2C, Ti_3C_2, and Ti_4C_3) to demonstrate their bending resistance under applied bending load [60]. Ti_2C was found to possess higher resistance for bending than atomically thin graphene due to its larger thickness. In contrast, the bending strength of Ti_2C is lower than that of MoS_2 due to different atomic arrangements and larger thickness in MoS_2 compared to Ti_2C [60].

DFT calculations have shown that the in-plane stiffness (C) and out-of-plane bending rigidity (D) of Ti_2CT_x, $Ti_3C_2T_x$, Nb_2CT_x, and $Nb_4C_3T_x$ (T = O, OH, and F) are highly dependent on the layer thickness of $[M_{n+1}X_n]$ and functionalization groups [61]. As the $[M_{n+1}X_n]$ layer thickness increases, the in-plane stiffness increases (Figure 8a) due to the increase in the number of M-C bonds [61]. Nb_2CT_2 and Ti_2CT_2 have relatively low in-plane stiffness due to having only a three-atomic layer in $[M_2X]$, compared to seven-atomic-thick $[Nb_4C_3]$ layer in $Nb_4C_3T_2$ with the largest in-plane stiffness. Moreover, the surface terminations in MXenes significantly increased the stiffness (Figure 8a) [61]. The O-functionalized MXenes were found to have higher in-plane stiffness than that of bare MXenes due to the strong O−M bonding. However, similar in-plane stiffness was noticed for OH, and F terminated Mxenes [61]. Figure 8b depicts the bending rigidities of the four MXenes and their functional groups, showing that Ti_2C has a D value of 4.47 eV [61]. Compared to MXenes, the surface terminations groups decreased their stiffness of graphene and graphene oxides [62,63]. Additionally, the measured bending rigidities of Ti_2C with surface functionalities (4.47 eV) were relatively lower than the previously reported value by MD calculations (5.21 eV) [61], but was higher than that of a graphene monolayer (1.2 eV) [64]. Meanwhile, three-atom-thick Ti_2C and Nb_2C revealed superior flexibility (observed by Foppl-von Karman number per unit area γ) and higher in-plane stiffness, compared to three-atom-thick MoS_2 (9.14 eV) [61]. Therefore, increasing the layer thickness decreases the flexibility of MXenes; however, better flexibility could be observed in MXenes with OH terminations, and the thinnest MXenes with a noticeable decrease in the in-plane stiffness (requires milder exfoliation techniques) [61]. As observed in Figure 8c, with increasing the layer thickness of MXenes, the in-plane stiffness, and out-of-plane bending rigidity increases [61]. Lastly, the bending rigidity increases with effective thickness t_s in a cubic manner, as presented in Figure 8d where C/D ratios were plotted and γ and D considered as a function of effective thickness for 2D materials [61].

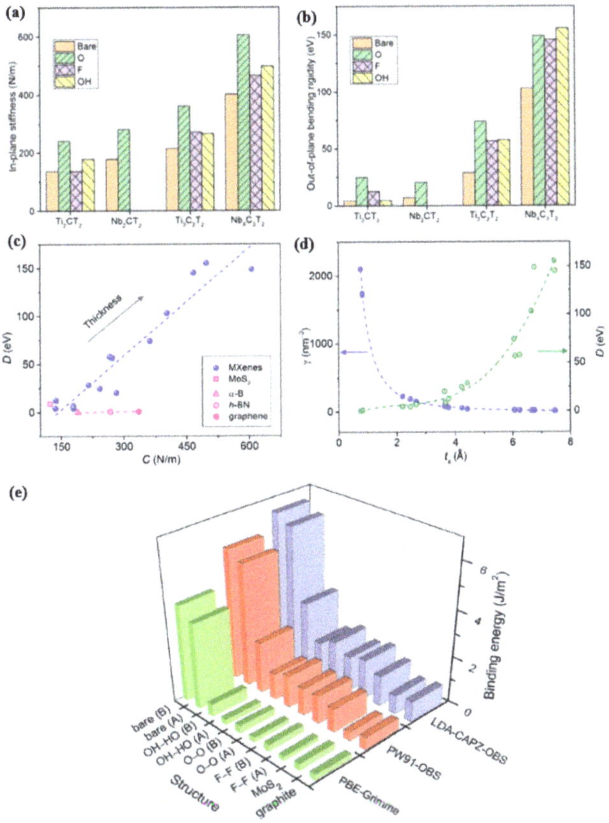

Figure 8. (**a**) In-plane stiffness and (**b**) out-of-plane bending rigidity MXenes. (**c**) Rigidity "D" vs. stiffness, (**d**) *D* vs. *thickness t_s*. Reproduced with permission from [61]. Copyright RSC, 2020. (**e**) Comparison of the theoretical binding energies of $Ti_3C_2T_2$, T = OH, F, and O relative to different 2D materials. Reproduced with permission from [65]. Copyright RSC, 2016.

2.3. Interlayer Adhesion and Sliding

The interlayer adhesion energy and sliding resistance are two critical characteristics of MXenes that are affected by several parameters such as composition, shape, adhesion, and functional species. For instance, understanding the adhesion between MXenes and various substrates is crucial for MXene device fabrication and performance. DFT calculations demonstrated that the surface functionalities (T = OH, F, and O) weaken interlayer coupling of $Ti_3C_2T_2$, relative to the bare counterparts as well as other different 2D materials (Figure 8e) [65]. The binding energies of stacked $Ti_{n+1}C_nT_2$ were found to be about 2- to 6-fold those of 2D graphite and MoS_2 materials with weak interlayer coupling. The interlayer coupling of $Ti_3C_2T_2$ depends on the surface functionalities, which decrease the interlayer coupling, resulting in exfoliation of the stacked $Ti_3C_2T_2$ into monolayers with outstanding mechanical properties compared to other 2D materials. The OH-containing functionalities were the most strongly coupled $Ti_3C_2T_2$ with the highest mechanical properties. The determined Young's moduli normal to the layer plane was 226 GPa for Bernal-$Ti_3C_2(OH)_2$, which is more energetically preferred and also higher than that of highly oriented pyrolytic graphite (about 34 GPa). Furthermore, another study investigated the effect of surface functionalization on the sliding resistance of M_2CO_2 compared to bare counterparts using DFT, along with exploring the strain effect on the sliding resistance [66]. At equilibrium, the layers can easily slide due to the smaller binding energy as a consequence of larger interlayer distance. Due to the oxygen hollow at the surface of oxygen functionalized MXenes, the sliding resistance is increased. However, due to the strong metallic interactions between the stacked M_2C layers, the sliding resistance is much higher than that of M_2CO_2. Another finding is that the relation between the gap and the energy barrier is not linear, whereas as the strain increases, the gap first starts increasing until it reaches maximum value then starts decreasing again. Comparing different stacking configurations, the mirror stacked M_2CO_2-II possesses a better lubricant property than the parallel stacked M_2CO_2-I because its sliding energy barrier is much lower. In addition, the sliding barrier can be significantly enhanced by normal compression. Whereas, the interlayer sliding, owing to the transfer of different charges from M to O atom, may effectively be hindered by the in-plane biaxial tension. The minimum energy pathway can be modified entirely by the uniaxial tension strain due to anisotropic expansion of the surface electronic state. The functionalized MXenes with strain-controllable frictional properties promise lubricating materials due to their lower sliding resistance and superior mechanical properties.

Another study [67] investigated the effect of point defects on the friction coefficients using DFT calculations and classical MD simulations with reactive force-field (ReaxFF) potentials. The results revealed that the sliding pathways are with low energy barriers in all $Ti_{n+1}C_n$ (n = 1, 2, and 3) systems. For these systems, both DFT and ReaxFF methods predicted friction coefficients for interlayer sliding, for normal loads below 1.2 GPa, to be between 0.24 and 0.273. It was found that titanium (Ti) vacancies in sublayers and terminal oxygen (O) vacancies at surfaces increased the friction coefficients, reaching almost 0.31. That is because the surface roughness increased, resulting in additional attractive forces between adjacent layers. Thereby, Ti_3C_2 with surfaces functionalized with –OH and –OCH$_3$ groups were studied and found to be able to reduce the friction coefficient to 0.10 and 0.14, respectively.

Understanding of the adhesion among MXenes and different substrates is crucial for the fabrication of MXene devices. In this regard, the adhesion of $Ti_3C_2T_x$ and Ti_2CT_x) with a SiO_2-coated spherical Si tip was benchmarked compared to graphene (mono-, bi-, and tri-layer) and SiO_2-coated Si tip substrate using direct AFM measurements [68]. This is based on using the Maugis-Dugdale theory for conversion of the adhesion force measured by the AFM to adhesion energy with consideration of the surface roughness [68]. The average adhesion energies of $Ti_3C_2T_x$ (0.90 ± 0.03 J m^{-2}) was higher than that of Ti_2CT_x (0.40 ± 0.02 J m^{-2}) and was in the range of adhesion between graphene and SiO_2. The superior adhesion energy between SiO_2 and $Ti_3C_2T_x$ is due to its thicker monolayer relative to Ti_2CT_x. Another observation was that the adhesion energy of multilayer MXene stacks is dependent on the number of monolayers, in contrast to graphene, which is attributed to the larger interlayer spacing and monolayer thickness of the MXenes.

3. Self-Standing MXene as Electrode for Supercapacitors

Supercapacitors are highly efficient energy storage devices, owing to their excellent power density, fast charge propagation, and long-term durability. The capacitance performance can be calculated using the maximum stored energy (E) and this equation $E = 1/2\ CV^2$, where C is the total capacitance, and V is the working voltage. Meanwhile, the power delivery (P) can be calculated using this equation $P = V^2/4R$, where R is the equivalent series resistance of the supercapacitor. Self-standing MXenes are among the most promising materials for supercapacitors due to their excellent electrical conductivity, mechanical flexibility, high surface area, and high capacitance [24,69,70]. Thereby, few reviews emphasized the utilization of MXenes as supercapacitors, which showed that self-standing $Ti_3C_2T_x$ is the most studied MXenes [71–73]. Several factors determine the capacitance performance of MXenes, such as their morphology, surface area, composition, preparation approaches, as well as the type of electrolytes. Table 2 shows the utilization of self-standing $Ti_3C_2T_x$ prepared by various approaches as efficient supercapacitors, which showed comparable or better performance than that of hybrid $Ti_3C_2T_x$ (i.e., combined with PPy, rGO, and CNTs) [74–76]. Table 3 shows the supercapacitance performance of self-standing $Ti_3C_2T_x$ and hybrid $Ti_3C_2T_x$ in different electrolytes solutions. The performance of both self-standing and hybrid $Ti_3C_2T_x$ in acidic electrolytes (H_2SO_4) was significantly higher than that in alkaline or neutral electrolytes (Table 3). For example, $Ti_3C_2T_x$ showed capacitance performance of 70, 95, 245, and 450 Fg^{-1} in KOH, $MgSO_4$, 1 M H_2SO_4, and 3 M H_2SO_4, respectively [76–78]. The same phenomenon was observed in self-standing V_2CT_x, which showed the capacitance performance of (487 F g^{-1}) in H_2SO_4 compared to 225 Fg^{-1} in $MgSO_4$ and 184 Fg^{-1} in KOH [79]. Interestingly, the capacitance of self-standing V_2CT_x in H_2SO_4 electrolyte (487 F g^{-1}) [79] was superior to $Ti_3C_2T_x$, $Mo2CT_x$, $Mo_{1.33}CT_x$ 245, 196, and 339 F g^{-1}, respectively [74,75,80]. The superior capacitances performance of self-standing MXenes in acidic electrolytes compared to in neutral or alkaline electrolytes is owing to the pseudocapacitive performance with surface redox reactions in acidic electrolytes relative to compared to the ion-intercalation capacitance in neutral and alkaline electrolytes [24]. The surface functionalities (i.e., O_2, OH, and F) have a significant effect on the capacity of the H_2SO_4 electrolyte; The increase of O and decrease of F ions termination in $Ti_3C_2T_x$ increases the capacitance [81].

The pseudocapacitance characteristics and internal mechanism of various MXenes as supercapacitors deeply studied in H_2SO_4 electrolyte, in addition to the factors determine the capacitance effect via DFT calculations [81]. This is included various MXenes ($M_{n+1}X_nT_x$), where M = Sc, Ti, V, Zr, Nb, Mo; X = C, N; T = O, OH; n = 1–3) in H_2SO_4 electrolyte [82]. The predicted capacitance performance of $Ti_3C_2T_x$, Mo_2CT_x, and V_2CT_x [82,83] were similar to the experimentally measured capacitances 235, 245, 90, and 380 F g^{-1}, respectively [77,79,82].

Evaluating the descriptors for the capacitance trends, we find that more positive hydrogen adsorption free energy (weak binding to H) and smaller change of the potential at the point of zero charge after H binding lead to higher capacitance. Interestingly, the pseudocapacitive performance of nitride MXenes electrodes outperformed carbide MXenes. Mainly, Ti_2NT_x is expected to possess a high gravimetric capacitance under any applied voltage in H_2SO_4, owing to the low atomic weight and favorable redox chemistry of Ti. Meanwhile, $Zr_{n+1}N_nT_x$ is anticipated to possess the best areal capacitive performance [82]. The higher capacitance performance is attributed to the higher adsorption free energy and lower change of the potential at the point of zero charge after H binding [82]. The relationship between the charge storage of nitride and carbide MXenes against the shift in the point-of-zero-charge (V_{PZC}) and H_2 adsorption free energy (ΔG_H) displayed that the large ΔGH and the low Δpzc lead to higher charge storage per unit of formula (Figure 9) [82]. Thereby, Zr-based nitride MXenes (Zr_2N, Zr_3N_2, and Zr_4N_3) reveal the highest charge storage under an applied potential range from −1 to 1 V vs. standard Hydrogen Electrode (SHE) [82].

Although the tremendous progress in the capacitance performance MXenes, some remaining gaps exist among the theoretical calculations and experiments, such as inaccurate consideration of the multilayered structures of MXenes along with ignoring the F-rich MXenes surface [84,85].

Table 2. Freestanding $Ti_3C_2T_x$ MXenes prepared by various approaches as efficient supercapacitors compared to some $Ti_3C_2T_x$ composites as a function of preparation method. Abbreviations: PPy = polypyrrole, rGO = reduced graphene oxide, CNT = carbon nanotubes and EG = electrochemically exfoliated graphene.

MAX Phase MXene-Hybrid	Material Composition	Synthesis/Characterization Methods	Morphology	Performance	Ref.
Ti_3AlC_2	$Ti_3C_2T_x$	HF etching/TEM, SEM, CA, XRD, EIS	Nanosheets	517 F/g at 1 A/g	[16]
Ti_3AlC_2	$Ti_3C_2T_x$	HF etching/EIS, XRD	Paper	340 F/cm^3 at 1 A/g	[76]
Ti_3AlC_2	$Ti_3C_2T_x$-P	HCl–LiF/SEM, TEM, FTIR	Paper	416 F/g at 5 mV/s	[84]
Ti_3AlC_2	$Ti_3C_2T_x$	HCl–LiF/XRD, TEM, SEM	Nanosheets	900 F/cm^3 at 2 mV/s	[77]
Ti_3AlC_2	$Ti_3C_2T_x$-EG	HF etching/XRD, TEM, SEM	Nanosheets	33 F/cm^2 at 2 mV/s	[86]
Ti_3AlC_2	$Ti_3C_2T_x$	HF etching/NA	Film	528 F/cm^3 at 2 mV/s	[87]
Ti_3AlC_2	$Ti_3C_2T_x$	NH$_4$F-hydrothermal/XRD, SEM, Raman, XPS	Nanosheets	141 F/g at 2 A/g	[88]
Ti_3AlC_2	$Ti_3C_2T_x$-rGO	HCl–LiF/XRD, SEM	Nanosheets	8.6 mWh/cm^3 at 0.2 W/cm^3	[89]
Ti_3AlC_2	$Ti_3C_2T_x$-CNT	HCl–LiF/XRD, SEM, TEM	Nanosheets	314 F/cm^3 at 1.7 mg/cm^2	[90]
Ti_3AlC_2	$Ti_3C_2T_x$	HF etching/NA	Nanosheets	2.8 mWh/cm^3 at 0.225 W/cm^3	[91]
Ti_3AlC_2	BiOCl-$Ti_3C_2T_x$	HF etching/XRD, SEM, TEM, XPS	Nanosheets	397 F/cm^3 at 1 A/g	[92]
Ti_3AlCN	$Ti_3C_2T_x$	HCl–LiF/TEM, AFM, SEM	Nanosheets	61 mF/cm^2 at 5 µA/cm^2	[93]

Table 3. Freestanding $Ti_3C_2T_x$ and Ti_2CT_x MXenes supercapacitors compared to some $Ti_3C_2T_x$ composites as a function of electrolyte and scan rate/current density. Abbreviations: PPy = polypyrrole, rGO = reduced graphene oxide, CNT = carbon nanotubes, PVA = polyvinyl alcohol, and SWCNT = single-walled carbon nanotubes.

		Freestanding $Ti_3C_2T_x$ and Ti_2CT_x MXenes				
Electrode	Electrolyte	Scan Rate/Current Density	Initial Capacitance (IC)	Cycle Number (CN)	Capacity After Cycles (AC)	Ref.
Ti_2CT_x	30 wt % KOH	$10\ A\ g^{-1}$	$51\ F\ g^{-1}$	6000	93%	[74]
$Ti_3C_2T_x$	1 M KOH	$1\ A\ g^{-1}$	$350\ F\ cm^{-3}$	10,000	~94%	[76]
$Ti_3C_2T_x$	1 M H_2SO_4	$5\ A\ g^{-1}$	$415\ F\ cm^{-3}$	10,000	~100%	[81]
$Ti_3C_2T_x$	1 M H_2SO_4	$10\ A\ g^{-1}$	$900\ F\ cm^{-3}$	10,000	~100%	[77]
$Ti_3C_2T_x$	1 M H_2SO_4	$10\ A\ g^{-1}$	$499\ F\ g^{-1}$	10,000	~100%	[94]
$Ti_3C_2T_x$	6 M KOH	$5\ A\ g^{-1}$	$118\ F\ g^{-1}$	5000	~100%	[95]
$Ti_3C_2T_x$	1 M H_2SO_4	$5\ A\ g^{-1}$	$215\ F\ g^{-1}$	10,000	~100%	[96]
$Ti_3C_2T_x$	1 M H_2SO_4	$5\ A\ g^{-1}$	$892\ F\ g^{-1}$	10,000	~100%	[97]
$N\text{-}Ti_3C_2T_x$	1 M H_2SO_4	$50\ mv\ s^{-1}$	$192\ F\ g^{-1}$	10,000	92%	[98]
$Ti_3C_2T_x/paper$	1 M H_2SO_4	$2\ mA\ cm^{-2}$	$25\ mF\ cm^{-2}$	10,000	92%	[99]
$Ti_3C_2T_x$/3D porous layered double hydroxide	6 M KOH	$1\ A\ g^{-1}$	$1061\ F\ g^{-1}$	4000	70%	[100]
$400\text{-}KOH\text{-}Ti_3C_2T_x$	1 M H_2SO_4	$1\ A\ g^{-1}$	$517\ F\ g^{-1}$	10,000	>99%	[101]
		$Ti_3C_2T_x$ MXenes Hybrid Composites				
Electrode	Electrolyte	Scan Rate/Current Density	Initial Capacitance (IC)	Cycle Number (CN)	Capacity After Cycles (AC)	Ref.
$Ti_3C_2T_x/PVA$	1 M KOH	$5\ A\ g^{-1}$	$\sim370\ F\ cm^{-3}$	10,000	~85%	[16]
$PPy/Ti_3C_2T_x$	1 M H_2SO_4	$100\ mV\ s^{-1}$	$\sim250\ F\ g^{-1}$	25,000	92%	[84]
$Ti_3C_2T_x/SWCNT$	1 M $MgSO_4$	$5\ A\ g^{-1}$	$345\ F\ cm^{-3}$	10,000	~100%	[102]
$Ti_3C_2T_x/rGO$	3 M H_2SO_4	$100\ mV\ s^{-1}$	$777\ F\ cm^{-3}$	20,000	~100%	[103]
$Ti_3C_2T_x/CNT$	1 M EMITFSI	$1\ A\ g^{-1}$	$\sim80\ F\ g^{-1}$	1000	~90%	[104]
$Ti_3C_2T_x/CNT$	6 M KOH	$10\ mv\ s^{-1}$	$\sim384\ F\ g^{-1}$	10,000	~100%	[105]
$TiO_2/Ti_3C_2T_x$	6 M KOH	$5\ mV\ s^{-1}$	$143\ F\ g^{-1}$	3000	~96%	[106]
$MnO_x/Ti_3C_2T_x$	1 M Li_2SO_4	$2\ mV\ s^{-1}$	$602\ F\ cm^{-3}$	10,000	89.8%	[107]
$PPy/Ti_3C_2T_x$	0.5 M H_2SO_4	$1\ mA\ cm^{-2}$	$406\ F\ cm^{-3}$	20,000	~96%	[108]

Increasing the specific surface areas (SSA) and the active redox sites of MXenes can enhance their capacitance performance. Using these strategies, the capacitances of macroporous $Ti_3C_2T_x$ and $Ti_3C_2T_x$ hydrogels reached 210 and 380 F g^{-1}, respectively, owing to their abundance of active sites resulted from the high SSA [84,109]. Moreover, macroporous $Ti_3C_2T_x$ shows capacitances of 310, 210, and 100 F g^{-1} at scan rates of 0.01, 10, and 40 V s^{-1}, respectively [78]. This indicated the direct correlation between the current density peak (*i*) current and scan rate (*v*), which can be an indicator for the inherent charge storage kinetics as can be calculated using this equation $i = av^b$, where *a* and *b* are constants. Electrodes of supercapacitors usually possess a linear relationship between *v* and *i*, i.e., (i~v).

To this end, macroporous $Ti_3C_2T_x$ in H_2SO_4 electrolyte showed a pseudocapacitive behavior as found in the linear dependence of log *i* vs. log *v*, i.e., $b \approx 1$ [78].

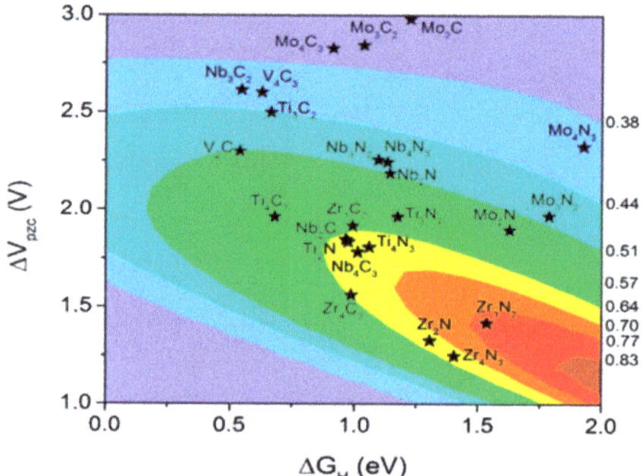

Figure 9. Color map of the relationship between the charge storage per formula unit against the shift in the V_{PZC} and H_2 adsorption free energy (ΔG_H). The applied potential ranged from −1 to 1 V vs. standard Hydrogen Electrode (SHE). Reproduced with permission from [82]. Copyright ACS, 2019.

Another factor for improvement of the capacitance performance of MXenes is the selection of appropriate electrolyte, due to the difference in the ionic conductivity, operation voltage, the temperature of different electrolytes that subsequently tailor the capacitance performance. Thereby, aqueous electrolytes with their outstanding ionic conductivity are preferred compared to organic or ionic liquids electrolytes, although the ionic liquids electrolytes have the largest potential window and feasible for high working temperatures. This finding was achieved in the superior capacitance of $Ti_3C_2T_x$ (325 Fg^{-1}) in H_2SO_4 [81], compared to (70 F g^{-1}) in ionic liquid electrolytes [110] and (32 F g^{-1}) in organic electrolytes [104]. The capacitance performance of $Ti_3C_2T_x$ in propylene carbonate (PC) organic electrolyte with higher ionic conductivity was higher than that in acetonitrile and dimethyl sulfoxide (DMSO), with lower ionic conductivity [111]. The preparation method of MXenes is also an essential factor for boosting the mechanical properties and electrochemical or capacitance performance. Investigation of the mechanical evolution during the intercalation/deintercalation of MXenes revealed that Li-ion intercalation increases the out-of-plane stiffness (elastic properties) in aqueous electrolytes [59]. This is achieved by proposing a theoretical correlation among the cation content and the out-of-plane elastic properties during electrochemical reactions.

Although MXenes were reported to be promising for energy storage applications, their restacking issue, low intrinsic electronic and ionic conductivity, and low specific capacity hinder their use in the practical applications [24]. Besides, the underlying mechanism of the use of MXenes for supercapacitors still needs to be clarified, and that requires further in-depth theoretical and experimental investigations. Furthermore, due to the remarkable influence of the electrolytes on the MXene supercapacitors, more studies are needed for electrolytes optimization.

4. Mechanical of Self-Standing MXenes vs. Hybrid MXenes

MXenes, especially $Ti_3C_2T_x$, was found to be a promising candidate for enhancing the mechanical properties of various polymers, metals, and carbon materials. This is owing to the multilayered 2D structure and outstanding Young's modulus of $Ti_3C_2T_x$ monolayer (0.33 ± 0.03 TPa), measured via the nanoindentation experiments [2]. For instance, the mechanical properties of polyvinyl alcohol (PVA) nanofibers were significantly enhanced via using $Ti_3C_2T_x$ and cellulose nanocrystals (CNC) fillers (denoted as PVA/CNC/$Ti_3C_2T_x$) compared to pristine PVA [112]. Notably, PVA nanofibers containing 0.07 wt.% of both CNC and $Ti_3C_2T_x$ displayed more than 100% enhancement of the storage modulus relative to PVA nanofibers. In comparison, PVA nanofibers with 3 wt.% nanocellulose (PVA/CNC) revealed a 74% increase in storage modulus of PVA at 25 °C [112]. Additionally, the elastic modulus of PVA/CNC/$Ti_3C_2T_x$ nanofibers (855 MPa) was 2.1 times higher than that of PVA nanofibers (392 MPa). The Young's modulus of PVA/CNC/$Ti_3C_2T_x$ nanofibers (293 ± 59 MPa.) was higher than that of PVA/CNC (241 ± 51 MPa), PVA/$Ti_3C_2T_x$ (283 ± 60 MPa), and PVA nanofibers (221 ± 51 MPa) [112]. Likewise, polyimide/$Ti_3C_2T_x$ aerogel prepared via the freeze-drying of and annealing to form a robust, lightweight, and hydrophobic aerogel (Figure 10a) with three-dimensional "house of cards" structure (Figure 10b) [113]. The compressive strength at 80% strain and Young's modulus of elasticity for PI/$Ti_3C_2T_x$ aerogel increased significantly with decreasing the $Ti_3C_2T_x$ concentration. This is owing to greater porosity and lower density of PI/MXene aerogels with the increase of the $Ti_3C_2T_x$ amount [113]. Interestingly, the elastic properties, PI/MXene-3 with a ratio of 5.2:1, respectively, showed impressive stress−strain repeatability after 50 cycles of compression-release (Figure 10c), attributed to the strong interactions between PI chains and $Ti_3C_2T_x$ nanosheets in the hybrid aerogel [113]. Silver nanowires, combined with $Ti_3C_2T_x$ (AgNWs-$Ti_3C_2T_x$) transparent conductive electrode, displayed a higher conductivity, chemical stability, and mechanical stability than that of pristine AgNW electrode [114].

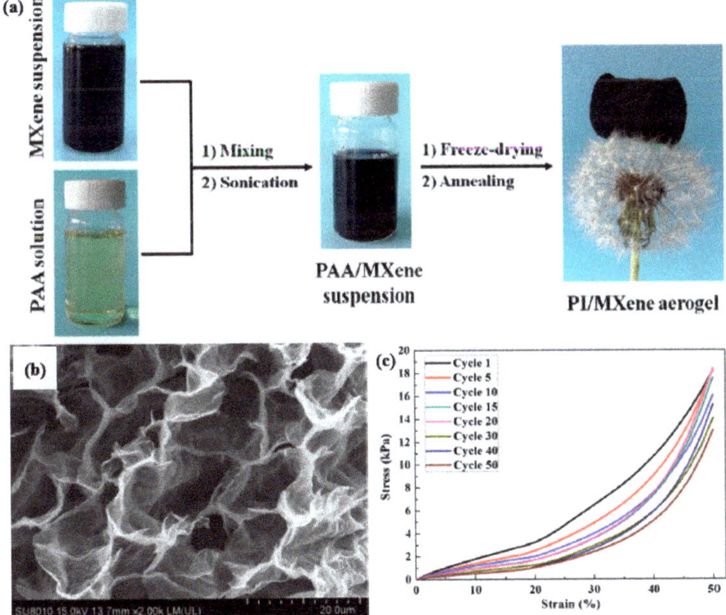

Figure 10. (a) Preparation route of PI/$Ti_3C_2T_x$ aerogels (b) Internal structure of PI/$Ti_3C_2T_x$-3 aerogels (c) Stress−strain curves for 50 repeated compression cycles on the PI/MXene-3 aerogel at 50% strain. Adapted and reproduced with permission from [113]. Copyright ACS, 2019.

Ti$_3$C$_2$T$_x$/carbon nanotube (CNT) 3D porous aerogel (denoted as MXCNT) was synthesized using the bidirectional freezing approach (Figure 11a) [115]. Figure 11b displays the compressive stress-strain curves for Ti$_3$C$_2$T$_x$ and MXCNT aerogels measured under compression at a displacement rate of 1 mm/min up to 50% strain. The compressive strength of MXCNT was substantially higher than that of Ti$_3$C$_2$T$_x$ Figure 11b. Also, the compressive strength of MXCNT increased with increasing CNT concentration to reach the maximum value of 25,000 Pa using a ratio of 1/3 of Ti$_3$C$_2$T$_x$/CNT, respectively. This is originated from the uniform distribution of Ti$_3$C$_2$T$_x$ multilayered sheets with CNT in the direction of the compressive force resulting in a uniform aerogel, as shown in (Figure 11c). Interestingly, the as-formed MXCNT aerogel can afford more than 500 times (Figure 11d) and more than 2100 times (Figure 11d) of its weight without collapsing along with recovery of 12.1 % strain after eliminating the applied load. The significant enhancement in the compressive strength of MXCNT is ascribed to the ordered porous framework supported by vertical pillars, that warrants the cell walls deformation on compression rather than sliding between the walls [115]. The MXCNT aerogel is highly promising for electromagnetic interference (EMI) shielding applications.

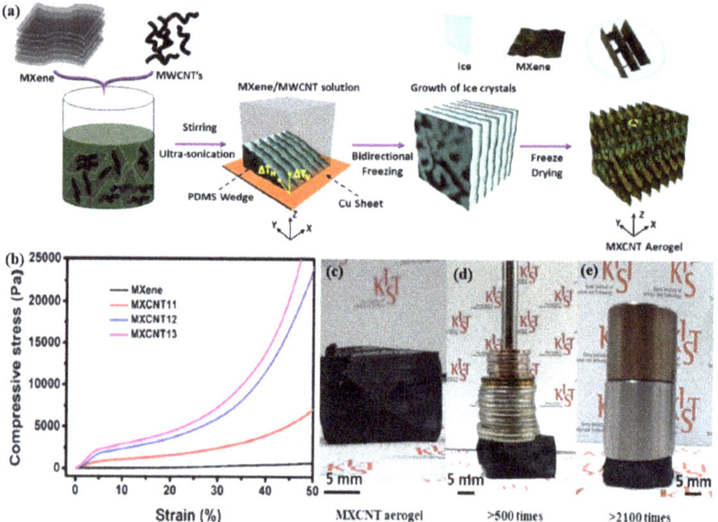

Figure 11. (**a**) Preparation of MXene/CNT hybrid aerogels through bidirectional freezing approach (**b**) Stress-strain curves of pristine MXene aerogels and MXene/CNT hybrid aerogels samples (**c–e**) MXene/CNT hybrid aerogels supporting more than >500 and >2100 times of its weight with no obvious collapsing. Adapted and reproduced with permission from [115]. Copyright ACS, 2019.

5. Summary and Perspectives

In summary, this review emphasized the recent advances in the mechanical properties of self-standing MXenes, including elastic properties, bending rigidity, and adhesion and sliding resistance from the experimental and theoretical views. This is, besides, to compare the mechanical properties of self-standing MXenes with hybrid MXenes along with their utilization as supercapacitors. Both experimental and theoretical calculations implied the significant effect of shape (i.e., layer thickness, interlayer spacing, dimensional, and porosity), preparation method, type (i.e., carbides or nitrides), composition (i.e., mono-/binary/multi-metals, doping, defects, and decoration with nanoparticles or single atoms), and functional groups (O, OH, and F) on enhancement the mechanical properties of MXenes. These features endowed the mechanical properties of MXenes are found to be closer to various 2D materials such as graphene, molybdenum disulfide, and boronitrene.

Despite the significant progress achieved in the rational design of self-standing MXenes, their mechanical properties are frequently investigated theoretically rather than experimentally. Additionally, the preparation approaches of MXenes entail multiple complicated steps, hazard chemicals, and without precise monitoring, shape, composition, and surface/bulk functionalities. However, the theoretical calculations predicted the synthesis of dozens of MXenes with outstanding mechanical merits coupled with electrical conductivity, high surface area, and ion adsorption/storage properties, which leaves extensive gates for the utilization of MXenes in various applications such as flexible devices, energy production/storage devices, and sensors. To this end, the capacitance performances of MXenes were enhanced significantly via their integration with conductive polymers, carbon-based materials (i.e., graphene, carbon nanotubes), and doping or functionalization metals (i.e., transition metals, noble metals, non-metals traces, semiconductors). Thereby, the mechanical properties of self-standing MXens and their mechanism should be highlighted experimentally rather than through theoretically. Also, the combination between MXenes and other carbon-based materials and novel metallic nano architectonics can lead to impressive properties and applications [116–119]. Thus, the presented review can provide a guided roadmap for the scientists to design novel MXenes for the coming generations of energy conversion and storage devices as well as smart sensors.

Author Contributions: All authors contributed equally to this work. Collecting the data and writing-original draft preparation, Y.I. and A.M. contributed equally; writing-review and editing, K.E.; review editing, A.M.A. (Ahmed M. Abdelgawad), supervision and project administration, A.M.A. (Aboubakr M. Abdullah) and A.E. All authors have read and agreed to the published version of the manuscript.

Funding: This publication was supported by Qatar University, internal grant number QUHI-CAS-19/20-1. The findings achieved herein are solely the responsibility of the authors.

Acknowledgments: The authors gratefully thank Hydro and Qatar Aluminium Limited (QSC).

Conflicts of Interest: The authors declare no conflict of interest.

References

1. Liu, M.; Zhang, R.; Chen, W. Graphene-supported nanoelectrocatalysts for fuel cells: Synthesis, properties, and applications. *Chem. Rev.* **2014**, *114*, 5117–5160. [CrossRef] [PubMed]
2. Lipatov, A.; Lu, H.; Alhabeb, M.; Anasori, B.; Gruverman, A.; Gogotsi, Y.; Sinitskii, A. Elastic properties of 2D Ti3C2Tx MXene monolayers and bilayers. *Sci. Adv.* **2018**, *4*, eaat0491. [CrossRef] [PubMed]
3. Eid, K.; Sliem, M.H.; Al-Kandari, H.; Sharaf, M.A.; Abdullah, A.M. Rational synthesis of porous graphitic-like carbon nitride nanotubes codoped with Au and Pd as an efficient catalyst for carbon monoxide oxidation. *Langmuir* **2019**, *35*, 3421–3431. [CrossRef] [PubMed]
4. Eid, K.; Sliem, M.H.; Abdullah, A.M. Unraveling template-free fabrication of carbon nitride nanorods codoped with Pt and Pd for efficient electrochemical and photoelectrochemical carbon monoxide oxidation at room temperature. *Nanoscale* **2019**, *11*, 11755–11764. [CrossRef] [PubMed]
5. Abdu, H.I.; Eid, K.; Abdullah, A.M.; Han, Z.; Ibrahim, M.H.; Shan, D.; Chen, J.; Elzatahry, A.A.; Lu, X. Unveiling one-pot scalable fabrication of reusable carboxylated heterogeneous carbon-based catalysts from eucalyptus plant with the assistance of dry ice for selective hydrolysis of eucalyptus biomass. *Renew. Energy* **2020**, *153*, 998–1004. [CrossRef]
6. Anasori, B.; Gogotsi, Y. *2D Metal. Carbides and Nitrides (MXenes): Structure, Properties and Applications*; Springer International Publishing: Berlin/Heidelberg, Germany, 2019; p. 534. [CrossRef]
7. Idris Abdu, H.; Eid, K.; Abdullah, A.M.; Sliem, M.H.; Elzatahry, A.; Lu, X. Dry ice-mediated rational synthesis of edge-carboxylated crumpled graphene nanosheets for selective and prompt hydrolysis of cellulose and eucalyptus lignocellulose under ambient reaction conditions. *Green Chem.* **2020**, *22*, 5437–5446. [CrossRef]
8. Deysher, G.; Shuck, C.E.; Hantanasirisakul, K.; Frey, N.C.; Foucher, A.C.; Maleski, K.; Sarycheva, A.; Shenoy, V.B.; Stach, E.A.; Anasori, B.; et al. Synthesis of Mo4VAlC4 MAX Phase and Two-Dimensional Mo4VC4 MXene with Five Atomic Layers of Transition Metals. *ACS Nano* **2020**, *14*, 204–217. [CrossRef]
9. Jun, B.M.; Her, N.; Park, C.M.; Yoon, Y. Effective removal of Pb(ii) from synthetic wastewater using Ti3C2T: X MXene. *Environ. Sci. Water Res. Technol.* **2020**, *6*, 173–180. [CrossRef]

10. Han, M.; Shuck, C.E.; Rakhmanov, R.; Parchment, D.; Anasori, B.; Koo, C.M.; Friedman, G.; Gogotsi, Y. Beyond $Ti_3C_2T_x$: MXenes for Electromagnetic Interference Shielding. *ACS Nano* **2020**, *14*, 5008–5016. [CrossRef]

11. Amiri, A.; Chen, Y.; Bee Teng, C.; Naraghi, M. Porous nitrogen-doped MXene-based electrodes for capacitive deionization. *Energy Storage Mater.* **2020**, *25*, 731–739. [CrossRef]

12. Rasool, K.; Pandey, R.P.; Rasheed, P.A.; Buczek, S.; Gogotsi, Y.; Mahmoud, K.A. Water treatment and environmental remediation applications of two-dimensional metal carbides (MXenes). *Mater. Today* **2019**, *30*, 80–102. [CrossRef]

13. Ibrahim, Y.; Kassab, A.; Eid, K.; Abdullah, A.M.; Ozoemena, K.I.; Elzatahry, A. Unveiling Fabrication and Environmental Remediation of MXene-Based Nanoarchitectures in Toxic Metals Removal from Wastewater: Strategy and Mechanism. *Nanomaterials* **2020**, *10*, 885. [CrossRef] [PubMed]

14. Barsoum, M.; El-Raghy, T. The MAX phases: Unique new carbide and nitride materials ternary ceramics turn out to be surprisingly soft and machinable, yet also heat-tolerant, strong and lightweight. *Am. Sci.* **2001**, *89*, 334–343. [CrossRef]

15. Barsoum, M.W.; Radovic, M. Elastic and Mechanical Properties of the MAX Phases. *Annu. Rev. Mater. Res.* **2011**, *41*, 195–227. [CrossRef]

16. Ling, Z.; Ren, C.E.; Zhao, M.-Q.; Yang, J.; Giammarco, J.M.; Qiu, J.; Barsoum, M.W.; Gogotsi, Y. Flexible and conductive MXene films and nanocomposites with high capacitance. *Proc. Natl. Acad. Sci. USA* **2014**, *111*, 16676. [CrossRef] [PubMed]

17. Zhan, X.; Si, C.; Zhou, J.; Sun, Z. MXene and MXene-based composites: Synthesis, properties and environment-related applications. *Nanoscale Horiz.* **2020**, *5*, 235–258. [CrossRef]

18. Li, Y.; Wang, J.; Liu, X.; Zhang, S. Towards a molecular understanding of cellulose dissolution in ionic liquids: Anion/cation effect, synergistic mechanism and physicochemical aspects. *Chem. Sci.* **2018**, *9*, 4027–4043. [CrossRef] [PubMed]

19. Chen, J.; Huang, Q.; Huang, H.; Mao, L.; Liu, M.; Zhang, X.; Wei, Y. Recent progress and advances in the environmental applications of MXene related materials. *Nanoscale* **2020**, *12*, 3574–3592. [CrossRef]

20. Li, J.; Li, X.; Van der Bruggen, B. An MXene-based membrane for molecular separation. *Environ. Sci. Nano* **2020**. [CrossRef]

21. Zhao, P.; Jian, M.; Zhang, Q.; Xu, R.; Liu, R.; Zhang, X.; Liu, H. A new paradigm of ultrathin 2D nanomaterial adsorbents in aqueous media: Graphene and GO, MoS_2, MXenes, and 2D MOFs. *J. Mater. Chem. A* **2019**, *7*, 16598–16621. [CrossRef]

22. Yang, B.; She, Y.; Zhang, C.; Kang, S.; Zhou, J.; Hu, W. Nitrogen Doped Intercalation $TiO_2/TiN/Ti_3C_2T_x$ Nanocomposite Electrodes with Enhanced Pseudocapacitance. *Nanomaterials* **2020**, *10*, 345. [CrossRef]

23. Yao, Z.; Sun, H.; Sui, H.; Liu, X. Construction of $BPQDs/Ti_3C_2@TiO_2$ Composites with Favorable Charge Transfer Channels for Enhanced Photocatalytic Activity under Visible Light Irradiation. *Nanomaterials* **2020**, *10*, 452. [CrossRef]

24. Fu, Z.; Wang, N.; Legut, D.; Si, C.; Zhang, Q.; Du, S.; Germann, T.C.; Francisco, J.S.; Zhang, R. Rational Design of Flexible Two-Dimensional MXenes with Multiple Functionalities. *Chem. Rev.* **2019**, *119*, 11980–12031. [CrossRef] [PubMed]

25. Anasori, B.; Lukatskaya, M.R.; Gogotsi, Y. 2D metal carbides and nitrides (MXenes) for energy storage. *Nat. Rev. Mater.* **2017**, *2*, 16098. [CrossRef]

26. Kumar, K.S.; Choudhary, N.; Jung, Y.; Thomas, J. Recent Advances in Two-Dimensional Nanomaterials for Supercapacitor Electrode Applications. *ACS Energy Lett.* **2018**, *3*, 482–495. [CrossRef]

27. Zha, X.H.; Luo, K.; Li, Q.; Huang, Q.; He, J.; Wen, X.; Du, S. Role of the surface effect on the structural, electronic and mechanical properties of the carbide MXenes. *EPL* **2015**, *111*, 26007. [CrossRef]

28. Fu, Z.; Zhang, S.; Legut, D.; Germann, T.C.; Si, C.; Du, S.; Francisco, J.S.; Zhang, R. A synergetic stabilization and strengthening strategy for two-dimensional ordered hybrid transition metal carbides. *Phys. Chem. Chem. Phys.* **2018**, *20*, 29684–29692. [CrossRef]

29. Yorulmaz, U.; Özden, A.; Perkgöz, N.K.; Ay, F.; Sevik, C. Vibrational and mechanical properties of single layer MXene structures: A first-principles investigation. *Nanotechnology* **2016**, *27*, 335702. [CrossRef]

30. Zhang, H.; Fu, Z.; Zhang, R.; Zhang, Q.; Tian, H.; Legut, D.; Germann, T.C.; Guo, Y.; Du, S.; Francisco, J.S. Designing flexible 2D transition metal carbides with strain-controllable lithium storage. *Proc. Natl. Acad. Sci. USA* **2017**, *114*, E11082–E11091. [CrossRef]

31. Si, C.; Duan, W.; Liu, Z.; Liu, F. Electronic strengthening of graphene by charge doping. *Phys. Rev. Lett.* **2012**, *109*, 226802. [CrossRef]

32. Plummer, G.; Anasori, B.; Gogotsi, Y.; Tucker, G.J. Nanoindentation of monolayer $Ti_{n+1}C_nT_x$ MXenes via atomistic simulations: The role of composition and defects on strength. *Comput. Mater. Sci.* **2019**, *157*, 168–174. [CrossRef]

33. Zhang, L.; Yu, J.; Yang, M.; Xie, Q.; Peng, H.; Liu, Z. Janus graphene from asymmetric two-dimensional chemistry. *Nat. Commun.* **2013**, *4*, 1443. [CrossRef] [PubMed]

34. Zha, X.-H.; Yin, J.; Zhou, Y.; Huang, Q.; Luo, K.; Lang, J.; Francisco, J.S.; He, J.; Du, S. Intrinsic Structural, Electrical, Thermal, and Mechanical Properties of the Promising Conductor Mo_2C MXene. *J. Phys. Chem. C* **2016**, *120*, 15082–15088. [CrossRef]

35. Peng, Q.; De, S. Outstanding mechanical properties of monolayer MoS_2 and its application in elastic energy storage. *Phys. Chem. Chem. Phys.* **2013**, *15*, 19427–19437. [CrossRef] [PubMed]

36. Wu, D.; Wang, S.; Yuan, J.; Yang, B.; Chen, H. Modulation of the electronic and mechanical properties of phagraphene via hydrogenation and fluorination. *Phys. Chem. Chem. Phys.* **2017**, *19*, 11771–11777. [CrossRef]

37. Guo, Z.; Zhou, J.; Si, C.; Sun, Z. Flexible two-dimensional $Ti_{n+1}C_n$ (n = 1, 2 and 3) and their functionalized MXenes predicted by density functional theories. *Phys. Chem. Chem. Phys.* **2015**, *17*, 15348–15354. [CrossRef] [PubMed]

38. Zhou, J.; Zha, X.; Chen, F.Y.; Ye, Q.; Eklund, P.; Du, S.; Huang, Q. A Two-Dimensional Zirconium Carbide by Selective Etching of Al_3C_3 from Nanolaminated $Zr_3Al_3C_5$. *Angew. Chem. Int. Ed.* **2016**, *55*, 5008–5013. [CrossRef] [PubMed]

39. Fu, Z.H.; Zhang, Q.F.; Legut, D.; Si, C.; Germann, T.C.; Lookman, T.; Du, S.Y.; Francisco, J.S.; Zhang, R.F. Stabilization and strengthening effects of functional groups in two-dimensional titanium carbide. *Phys. Rev. B* **2016**, *94*, 104103. [CrossRef]

40. Bai, Y.; Zhou, K.; Srikanth, N.; Pang, J.H.L.; He, X.; Wang, R. Dependence of elastic and optical properties on surface terminated groups in two-dimensional MXene monolayers: A first-principles study. *RSC Adv.* **2016**, *6*, 35731–35739. [CrossRef]

41. Kurtoglu, M.; Naguib, M.; Gogotsi, Y.; Barsoum, M. First Principles Study of Two-dimensional Early Transition Metal Carbides. *MRS Commun.* **2012**, *2*. [CrossRef]

42. Borysiuk, V.N.; Mochalin, V.N.; Gogotsi, Y. Molecular dynamic study of the mechanical properties of two-dimensional titanium carbides $Ti_{n+1}C_n$ (MXenes). *Nanotechnology* **2015**, *26*, 265705. [CrossRef] [PubMed]

43. Lee, C.; Wei, X.; Kysar, J.W.; Hone, J. Measurement of the Elastic Properties and Intrinsic Strength of Monolayer Graphene. *Science* **2008**, *321*, 385–388. [CrossRef] [PubMed]

44. Jin, W.; Wu, S.; Wang, Z. Structural, electronic and mechanical properties of two-dimensional Janus transition metal carbides and nitrides. *Phys. E Low Dimens. Syst. Nanostruct.* **2018**, *103*, 307–313. [CrossRef]

45. Khazaei, M.; Ranjbar, A.; Arai, M.; Sasaki, T.; Yunoki, S. Electronic properties and applications of MXenes: A theoretical review. *J. Mater. Chem. C* **2017**, *5*, 2488–2503. [CrossRef]

46. Çakır, D.; Peeters, F.M.; Sevik, C. Mechanical and thermal properties of h -MX2 (M = Cr, Mo, W.; X = O, S, Se, Te) monolayers: A comparative study. *Appl. Phys. Lett.* **2014**, *41*, 10891–10896. [CrossRef]

47. Duerloo, K.A.N.; Ong, M.T.; Reed, E.J. Intrinsic Piezoelectricity in Two-Dimensional Materials. *J. Phys. Chem. Lett.* **2012**, *3*, 2871–2876. [CrossRef]

48. Zhou, J.; Zha, X.; Zhou, X.; Chen, F.; Gao, G.; Wang, S.; Shen, C.; Chen, T.; Zhi, C.; Eklund, P.; et al. Synthesis and Electrochemical Properties of Two-Dimensional Hafnium Carbide. *ACS Nano* **2017**, *11*, 3841–3850. [CrossRef]

49. Holm, A.; Park, J.; Goodman, E.D.; Zhang, J.; Sinclair, R.; Cargnello, M.; Frank, C.W. Synthesis, Characterization, and Light-Induced Spatial Charge Separation in Janus Graphene Oxide. *Chem. Mater.* **2018**, *30*, 2084–2092. [CrossRef]

50. Xu, C.; Wang, L.; Liu, Z.; Chen, L.; Guo, J.; Kang, N.; Ma, X.-L.; Cheng, H.-M.; Ren, W. Large-area high-quality 2D ultrathin Mo2C superconducting crystals. *Nat. Mater.* **2015**, *14*, 1135–1141. [CrossRef]

51. Chakraborty, P.; Das, T.; Nafday, D.; Saha-Dasgupta, T. Manipulating the mechanical properties of Ti_2C MXene: Effect of substitutional doping. *Phys. Rev. B* **2017**, *95*, 184106. [CrossRef]

52. Andrew, R.C.; Mapasha, R.E.; Ukpong, A.M.; Chetty, N. Mechanical properties of graphene and boronitrene. *Phys. Rev. B* **2012**, *85*, 125428. [CrossRef]

53. Wu, D.; Wang, S.; Zhang, S.; Yuan, J.; Yang, B.; Chen, H. Highly negative Poisson's ratio in a flexible two-dimensional tungsten carbide monolayer. *Phys. Chem. Chem. Phys.* **2018**, *20*, 18924–18930. [CrossRef]

54. Isaacs, E.B.; Marianetti, C.A. Ideal strength and phonon instability of strained monolayer materials. *Phys. Rev. B* **2014**, *89*, 184111. [CrossRef]

55. Chen, S.; Fu, Z.; Zhang, H.; Legut, D.; Germann, T.C.; Zhang, Q.; Du, S.; Francisco, J.S.; Zhang, R. Surface Electrochemical Stability and Strain-Tunable Lithium Storage of Highly Flexible 2D Transition Metal Carbides. *Adv. Funct. Mater.* **2018**, *28*, 1804867. [CrossRef]

56. Radajewski, M.; Henschel, S.; Grützner, S.; Krüger, L.; Schimpf, C.; Chmelik, D.; Rafaja, D. Microstructure and mechanical properties of bulk TiN–AlN composites processed by FAST/SPS. *Ceram. Int.* **2016**, *42*, 10220–10227. [CrossRef]

57. Ivashchenko, V.I.; Turchi, P.E.A.; Gonis, A.; Ivashchenko, L.A.; Skrynskii, P.L. Electronic origin of elastic properties of titanium carbonitride alloys. *Met. Mater. Trans. A Phys. Met. Mater. Sci.* **2006**, *37*, 3391–3396. [CrossRef]

58. Yang, Q.; Lengauer, W.; Koch, T.; Scheerer, M.; Smid, I. Hardness and elastic properties of Ti(C_xN_{1-x}), Zr(C_xN_{1-x}) and Hf(C_xN_{1-x}). *J. Alloys Compd.* **2000**, *309*, L5–L9. [CrossRef]

59. Come, J.; Xie, Y.; Naguib, M.; Jesse, S.; Kalinin, S.V.; Gogotsi, Y.; Kent, P.R.C.; Balke, N. Nanoscale Elastic Changes in 2D $Ti_3C_2T_x$ (MXene) Pseudocapacitive Electrodes. *Adv. Energy Mater.* **2016**, *6*, 1–9. [CrossRef]

60. Borysiuk, V.N.; Mochalin, V.N.; Gogotsi, Y. Bending rigidity of two-dimensional titanium carbide (MXene) nanoribbons: A molecular dynamics study. *Comput. Mater. Sci.* **2018**, *143*, 418–424. [CrossRef]

61. Hu, T.; Yang, J.; Li, W.; Wang, X.; Li, C. Quantifying the rigidity of 2D carbides (MXenes). *Phys. Chem. Chem. Phys.* **2020**, *22*, 2115–2121. [CrossRef]

62. Suk, J.W.; Piner, R.D.; An, J.; Ruoff, R.S. Mechanical properties of monolayer graphene oxide. *ACS Nano* **2010**, *4*, 6557–6564. [CrossRef] [PubMed]

63. Liu, L.; Zhang, J.; Zhao, J.; Liu, F. Mechanical properties of graphene oxides. *Nanoscale* **2012**, *4*, 5910–5916. [CrossRef] [PubMed]

64. Wang, S.; Li, J.-X.; Du, Y.; Cui, C. First-principles study on structural, electronic and elastic properties of graphene-like hexagonal Ti_2C monolayer. *Comput. Mater. Sci.* **2014**, *83*, 290–293. [CrossRef]

65. Hu, T.; Hu, M.; Li, Z.; Zhang, H.; Zhang, C.; Wang, J.; Wang, X. Interlayer coupling in two-dimensional titanium carbide MXenes. *Phys. Chem. Chem. Phys.* **2016**, *18*, 20256–20260. [CrossRef]

66. Zhang, H.; Fu, Z.H.; Legut, D.; Germann, T.C.; Zhang, R.F. Stacking stability and sliding mechanism in weakly bonded 2D transition metal carbides by van der Waals force. *RSC Adv.* **2017**, *7*, 55912–55919. [CrossRef]

67. Zhang, D.; Ashton, M.; Ostadhossein, A.; Van Duin, A.C.T.; Hennig, R.G.; Sinnott, S.B. Computational study of low interlayer friction in $Ti_{n+1}C_n$ (n = 1, 2, and 3) MXene. *ACS Appl. Mater. Interfaces* **2017**, *9*, 34467–34479. [CrossRef]

68. Li, Y.; Huang, S.; Wei, C.; Wu, C.; Mochalin, V.N. Adhesion of two-dimensional titanium carbides (MXenes) and graphene to silicon. *Nat. Commun.* **2019**, *10*, 1–8. [CrossRef]

69. Wu, Z.-S.; Feng, X.; Cheng, H.-M. Recent advances in graphene-based planar micro-supercapacitors for on-chip energy storage. *Natl. Sci. Rev.* **2014**, *1*, 277. [CrossRef]

70. Zhang, C.; Ma, Y.; Zhang, X.; Abdolhosseinzadeh, S.; Sheng, H.; Lan, W.; Pakdel, A.; Heier, J.; Nüesch, F. Two-Dimensional Transition Metal Carbides and Nitrides (MXenes): Synthesis, Properties, and Electrochemical Energy Storage Applications. *Energy Environ. Mater.* **2020**, *3*, 29–55. [CrossRef]

71. Zhang, C.; Nicolosi, V. Graphene and MXene-based transparent conductive electrodes and supercapacitors. *Energy Storage Mater.* **2019**, *16*, 102–125. [CrossRef]

72. Jiang, Q.; Lei, Y.; Liang, H.; Xi, K.; Xia, C.; Alshareef, H.N. Review of MXene electrochemical microsupercapacitors. *Energy Storage Mater.* **2020**, *27*, 78–95. [CrossRef]

73. Wang, K.; Zheng, B.; Mackinder, M.; Baule, N.; Qiao, H.; Jin, H.; Schuelke, T.; Fan, Q.H. Graphene wrapped MXene via plasma exfoliation for all-solid-state flexible supercapacitors. *Energy Storage Mater.* **2019**, *20*, 299–306. [CrossRef]

74. Rakhi, R.B.; Ahmed, B.; Hedhili, M.N.; Anjum, D.H.; Alshareef, H.N. Effect of Postetch Annealing Gas Composition on the Structural and Electrochemical Properties of Ti_2CTx MXene Electrodes for Supercapacitor Applications. *Chem. Mater.* **2015**, *27*, 5314–5323. [CrossRef]

75. Tao, Q.; Dahlqvist, M.; Lu, J.; Kota, S.; Meshkian, R.; Halim, J.; Palisaitis, J.; Hultman, L.; Barsoum, M.W.; Persson, P.O.Å.; et al. Two-dimensional Mo1.33C MXene with divacancy ordering prepared from parent 3D laminate with in-plane chemical ordering. *Nat. Commun.* **2017**, *8*, 14949. [CrossRef] [PubMed]

76. Lukatskaya, M.R.; Mashtalir, O.; Ren, C.E.; Dall'Agnese, Y.; Rozier, P.; Taberna, P.L.; Naguib, M.; Simon, P.; Barsoum, M.W.; Gogotsi, Y. Cation Intercalation and High Volumetric Capacitance of Two-Dimensional Titanium Carbide. *Science* **2013**, *341*, 1502–1505. [CrossRef]

77. Ghidiu, M.; Lukatskaya, M.R.; Zhao, M.-Q.; Gogotsi, Y.; Barsoum, M.W. Conductive two-dimensional titanium carbide 'clay' with high volumetric capacitance. *Nature* **2014**, *516*, 78–81. [CrossRef] [PubMed]

78. Lukatskaya, M.R.; Kota, S.; Lin, Z.; Zhao, M.-Q.; Shpigel, N.; Levi, M.D.; Halim, J.; Taberna, P.-L.; Barsoum, M.W.; Simon, P.; et al. Ultra-high-rate pseudocapacitive energy storage in two-dimensional transition metal carbides. *Nat. Energy* **2017**, *2*, 1–6. [CrossRef]

79. Shan, Q.; Mu, X.; Alhabeb, M.; Shuck, C.E.; Pang, D.; Zhao, X.; Chu, X.-F.; Wei, Y.; Du, F.; Chen, G.; et al. Two-dimensional vanadium carbide (V2C) MXene as electrode for supercapacitors with aqueous electrolytes. *Electrochem. Commun.* **2018**, *96*, 103–107. [CrossRef]

80. Halim, J.; Kota, S.; Lukatskaya, M.R.; Naguib, M.; Zhao, M.-Q.; Moon, E.J.; Pitock, J.; Nanda, J.; May, S.J.; Gogotsi, Y.; et al. Synthesis and Characterization of 2D Molybdenum Carbide (MXene). *Adv. Funct. Mater.* **2016**, *26*, 3118–3127. [CrossRef]

81. Dall'Agnese, Y.; Lukatskaya, M.R.; Cook, K.M.; Taberna, P.-L.; Gogotsi, Y.; Simon, P. High capacitance of surface-modified 2D titanium carbide in acidic electrolyte. *Electrochem. Commun.* **2014**, *48*, 118–122. [CrossRef]

82. Zhan, C.; Sun, W.; Kent, P.R.C.; Naguib, M.; Gogotsi, Y.; Jiang, D.-E. Computational Screening of MXene Electrodes for Pseudocapacitive Energy Storage. *J. Phys. Chem. C* **2019**, *123*, 315–321. [CrossRef]

83. Zhan, C.; Naguib, M.; Lukatskaya, M.; Kent, P.R.; Gogotsi, Y.; Jiang, D.-E. Understanding the MXene pseudocapacitance. *J. Phys. Chem. Lett.* **2018**, *9*, 1223–1228. [CrossRef] [PubMed]

84. Boota, M.; Anasori, B.; Voigt, C.; Zhao, M.Q.; Barsoum, M.W.; Gogotsi, Y. Pseudocapacitive Electrodes Produced by Oxidant-Free Polymerization of Pyrrole between the Layers of 2D Titanium Carbide (MXene). *Adv. Mater. (Deerfield Beach, Fla)* **2016**, *28*, 1517–1522. [CrossRef]

85. Wang, X.; Kajiyama, S.; Iinuma, H.; Hosono, E.; Oro, S.; Moriguchi, I.; Okubo, M.; Yamada, A. Pseudocapacitance of MXene nanosheets for high-power sodium-ion hybrid capacitors. *Nat. Commun.* **2015**, *6*, 6544. [CrossRef]

86. Li, H.; Hou, Y.; Wang, F.; Lohe, M.R.; Zhuang, X.; Niu, L.; Feng, X. Flexible All-Solid-State Supercapacitors with High Volumetric Capacitances Boosted by Solution Processable MXene and Electrochemically Exfoliated Graphene. *Adv. Energy Mater.* **2017**, *7*, 1601847. [CrossRef]

87. Tang, Q.; Zhou, Z.; Shen, P. Are MXenes Promising Anode Materials for Li Ion Batteries? Computational Studies on Electronic Properties and Li Storage Capability of Ti_3C_2 and $Ti_3C_2X_2$ (X = F, OH) Monolayer. *J. Am. Chem. Soc.* **2012**, *134*, 16909–16916. [CrossRef] [PubMed]

88. Wang, L.; Zhang, H.; Wang, B.; Shen, C.; Zhang, C.; Hu, Q.; Zhou, A.; Liu, B. Synthesis and electrochemical performance of $Ti_3C_2T_x$ with hydrothermal process. *Electron. Mater. Lett.* **2016**, *12*, 702–710. [CrossRef]

89. Couly, C.; Alhabeb, M.; Van Aken, K.L.; Kurra, N.; Gomes, L.; Navarro-Suárez, A.M.; Anasori, B.; Alshareef, H.N.; Gogotsi, Y. Asymmetric Flexible MXene-Reduced Graphene Oxide Micro-Supercapacitor. *Adv. Electron. Mater.* **2018**, *4*, 1700339. [CrossRef]

90. Fu, Q.; Wang, X.; Zhang, N.; Wen, J.; Li, L.; Gao, H.; Zhang, X. Self-assembled $Ti_3C_2T_x$/SCNT composite electrode with improved electrochemical performance for supercapacitor. *J. Colloid Interface Sci.* **2018**, *511*, 128–134. [CrossRef]

91. Jiang, Q.; Wu, C.; Wang, Z.; Wang, A.C.; He, J.-H.; Wang, Z.L.; Alshareef, H.N. MXene electrochemical microsupercapacitor integrated with triboelectric nanogenerator as a wearable self-charging power unit. *Nano Energy* **2018**, *45*, 266–272. [CrossRef]

92. Xia, Q.X.; Shinde, N.M.; Yun, J.M.; Zhang, T.; Mane, R.S.; Mathur, S.; Kim, K.H. Bismuth Oxychloride/MXene symmetric supercapacitor with high volumetric energy density. *Electrochim. Acta* **2018**, *271*, 351–360. [CrossRef]

93. Zhang, C.; Kremer, M.P.; Seral-Ascaso, A.; Park, S.-H.; McEvoy, N.; Anasori, B.; Gogotsi, Y.; Nicolosi, V. Stamping of Flexible, Coplanar Micro-Supercapacitors Using MXene Inks. *Adv. Funct. Mater.* **2018**, *28*, 1705506. [CrossRef]

94. Hu, M.; Li, Z.; Zhang, H.; Hu, T.; Zhang, C.; Wu, Z.; Wang, X. Self-assembled Ti3C2Tx MXene film with high gravimetric capacitance. *Chem. Commun.* **2015**, *51*, 13531–13533. [CrossRef]

95. Tang, Y.; Zhu, J.; Yang, C.; Wang, F. Enhanced Capacitive Performance Based on Diverse Layered Structure of Two-Dimensional Ti3C2MXene with Long Etching Time. *J. Electrochem. Soc.* **2016**, *163*, A1975–A1982. [CrossRef]

96. Mashtalir, O.; Lukatskaya, M.R.; Kolesnikov, A.I.; Raymundo-Piñero, E.; Naguib, M.; Barsoum, M.W.; Gogotsi, Y. The effect of hydrazine intercalation on the structure and capacitance of 2D titanium carbide (MXene). *Nanoscale* **2016**, *8*, 9128–9133. [CrossRef] [PubMed]

97. Fu, Q.; Wen, J.; Zhang, N.; Wu, L.; Zhang, M.; Lin, S.; Gao, H.; Zhang, X. Free-standing $Ti_3C_2T_x$ electrode with ultrahigh volumetric capacitance. *RSC Adv.* **2017**, *7*, 11998–12005. [CrossRef]

98. Wen, Y.; Rufford, T.E.; Chen, X.; Li, N.; Lyu, M.; Dai, L.; Wang, L. Nitrogen-doped $Ti_3C_2T_x$ MXene electrodes for high-performance supercapacitors. *Nano Energy* **2017**, *38*, 368–376. [CrossRef]

99. Kurra, N.; Ahmed, B.; Gogotsi, Y.; Alshareef, H.N. MXene-on-Paper Coplanar Microsupercapacitors. *Adv. Energy Mater.* **2016**, *6*, 1601372. [CrossRef]

100. Wang, Y.; Dou, H.; Wang, J.; Ding, B.; Xu, Y.; Chang, Z.; Hao, X. Three-dimensional porous MXene/layered double hydroxide composite for high performance supercapacitors. *J. Power Sources* **2016**, *327*, 221–228. [CrossRef]

101. Li, J.; Yuan, X.; Lin, C.; Yang, Y.; Xu, L.; Du, X.; Xie, J.; Lin, J.; Sun, J. Achieving High Pseudocapacitance of 2D Titanium Carbide (MXene) by Cation Intercalation and Surface Modification. *Adv. Energy Mater.* **2017**, *7*, 1602725. [CrossRef]

102. Zhao, M.-Q.; Ren, C.E.; Ling, Z.; Lukatskaya, M.R.; Zhang, C.; Van Aken, K.L.; Barsoum, M.W.; Gogotsi, Y. Flexible MXene/Carbon Nanotube Composite Paper with High Volumetric Capacitance. *Adv. Mater.* **2015**, *27*, 339–345. [CrossRef] [PubMed]

103. Yan, J.; Ren, C.E.; Maleski, K.; Hatter, C.B.; Anasori, B.; Urbankowski, P.; Sarycheva, A.; Gogotsi, Y. Flexible MXene/Graphene Films for Ultrafast Supercapacitors with Outstanding Volumetric Capacitance. *Adv. Funct. Mater.* **2017**, *27*, 1701264. [CrossRef]

104. Dall'Agnese, Y.; Rozier, P.; Taberna, P.-L.; Gogotsi, Y.; Simon, P. Capacitance of two-dimensional titanium carbide (MXene) and MXene/carbon nanotube composites in organic electrolytes. *J. Power Sources* **2016**, *306*, 510–515. [CrossRef]

105. Yan, P.; Zhang, R.; Jia, J.; Wu, C.; Zhou, A.; Xu, J.; Zhang, X. Enhanced supercapacitive performance of delaminated two-dimensional titanium carbide/carbon nanotube composites in alkaline electrolyte. *J. Power Sources* **2015**, *284*, 38–43. [CrossRef]

106. Zhu, J.; Tang, Y.; Yang, C.; Wang, F.; Cao, M. Composites of TiO_2 Nanoparticles Deposited on Ti_3C_2 MXene Nanosheets with Enhanced Electrochemical Performance. *J. Electrochem. Soc.* **2016**, *163*, A785–A791. [CrossRef]

107. Tian, Y.; Yang, C.; Que, W.; Liu, X.; Yin, X.; Kong, L.B. Flexible and freestanding 2D titanium carbide film decorated with manganese oxide nanoparticles as a high volumetric capacity electrode for supercapacitor. *J. Power Sources* **2017**, *359*, 332–339. [CrossRef]

108. Zhu, M.; Huang, Y.; Deng, Q.; Zhou, J.; Pei, Z.; Xue, Q.; Huang, Y.; Wang, Z.; Li, H.; Huang, Q.; et al. Highly Flexible, Freestanding Supercapacitor Electrode with Enhanced Performance Obtained by Hybridizing Polypyrrole Chains with MXene. *Adv. Energy Mater.* **2016**, *6*, 1600969. [CrossRef]

109. Luo, J.; Zhang, W.; Yuan, H.; Jin, C.; Zhang, L.; Huang, H.; Liang, C.; Xia, Y.; Zhang, J.; Gan, Y.; et al. Pillared Structure Design of MXene with Ultralarge Interlayer Spacing for High-Performance Lithium-Ion Capacitors. *ACS Nano* **2017**, *11*, 2459–2469. [CrossRef]

110. Lin, Z.; Barbara, D.; Taberna, P.-L.; Van Aken, K.L.; Anasori, B.; Gogotsi, Y.; Simon, P. Capacitance of $Ti_3C_2T_x$ MXene in ionic liquid electrolyte. *J. Power Sources* **2016**, *326*, 575–579. [CrossRef]

111. Wang, X.; Mathis, T.S.; Li, K.; Lin, Z.; Vlcek, L.; Torita, T.; Osti, N.C.; Hatter, C.; Urbankowski, P.; Sarycheva, A.; et al. Influences from solvents on charge storage in titanium carbide MXenes. *Nat. Energy* **2019**, *4*, 241–248. [CrossRef]

112. Sobolčiak, P.; Ali, A.; Hassan, M.K.; Helal, M.I.; Tanvir, A.; Popelka, A.; Al-Maadeed, M.A.; Krupa, I.; Mahmoud, K.A. 2D $Ti_3C_2T_x$ (MXene)-reinforced polyvinyl alcohol (PVA) nanofibers with enhanced mechanical and electrical properties. *PLoS ONE* **2017**, *12*, e0183705. [CrossRef] [PubMed]

113. Wang, N.-N.; Wang, H.; Wang, Y.-Y.; Wei, Y.-H.; Si, J.-Y.; Yuen, A.C.Y.; Xie, J.-S.; Yu, B.; Zhu, S.-E.; Lu, H.-D.; et al. Robust, Lightweight, Hydrophobic, and Fire-Retarded Polyimide/MXene Aerogels for Effective Oil/Water Separation. *ACS Appl. Mater. Interfaces* **2019**, *11*, 40512–40523. [CrossRef] [PubMed]

114. Liu, J.; Zhang, L.; Li, C. Highly Stable, Transparent, and Conductive Electrode of Solution-Processed Silver Nanowire-Mxene for Flexible Alternating-Current Electroluminescent Devices. *Ind. Eng. Chem. Res.* **2019**, *58*, 21485–21492. [CrossRef]

115. Sambyal, P.; Iqbal, A.; Hong, J.; Kim, H.; Kim, M.-K.; Hong, S.M.; Han, M.; Gogotsi, Y.; Koo, C.M. Ultralight and Mechanically Robust $Ti_3C_2T_x$ Hybrid Aerogel Reinforced by Carbon Nanotubes for Electromagnetic Interference Shielding. *ACS Appl. Mater. Interfaces* **2019**, *11*, 38046–38054. [CrossRef] [PubMed]

116. Abualrejal, M.M.; Eid, K.; Tian, R.; Liu, L.; Chen, H.; Abdullah, A.M.; Wang, Z. Rational synthesis of three-dimensional core–double shell upconversion nanodendrites with ultrabright luminescence for bioimaging application. *Chem. Sci.* **2019**, *10*, 7591–7599. [CrossRef]

117. Abualrejal, M.M.; Eid, K.; Abdullah, A.M.; Numan, A.A.; Chen, H.; Zhang, H.; Wang, Z. Smart design of exquisite multidimensional multilayered sand-clock-like upconversion nanostructures with ultrabright luminescence as efficient luminescence probes for bioimaging application. *Microchim. Acta* **2020**, *187*, 1–13. [CrossRef]

118. Wu, F.; Eid, K.; Abdullah, A.M.; Niu, W.; Wang, C.; Lan, Y.; Elzatahry, A.A.; Xu, G. Unveiling One-Pot Template-Free Fabrication of Exquisite Multidimensional PtNi Multicube Nanoarchitectonics for the Efficient Electrochemical Oxidation of Ethanol and Methanol with a Great Tolerance for CO. *ACS Appl. Mater. Interfaces* **2020**, *12*, 31309–31318. [CrossRef]

119. Eid, K.; Sliem, M.H.; Jlassi, K.; Eldesoky, A.S.; Abdo, G.G.; Al-Qaradawi, S.Y.; Sharaf, M.A.; Abdullah, A.M.; Elzatahry, A.A. Precise fabrication of porous one-dimensional gC3N4 nanotubes doped with Pd and Cu atoms for efficient CO oxidation and CO_2 reduction. *Inorg. Chem. Commun.* **2019**, *107*, 107460. [CrossRef]

Review

Advances of RRAM Devices: Resistive Switching Mechanisms, Materials and Bionic Synaptic Application

Zongjie Shen [1,2], Chun Zhao [1,2,*], Yanfei Qi [1,3], Wangying Xu [4], Yina Liu [5], Ivona Z. Mitrovic [2], Li Yang [6] and Cezhou Zhao [1,2]

[1] Department of Electrical and Electronic Engineering, Xi'an Jiaotong-Liverpool University, Suzhou 215123, China; Zongjie.Shen@xjtlu.edu.cn (Z.S.); Yanfei.Qi01@xjtlu.edu.cn (Y.Q.); Cezhou.Zhao@xjtlu.edu.cn (C.Z.)

[2] Department of Electrical Engineering and Electronics, University of Liverpool, Liverpool L69 3BX, UK; Ivona@liverpool.ac.uk

[3] School of Electronic and Information Engineering, Xi'an Jiaotong University, Xi'an 710061, China

[4] College of Materials Science and Engineering, Shenzhen University, Shenzhen 518060, China; wyxu@szu.edu.cn

[5] Department of Mathematical Sciences, Xi'an Jiaotong-Liverpool University, Suzhou 215123, China; Yina.Liu@xjtlu.edu.cn

[6] Department of Chemistry, Xi'an Jiaotong-Liverpool University, Suzhou 215123, China; Li.Yang@xjtlu.edu.cn

* Correspondence: chun.zhao@xjtlu.edu.cn; Tel.: +86-(0)512-8816-1402

Received: 31 May 2020; Accepted: 19 July 2020; Published: 23 July 2020

Abstract: Resistive random access memory (RRAM) devices are receiving increasing extensive attention due to their enhanced properties such as fast operation speed, simple device structure, low power consumption, good scalability potential and so on, and are currently considered to be one of the next-generation alternatives to traditional memory. In this review, an overview of RRAM devices is demonstrated in terms of thin film materials investigation on electrode and function layer, switching mechanisms and artificial intelligence applications. Compared with the well-developed application of inorganic thin film materials (oxides, solid electrolyte and two-dimensional (2D) materials) in RRAM devices, organic thin film materials (biological and polymer materials) application is considered to be the candidate with significant potential. The performance of RRAM devices is closely related to the investigation of switching mechanisms in this review, including thermal-chemical mechanism (TCM), valance change mechanism (VCM) and electrochemical metallization (ECM). Finally, the bionic synaptic application of RRAM devices is under intensive consideration, its main characteristics such as potentiation/depression response, short-/long-term plasticity (STP/LTP), transition from short-term memory to long-term memory (STM to LTM) and spike-time-dependent plasticity (STDP) reveal the great potential of RRAM devices in the field of neuromorphic application.

Keywords: artificial intelligence; thin film; 2D materials; switching mechanisms; bionic synaptic application; RRAM

---

1. Introduction

In the neuromorphic system of the human brain, neuromorphic synapses are believed to be responsible for transmitting biological information. As the most typical and exquisite representative of the biological memory system, the human brain can store and process massively biological information with the adjustment of synaptic connection strength (synaptic weight) [1]. Meanwhile, they also provide complicated orthosympathetic space and energy balance [1,2]. In general, a biological synapse with the structure of dendrite, axon terminal and synaptic cleft is identified as a neuron linker that can permit a neuron to transmit neurotransmitters to another adjacent neuron [2]. For now, artificial synapse has attracted extensive interests and attention as the research focus in the development of artificial intelligence (AI) industry, which mainly concentrates on biomimetic synaptic functions simulated by memory functional devices in the computer industry [2]. As one of the most critical carriers for the inheritance of human civilization and the development of information technology, volatile and non-volatile memory (NVM) devices have always acted as dominating components of the development of memory devices. In the trend of miniaturization of electronic equipment, the demand for memory devices with small size, low voltage, low power consumption, and superior performance has been under extensive consideration. Currently, silicon-based flash memory, which dominates the market of data storage devices, has been difficult in meeting the needs of future development of data storage devices due to its physical and technological limitations, such as high operation voltage, high power consumption and low retention capacity [3–6]. As one of the emerging technologies of NVM, resistive random access memory (RRAM) device has been given attention as one of the next-generation memory devices. Simmons et al. reported a resistive switching (RS) characteristic in the memory device with the structure of Au/SiO$_2$/Al as early as in 1967, which provided the theoretical and experimental foundation of RRAM [7]. They demonstrated the new type of memory with the simple sandwiched structure, including top electrode (TE) layer, bottom electrode (BE) layer and an intermediate functional thin film layer (RS layer), which can be observed in Figure 1. This simple structure consisting of conductor/semiconductor or insulator/conductor makes RRAM cells be integrated into the passive crossbar array easily, with the size as small as 4F^2 (F—the minimum feature size), which can be evenly divided into n parts (4F^2/n) in the vertically stacked three-dimensional (3-D) architectures (n means the stacking layer number of the crossbar array) [1–3].

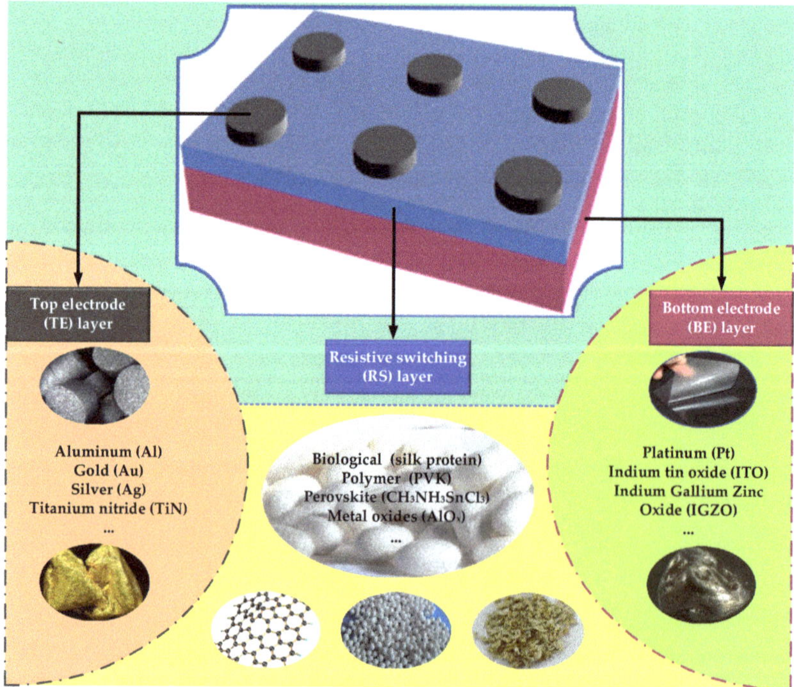

Figure 1. Illustration of sandwich structure for RRAM devices.

The simple sandwich structure of RRAM device indicates the significance on thin film materials investigation and application. In general, performance of a RRAM device largely depends on characteristics of thin film materials for electrode and RS layers. As the main conducting medium, top and bottom electrodes are considered to enhance the electrical conductivity of RRAM devices [2]. Therefore, some thin film materials with good electrical conductivity are then chosen as candidates of the electrode, such as pure metal, semiconductor and graphene materials [1–3]. For most bipolar RRAM devices, reactive metal like titanium (Ti) [8], nickel (Ni) [9], copper (Cu) [10] and silver (Ag) [11] are always used as TE due to their high metal activity. Compared with reactive metal, where Platinum (Pt) [12] always acts as BE in order to provide activity variation of RRAM devices. In addition, semiconductor materials like heavily doped silicon and indium tin oxide (ITO) are always chosen as electrodes due to their high electrical conductivity [13,14]. Currently, 2D thin film materials such as graphene and graphene oxide (GO) are also attracting increasing attention as candidates of electrode because of the excellent mobility of carriers and high thermal/electrical conductivity [15–17]. Compared with electrode thin film materials, more researchers focus on thin film materials applicated on RS medium, and inorganic materials are the main investigation objectives, including oxides and solid electrolyte, which will be discussed in Chapter 3 [11,18–27].

The performances of thin-film-material-based RS layers have a decisive influence on the performance of RRAM devices, which indicates that the fabrication methods of RS layers or synthesis technologies of thin film materials cannot be neglected [1–6]. Currently, several main fabrication technologies for RS layers have received extensive recognition by researchers, such as atomic layer deposition (ALD), magnetron sputtering, chemical vapor deposition (CVD) and solution-processed deposition. For most inorganic thin film materials used on RRAM devices (such as metal oxides and solid electrolyte), ALD and sputtering are two of the most advanced technologies due to stable performance of RS layers fabricated accordingly [5–11]. In addition, our previous study indicated that some other fabrication techniques for 2D thin film materials (such as graphene, GO and hexagonal boron nitride (h-BN), h-BN is considered as one of the most promising 2D materials with the function as 2D insulating template for high performance 2D electronic and photonic devices) have also been under intensive consideration, including nucleation and growth, liquid phase exfoliation and electrochemical exfoliation [21].

Apart from its typical physical structure, two basic switching states related to the conductive filament (CF) of RRAM devices, OFF and ON states, are commonly electrically characterized, which are also referred to as high-resistance-state (HRS) and low-resistance-state (LRS), respectively. With the transformation of resistance states, RRAM devices can complete the data storage process based on '0' or '1'. HRS value of the device shows the low conductance state while the device demonstrates a high conductance state with LRS value. ON/OFF ratio is determined by the ratio between HRS and LRS. With the applied voltage bias, the SET operation is defined from HRS to LRS and the RESET operation is the transition from LRS to HRS. Stop voltages of SET and RESET process are defined as V_{SET} and V_{RESET}. In general, two different switching types are defined as unipolar and bipolar [28–30], as illustrated in Figure 2. The unipolar switching mode is defined by the amplitude of the applied voltage bias while bipolar switching depends on the polarity of the applied voltage bias. In addition, as one of NVM devices, the endurance and retention properties of RRAM are also the presence of device reliability [4–7,28–30].

Compared with conventional silicon-based memory devices like flash memory, it is noted that RRAM devices have demonstrated a series of advantages such as low operation voltage, low power consumption, high density, and enhanced compatibility with traditional complementary metal oxide semiconductor (CMOS) technology [31–33]. In addition, with the deepening of research on artificial intelligence (AI) hardware equipment, biomimetic synapse behaviors of RRAM devices have also received extensive attention, which has non-negligible influence in the investigation of electrical artificial synapse [15,17,34–38]. However, some other limitations and challenges of RRAM devices cannot be neglected, such as synthesis methods of RS materials, stability of device performance and storage mechanism of devices with different materials.

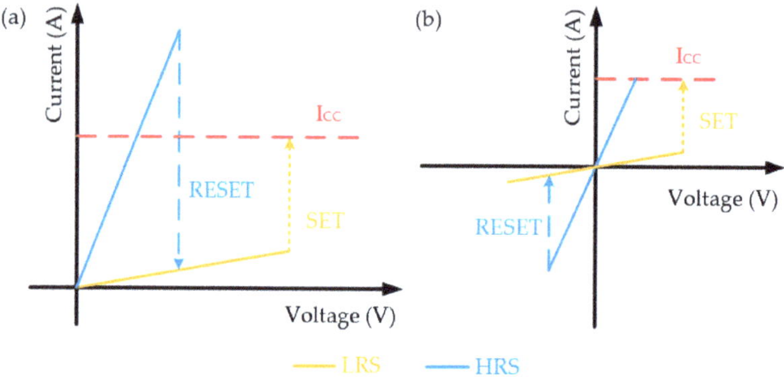

Figure 2. (**a**) Unipolar and (**b**) bipolar Modes for RRAM devices.

In this work, we provide a review of research for several main aspects, including exploration on switching mechanisms, investigation of thin film materials of RRAM devices and bionic synaptic application of RRAM devices. Section 1 provides an overview of background induction for RRAM devices. Section 2 will demonstrate a detailed discussion of different switching mechanisms. Section 3 will focus on various thin film materials applied to RRAM devices including RS medium materials and electrode materials. Section 4 will show an investigation on the neuromorphic application for RRAM devices, and Section 5 will be the conclusion of this review.

2. Resistive Switching (RS) Mechanisms of RRAM Devices

To present a comprehensive overview of RRAM devices, firstly, it is necessary to implement in-depth survey on different RS mechanisms of RRAM devices, which is still a controversial issue. The current investigation of RS mechanisms for sandwich structure RRAM devices is related to not only materials selection of electrode/RS medium but also utilized operation modes. For now, the most widely recognized switching mechanism is based on conductive filaments (CFs). However, there is no uniform and standard answer for some important issues about CFs with microscope chemical composition, physical morphology and the formation/rupture process, which have an intensive relationship to performance and working principle of RRAM devices. In this work, we will focus on the research of several working mechanisms related to anion/cation migration and thermal-chemical reaction, including thermal-chemical mechanism (TCM), valance change mechanism (VCM) and electrochemical metallization (ECM).

2.1. Thermal-Chemical Mechanism (TCM)

Theory about TCM can be applied to explain the formation and fracture of CFs resulted from ions (oxygen ion or metal ion) migration induced by thermal-chemical reaction (Joule heating), which is independent of the switching mode (unipolar and bipolar) for RRAM devices [39–41]. Zhang et al. explained the working principle of their Pt/Al/AlO$_x$/ITO RRAM device with TCM theory [39]. As illustrated in Figure 3, oxygen ions driven by Joule heating effect drifted to TE and left oxygen vacancies in the AlO$_x$ layer; consequent CFs based on the accumulation of oxygen vacancies set the device to LRS. For the RESET process of unipolar device, the current steadily increased with the increasing positive voltage bias, the formed CFs finally broke when it reached the critical temperature induced by Joule heating, which made the device switch back to HRS. Similarly, for the RESET process of bipolar device, the oxygen ions drifted back to the AlO$_x$ layer due to the melting of CFs and switched the device to HRS again. TCM based on Joule heating reaction is related to the formation and rupture of CFs. Tsuruoka et al. also proposed a research report on Ag/Ta$_2$O$_5$/Pt RRAM device with TCM based on the Joule heating effect [40]. The filament based on metal Ag played a dominating role during the

RS process. The formation of Ag CF made the device from HRS to LRS during the SET process. Due to the Joule-heating-based oxidation, the Ag CF ruptured by the thermal dissolution and completed the RESET operation. In general, with the SET/Forming operation, the thermal decomposition process that occurs in the RS medium generates the ions migration in the RRAM device and the resulting formation process of CFs transforms the device from HRS to LRS. With the reversed voltage bias applied onto the electrode, the existing CFs rupture due to the thermal melting reaction, which transforms the device back to HRS and completes the RESET process accordingly.

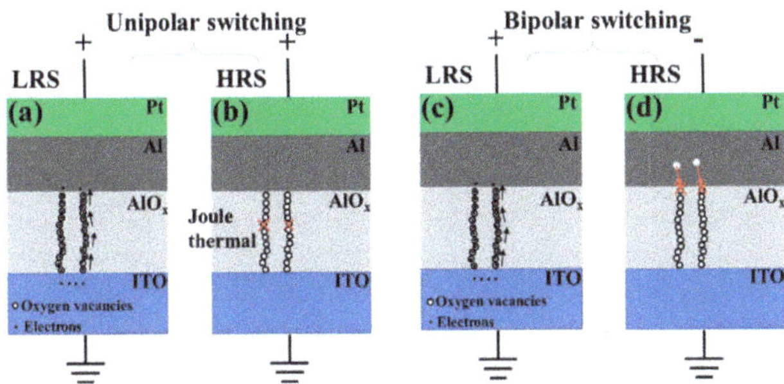

Figure 3. Switching mechanism of unipolar (**a**,**b**) and bipolar (**c**,**d**) AlO_x-based RRAM devices, reproduced from [39], with permission from Springer Nature, 2020.

2.2. Valance Change Mechanism (VCM)

Unlike the TCM mentioned above, VCM has oxygen-related defects/vacancies and their electrochemical reaction occurred in the RS medium [42–47]. In addition, it is not necessary for an RRAM device to operate with the structure that consists of an active electrode and an inert electrode, namely, the activity difference between TE and BE is not required [48]. Chen et al. researched the unipolar performance of $Pt/SiO_x/Pt$ RRAM device with VCM and dangling bond (DB) [43]. As illustrated in Figure 4, with the effect of external electric field, the strength of the polar covalent Si–O bond was weakened and finally broken. With the much higher concentration of DB near the middle of the silicon band gap, the hopping process could make the transportation of the electron through the discontinuation of DB, which was similar to the initial state (HRS) of their devices. If the DB concentration arose up to the threshold value of the percolation path, the electron transport could occur in the mini-band of DBs and the device switched into LRS, which accordingly indicated the SET process. Munjal also initiated the analysis process of $CoFe_2O_4$-based RRAM devices with the VCM theory [47].

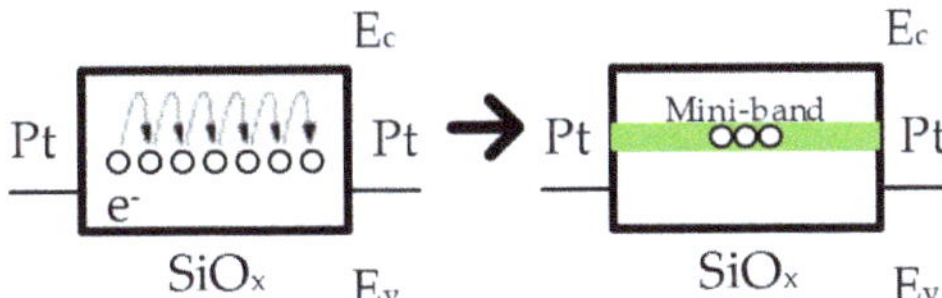

Figure 4. Schematic of discontinuous and continuous states of mini-band DBs in the middle of silicon oxide band gap.

In most cases, for VCM RRAM devices, the resistance change performance is attributed to the formation and rupture process of CF based on oxygen vacancies in the RS layer [42–47]. With the positive voltage bias applied onto the inert electrode, oxygen ions drift from where they stayed before with the effect of external electric field and oxygen vacancies left in the RS medium. The consequent CF path made up of leftover oxygen vacancies connects TE and BE through the functional layer, which increases the electric conductivity of the RS thin film and switches the device from HRS to LRS. Whereas, with the reversed voltage bias onto the same electrode, oxygen ions drift back to the RS layer and result in the rupture of formed CF, which makes the device switch back to HRS again. Therefore, oxygen defects/vacancies and oxygen may be the dominating aspect during the growth and destruction of CF in the functional layer.

2.3. Electrochemical Metallization (ECM)

Compared with TCM and VCM, ECM based on electrochemical reaction and cation migration, as the most recognized mechanism, is always used to explain the working principle of RRAM device with an active electrode, which is similar to VCM [49–55]. Generally, most active electrodes for ECM devices are active metal such as Cu [50–52] and Ag [49,53–55]. Tsuruoka et al. investigated $Cu/Ta_2O_5/Pt$ RRAM device based on Cu filaments [50]. With a positive voltage bias applied onto Cu TE, Cu atoms near the interface between Cu layer and Ta_2O_5 layer were dissolved into Cu ions (Cu^{2+}) and electron (e^-) due to the electrochemical reaction. These Cu^{2+} ions drifted towards the RS layer with the effect of external electric field, which induced the Cu^{2+} ions supersaturation near the Ta_2O_5/Pt interface. Then continuous cathodic deposition reaction occurred between Cu^{2+} and e^- led the formation of Cu-based filament and switched the device into ON state.

Yu et al. also confirmed that multiple Ag filaments attributed to the multilevel RESET behavior of RRAM device with a switching layer based on nonmetal materials ($Ag/SiO_2/Pt$) [49]. With a small negative voltage bias, the $Ag/SiO_2/Pt$ device exhibited gradual resistance increasement. When the voltage bias continued to increase beyond a threshold value, the resistance of device was increased to a higher state sharply, which suggested that multiple Ag filaments were effective as predicted. As demonstrated in Figure 5, in the SET process, Ag CFs with different sizes existed under a big CC after several switching cycles. During the RESET process, Ag from Ag CFs transferred into Ag^+ due to the dissolution reaction, which resulted in a gradual resistance increase of device. When these smaller CFs were broken, the resistance changed significantly. After that, CFs with larger sizes were getting thinner until they ruptured, which further induced the multilevel performance of RESET process. Long et al. also used ECM based on Ag filament to explain the switching mechanism of Ag/ZrO_2: Cu/Pt RRAM device [53]. With the effect of external electric field induced by voltage bias applied onto TE Ag, the oxidation process occurred on Ag atoms and Ag atoms transferred into Ag ions ($Ag \rightarrow Ag^+ + e^-$). Then Ag^+ migrated gradually to BE Pt as the electric field increased in the ZrO_2 thin film and ions were reduced back to atoms ($Ag^+ + e^- \rightarrow Ag$). Finally, the formed Ag filament switched the device into LRS when the voltage reached V_{SET}, which showed the related metallic transportation behavior. However, for the RESET process, when the voltage bias with reversed polarity was applied onto the active electrode, the existing Ag filament was broken within the oxide layer due to the electrochemical reaction w/o Joule heat assistance. Similarly, research reported by Tsuruoka et al. also presented the same perspective, which indicated that the RESET process related to formed metallic filaments might be related to electrochemical reaction w/o Joule heat assistance [50].

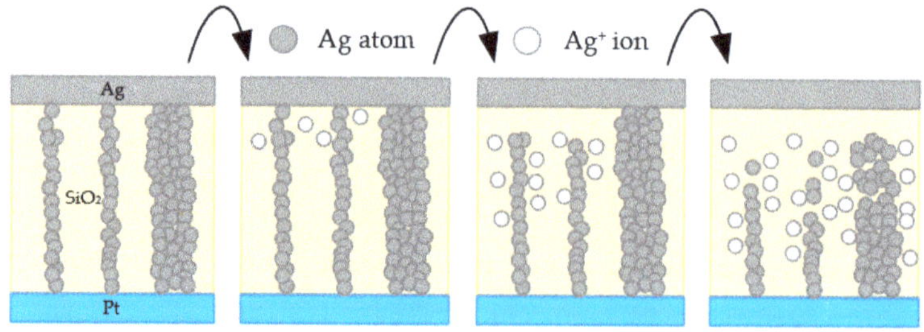

Figure 5. Illustration of multilevel RESET process of $Ag/SiO_2/Pt$ RRAM device.

2.4. Modeling Analysis for Switching Mechanisms

In order to provide convincing explanations about switching mechanisms, some researchers focused on establishing analytical models during the investigation process [41,56–61]. Ren et al. provided a specific explanation on the RS mechanism of the $Al/CH_3NH_3PbI_3/FTO$ RRAM device, which was mainly based on a physics-based analytical model with mathematical equations [56]. In this model, the RS mechanism was generally attributed to the migration of iodine vacancy (V_I) driven by electric field, which mainly includes operation phases such as electroforming, RESET and SET transitions. During the electroforming process, V_I nucleation occurred, induced by external electric field, and then CFs were formed in the perovskite film with the voltage bias applied onto the TE. They simplified the CF as a cylinder with radius (r) and height (h) and the obvious boundary existed between CF and outer region. In general, the region of the device where CF itself existed was the LRS region while other regions without CF indicated the HRS region. As illustrated in Figure 5, CFs themselves formed by Ag atoms (gray dots in the figure) represented LRS region while the pure wheat region without any gray dots indicated the HRS regions without CFs. With the combination of equations of Fick and Soret diffusion and related carrier drift theory, operations of RESET and SET were simulated during the switching process.

During the RESET process, V_I migrated from TE to BE due to the existence of Fick diffusion, which could be regulated by equations:

$$J_{Fick} = D\nabla n \tag{1}$$

$$D = 0.5\alpha^2 f \, exp(-E_a/kT) \tag{2}$$

where D was the diffusivity coefficient, ∇n was the concentration gradient of V_I; α, f, E_a, k and T were vacancy hopping distance, escape-attempt frequency, migration activation energy, Boltzmann constant and local temperature, respectively. The gradient direction was always from BE to TE due to the function of V_I reservoir for BE, thus the RESET process was expected to be confined by the Fick diffusion. Apart from Fick diffusion, another vital factor, the voltage-field activated V_I drift could not be neglected, which could be expressed by equations:

$$J_{drift} = -vn \tag{3}$$

$$v = af \, exp(-E_a/kT)sinh(qaE/2kT) \tag{4}$$

where v was the drift velocity, E was the electric field inside the CF, and q was elementary charge. With the combination of J_{Fick} and J_{drift}, the time evolution of CF size could be expressed by equation:

$$\frac{dg}{dt} = D\alpha_1 e^{-\frac{\beta_1}{0.5(L-g)}} - v \tag{5}$$

where L is the thickness of perovskite film, and g is the gap length. In addition, the Joule heating flow that occurred in the lateral direction of CF was also noted, which had a closed relationship with the length and diameter of CF. For the SET process, V_I migration occurred from BE to TE with the effect of reversed applied voltage bias resulting in the refill of the gap region. They explained the phenomenon with a set kinetics model based on Fick and Soret diffusion. Similar to Fick diffusion in the RESET process, the formation and growth of CFs were also suppressed by Soret diffusion in SET process, which was associated with the migration tendency of V_I towards the region with higher temperature.

According to their simulation results, the electric field inside CF decreased gradually when the depletion process occurred and then induced the slowing down of gap formation, which was used to explain why the growth of gap length mainly occurred in the positive part of voltage sweep. Apart from mathematical modeling, another main technique is modeling with software to justify the switching mechanisms. Software like COMSOL was also used to establish physics-based analytical models for further analysis [41,60]. Zhou et al. proposed a COMSOL-based model to research the switching mechanism of the Ti/HfO$_2$/TiN RRAM device [60]. They chose the 1st, 10th and 20th cycles as fitting analysis objectives, which all performed with Schottky emission. Based on the Schottky fitting formula, the gradual larger Schottky distance and the barrier had a relationship with a larger intercept and the slope of Ln(I)-(V)$^{1/2}$. According to the simulation results, the strongest electric field that existed in the tip of CF contacted the electrode in the initial SET process. After that, with more SET and RESET operations, oxygen ions accumulation occurred in the interface between TiN and HfO$_2$ layers, which made SET and RESET more difficult to achieve. Sun et al. also reported a similar mechanism on Cu/ZrO$_2$/Pt RRAM device [41].

3. Thin Film Materials of RRAM Devices

Apart from different RS switching mechanisms, most researches also put effort into the study of various thin film materials applied in RRAM devices. Before we present an investigation of materials application, it is necessary to provide brief induction of figures of merit (FoMs), which are mainly used to assess the performance of RRAM devices, including operation speed, power consumption, reliability, scalability, and cost [62]. Operation speed is defined by random-access time and effective time of write & erase (w & e) speed of a single RRAM cell; power consumption is affected by static and dynamic power consumption; reliability consists of endurance; retention properties are used to determine whether an RRAM device is reliable; scalability of an RRAM device determines whether RRAM devices can be developed in line with current trends of increasing device density; and the cost directly has an impact on marketization or mass-production of the proposed devices [62]. Apart from the main FoMs presented above, some other specific characteristics of FoMs such as SET/RESET voltage, ON/OFF switching ratio, distribution of operation voltage, and resistance cannot be neglected accordingly.

Thin film materials application of RRAM devices can be divided into two directions: materials for RS medium and materials for electrode (TE/BE). Materials of RS medium play a decisive role in the switching process, which has a direct and significant influence on the performance of RRAM devices such as ON/OFF ratio and device stability. On the other hand, electrode materials of RRAM devices more affect switching modes of RRAM devices, which should also be under further investigation [13,14,18,48,63,64].

3.1. Thin Film Materials of RS Medium

A lot of thin film materials have been researched as RS mediums of RRAM devices due to their exhibition of RS characteristics with the effect of external electric field. In general, organic materials and inorganic materials are two main categories of RS medium, as illustrated in Figure 6. Research of RS medium based on organic materials mainly focuses on biological materials (silk protein/fibroin, nanocellulose and albumen) [65–67], polymer materials (PVK (polyvinyl carbazole), PVA (polyvinyl alcohol), PDA (Polydiacetylene), PTH (polythiophene)) [68–71], and other materials. Chen et al. presented an RRAM device fabricated with spin-coated chicken egg albumen layer [67]. With the low SET/RESET voltage ~3 V, the reliable switching endurance was observed over 500 cycles with ~10^3 ON/OFF ratio. Similarly, Wang et al. reported an RRAM device with a structure of Au/Mg/fibroin/Mg and the fibroin-based RS layer was fabricated with drop-casting method [65]. Their device exhibited excellent performance with operation voltage lower than ~2 V, ON/OFF ratio higher than ~10^3 and retention time longer than ~10^4 s. In addition, they evaluated the transient behaviors of RRAM devices with the immersion process in DI water, which indicated the great potential of silk fibroin applied to transient and biocompatible electronics. Although the increasing number of research studies has demonstrated the feasibility of RRAM devices with organic materials, high power consumption under high operation voltage and dispersed voltage/resistance distribution induced by disappointing stability and reliability of device fabricated with organic materials indicated that it is necessary to further investigate perfect application of organic materials in the industry of memristive devices. In the future, most of these organic materials will be considered as probable candidates of application in flexible and wearable memristive devices with health diagnosis monitoring as the main function.

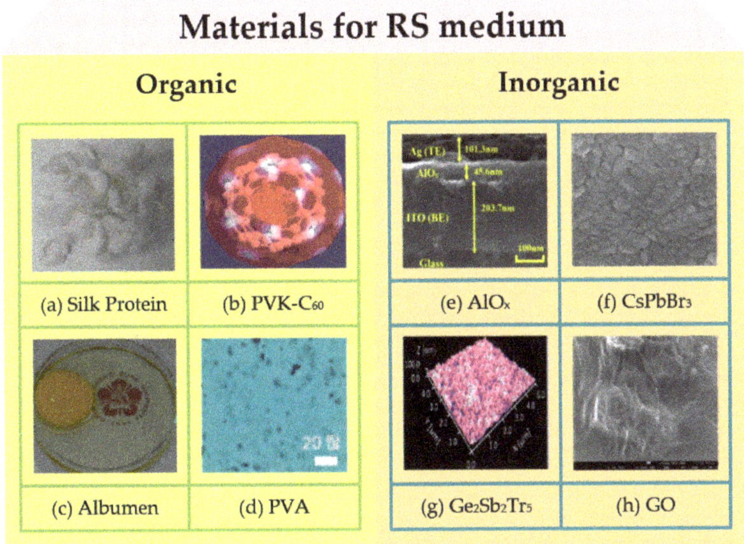

Figure 6. Various thin film materials of RS medium for RRAM devices. (**a**) Silk protein, reproduced from [65], with permission from John Wiley and Sons, 2016. (**b**) PVK-C_{60}, reproduced from [72], with permission from American Chemical Society, 2007. (**c**) Albumen, reproduced from [67], with permission from Springer Nature, 2015. (**d**) PVA, reproduced from [71], with permission from John Wiley and Sons, 2011. (**e**) AlO_x, reproduced from [19], with permission from Elsevier, 2020. (**f**) $CsPbBr_3$, reproduced from [73], with permission from John Wiley and Sons, 2019. (**g**) $Ge_2Sb_2Tr_5$, reproduced from [38], with permission from John Wiley and Sons, 2019. (**h**) GO, reproduced from [74], with permission from AIP Publishing, 2013.

Compared with RRAM devices fabricated with organic materials, better electrical performance of RRAM devices based on inorganic materials can be observed with more stable switching behavior, lower energy consumption and longer retention time. Inorganic materials are also being given extensive attention due to their simple manufacturing process and superior properties. In this review, we will present RRAM-related inorganic materials with oxides, solid electrolyte and two-dimensional (2D) materials. Table 1 demonstrated a performance comparison among RRAM devices with different oxide layers, including binary and complex oxide materials. As one of the most typical representatives of oxides, binary oxides have been explored for over half a century due to their simple composition, low manufacture cost, compatibility with traditional CMOS (complementary metal-oxide-semiconductor transistor) technology, and ease of fabrication and control. Among binary oxides, binary metal oxides such as Al_2O_3 [9,20,75], NiO [8,12,76], TiO_2 [21,77,78], HfO_2 [60,79,80], ZnO [81–83], and ZrO_2 [22,84,85] are always playing main roles in materials application of RS medium.

Table 1. Performance comparison among RRAM devices with various binary and complex oxides (metal and nonmetal materials).

Structure	Switching Mode	Thickness (nm)	$V_{Forming}$ (V)	V_{SET} (V)	V_{RESET} (V)	ON/OFF Ratio	Endurance (cycle)	Retention (s)	Ref.
Ni/AlO$_x$/Pt	bipolar	~40	Free	~1.0	~−1.0	~10^3	>150	>10^4	[9]
TaN/HfO$_2$/Al$_2$O$_3$/ITO	bipolar	~6	~4.5	~1.5	~−1.0	~10^2	>100	>2×10^3	[20]
Ti/IL-NiO/Pt	bipolar	~50	Free	~0.5	~−1.5	~10^3	>1300	>10^4	[8]
FeNi/Al$_2$O$_3$/NiO/Pt	bipolar	~180	~4.07	~6.0	~−5.0	~10^3	>100	>10^4	[12]
Au/TiO$_x$/TiO$_y$/Au	bipolar	~50	~5.62	~1.0	~−2.0	~10^2	N. A.	N. A.	[78]
Ni/SiGeO$_x$/TiO$_y$/TiN	bipolar	~25	Free	~3.0	~−2.5	~10^3	>10^4	>10^5	[21]
Ti/HfO$_2$/TiN	bipolar	~15	~6.5	~1.0	~−0.8	~10	N. A.	N. A.	[60]
Pt/Hf/HfO$_2$/TiN	bipolar	~20	Free	~0.8	~−1.5	~10^2	N. A.	>10^6	[80]
Pt/Ta/HfO$_2$/TiN	bipolar	~20	Free	~0.8	~−1.8	~10^2	N. A.	>10^4	[80]
Pt/Al:HfO$_2$/TiN	bipolar	~9	~2.3	~2.0	~−2.0	~10^4	>100	>10^4	[32]
TiN/ZnO/TiN	bipolar	~9	~4.2	~1.0	~−1.0	~10	240	N. A.	[82]
TiN/Al$_2$O$_3$/ZnO/Al$_2$O$_3$/TiN	bipolar	~15	~5.0	~1.0	~−1.0	~10^2	>10^4	>10^4	[82]
ITO/ZrO$_2$/Ag	bipolar	~50	N. A.	~5.0	~−15.0	~10^5	>100	>10^4	[22]
Pt/N:ZrO$_2$/TiN	bipolar	~25	~3.6	~0.5	~−1.0	~10^2	N. A.	N. A.	[85]
Ag/SiO$_2$/Pt	bipolar	~80	N. A.	~0.5	~−2.0	~10^6	>40	>2×10^3	[49]
ITO/LaAlO$_3$/ITO	bipolar	~30	~3.2	~3.0	~−3.0	~10^2	>100	N. A.	[86]
Cu/Cu:LaAlO$_3$/Pt	bipolar	~10	~7.0	~2.0	~−2.0	~10^3	>110	>10^4	[10]
GNR/SrTiO$_3$/GNR	bipolar	~50	N. A.	~2.0	~−3.0	~10	>200	>10^4	[87]
Pt/GO/PCMO/Pt	bipolar	~25	Free	~1.0	~−1.0	~10^2	>150	>10^4	[23]
Pt/BiFeO$_3$/Pt	unipolar	~200	N. A.	~5.0	~−15.0	N. A.	N. A.	N. A.	[88]
Ag/ZnO/BiFeO$_3$/ZnO/Ag	bipolar	~270	Free	~2.0	~−2.0	~10	N. A.	N. A.	[11]
Ag/Ag$_2$Se/MnO/Au	bipolar	~40	Free	~0.8	~−0.6	~10^2	>800	>10^4	[24]
TiN/SLG/HfO$_2$/Pt	bipolar	~35	~5.0	~2.0	~−3.0	~10^2	>120	>10^6	[89]
Ti/MoS$_2$-rGO/ITO	bipolar	~60	Free	~0.5	~−0.4	~10	>200	>10^4	[26]
Au/CsPbBr$_3$/ITO	bipolar	N, A.	Free	~1.0	~−1.0	~10^4	N. A.	>1200	[90]

The first report of RS performance in binary metal oxides was proposed by Hickmott in 1962 [91], which demonstrated the RS characteristics of Al/Al$_2$O$_3$/Al device under the effect of an electric field. With the development of fabrication methods for electronic devices with thin films, binary metal oxides thin films fabricated by sputtering, ALD (atomic layer deposition) and solution-processed methods have received more interest due to their superior performance. Also, it is indicated that some researchers not only focus on RS layer fabricated with conventional pure binary metal oxides, but also explore effective ways of optimization treatment on metal-oxides-based RS layers, such as stack layers, ionic liquid (IL) process and nanoparticle (NP) doping process, and other similar treatment methods [8,20,32].

Mahata et al. reported an RRAM device with ALD-based HfO$_2$/Al$_2$O$_3$ stack layers, which exhibited excellent performance with operation voltage lower than ~2 V and the ON/OFF ratio around ~10^3. The TaN/HfO$_2$/Al$_2$O$_3$/ITO RRAM device also presented neuromorphic synaptic behaviors with multi-level conductance properties by tuning the stop voltage in a DC sweep and the amplitude in pulse responses [20]. Shen et al. presented various RRAM devices with solution-processed AlO$_x$ layers annealed at different temperatures [9]. As illustrated in Figure 7, their research results on Ni/solution-processed AlO$_x$/Pt RRAM devices, indicating that the best performance can be found in the device with a solution-processed AlO$_x$ layer annealed at 250 °C, which exhibited operation voltage around ~1 V, ON/OFF ratio higher than ~10^3, endurance cycles over 100, and retention time longer than 10^4 s. Samanta et al. reported the threshold switching performance of their cross-point selector with Al$_2$O$_3$ and SiO$_x$ as bilayer dielectric [92]. Compared with the cross-point selector with only sputtering-deposited SiO$_x$ layer, the ALD-deposited Al$_2$O$_3$ layer in the bilayer platform controlled the dissolution gap of Ag filament and improved the uniformity of the device. The selectivity was larger than 5×10^7 and the rectifying ratio (RR) was over ~10^7. Other researchers like Banerjee, Knorr and Sleiman et al. also demonstrated their investigation on RS behaviors of AlO$_x$-based RRAM devices [93–97].

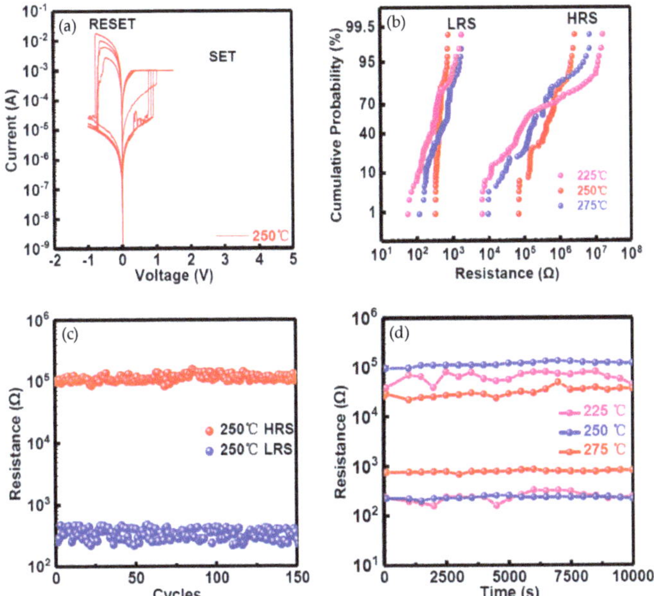

Figure 7. RS performance, including (**a**) bipolar I-V characteristic, (**b**) resistance distribution, (**c**) endurance and (**d**) retention performance, of Ni/solution-processed AlO$_x$/Pt RRAM devices annealed at different temperatures, reproduced from [9], with permission from MDPI (Basel, Switzerland), 2019.

Kang et al. proposed a Ti/IL-NiO/Pt RRAM device with IL-treated NiO layer, which made the RRAM device operated under only ~0.5 V voltage bias with ON/OFF ratio higher than ~10^3 [8] The IL treatment on NiO thin film created Ni^0-regions near the NiO/Pt interface and was helpful to the formation of oxygen vacancy filament, which improved device performance. The self-assembled memristive element based on NiO nanocrystal arrays was proposed by Kurnia et al. [98]. NiO nanocrystals were synthesized onto a SrRuO$_3$ substrate with PLD (pulsed laser deposition) using the phase separation method. Their devices exhibited memristive switching behavior with nonlinear bipolar characteristics under the ~5 V operation voltage, which was investigated via scanning probe microscopy, based on first-order reversal curve current–voltage spectroscopy. In addition, it was indicated that low electrical dissipation at the edge of the nanocrystals represented that less energy was consumed as heat, which enhanced the utilization efficiency during the nucleation process of CFs and reduced the energy consumption. Besides, further exploration on RS performance of NiO dielectric was also carried out by Yoshida, Russo, Cagli, and Ielmini et al. [99–103].

Kim et al. reported an RRAM device with a structure of Au/TiO$_x$/TiO$_y$/Au, and the TiO$_x$/TiO$_y$ layers were fabricated by the sputtering method [78]. During the experimental process, their modified operations on the gas environment made TiO$_x$ and TiO$_y$ layers deposited under a gas mixture of Ar and O$_2$ with flow ratios of 20:5 and 20:1, respectively. The Au/TiO$_x$/TiO$_y$/Au RRAM device operated under ~1.5 V with ~10^2 ON/OFF ratio, which also exhibited artificial synaptic characteristics such as long-term potentiation (LTP), long-term depression (LTD), and spike-timing-dependent plasticity (STDP). Mullani et al. reported the enhanced RS behavior of their devices based on hydrothermal-fabricated carbon nanotube/TiO$_2$ nanorods composite film through increasing oxygen vacancy reservoir [4]. The effect of concentration of TiO_2-fMWCNT (functionalized multiwalled carbon nanotube) nanocomposites was confirmed, which improved the RS performance of the device with forming-free and low operational voltage when the concentration of fMWCNT was 0.03 wt %. Sakellaropoulos et al. demonstrated a comparison among different devices with three kinds of dielectric structures such as HfO_x, TaO_y/HfO_x and $HfO_x/TaO_y/HfO_x$, which were corresponding to single-layer (SL), bilayer (BL) and triple-layer (TL) [104]. The forming-free sample TL exhibited enhanced RS behavior with ON/OFF ratio larger than $\sim10^2$, cycling variability smaller than 0.6 and number of endurance cycles over 10^6 under only $\sim$nW-level operation power in pulsing mode. Compared with sample SL, the RS performance of sample TL was confined due to the higher oxygen content and deeper oxygen vacancy levels of the TaO_y layer, which demonstrated analog switching characteristics and revealed great potential in artificial synaptic application. Besides, other researchers such as Nauenheim, Hermes, Salaoru and Otsuka et al. also provided a related investigation on TaO_x- and TiO_x-based RRAM devices in terms of electrical performance and physical characterization [105–109].

Wang et al. proposed an interface engineering method on ALD-based HfO_2 thin film with O_3, which improved the performance of $Pt/HfO_2/TiN$ RRAM device [79]. It is reported that the best stability could be observed in the HfO_2 RRAM device with 20 pulses of O_3 treatment. The TiON layer was observed at the interface between HfO_2 and TiN layers. With the voltage bias onto electrode, more abundant detects survived at the TiON layer due to the longer oxidation process, which had a positive influence on the migration efficiency of oxygen vacancy and formation of CFs. However, it also resulted in the obvious augmentation of the conductivity of the HfO_2 layer. In addition, their research on neuromorphic simulation based on potentiation, depression and STDP emulated the presynaptic and postsynaptic membranes of a biological synapse through applying the same pulses on both Pt and TiN electrodes, which revealed that oxide/metal interface engineering could have significant impact on RS characteristics of RRAM devices. Sharath et al. reported their RRAM devices based on RMBE (reactive molecular beam epitaxy)-deposited HfO_{2-x} layer with the operation voltage lower than 1 V and ON/OFF ratio higher than 10^2 [110]. With the Hard X-ray photoelectron spectroscopy, the presence of sub-stoichiometric hafnium oxide and defect states near the Fermi level were confirmed. Bipolar RS performance was also observed on forming-free RRAM devices with oxygen-deficient HfO_{2-x} thin films. Besides, other related research on RS behaviors of HfO_x dielectric thin films were also reported by Clima, Lanza and Stefano et al. [111–114].

Ha et al. demonstrated an $Ag/ZrO_2/ITO$ RRAM device with sol-gel-processed ZrO_2 thin film, which exhibited excellent performance with operation voltage around $\sim$2 V and ON/OFF ratio higher than 10^5 [22]. They investigated the effect of different annealing gas environments (air, vacuum and N_2) on the performance of RRAM devices. Sol-gel-processed ZrO_2 thin film annealed at vacuum showed enhanced performance with larger crystallinity and grain size, denser film, and a relatively small quantity of oxygen vacancies, which resulted in a decrease in the leakage current and an increase in the resistance ratio of HRS/ LRS and successfully improved non-volatile memory properties, such as endurance and retention characteristics. Abbas et al. proposed their investigation regarding $Ti/ZrO_x/Pt$ RRAM device with RTA (post-rapid thermal annealing) processing on ZrO_x dielectric layer [115]. Compared with samples based on an as-deposited ZrO_x layer without annealing process, RTA sample demonstrated improved RS performance such as lower operation voltage, higher RS ratio, longer retention time, and increasing number of endurance cycles. Particularly, with the annealing temperature of 700 °C during the RTA process, $Ti/ZrO_x/Pt$ RRAM device could operate with a voltage lower than 1 V and ON/OFF ratio higher than 10^3. Apart from Ha and Abbas, Verbakel, Awais, Kärkkänen and Ismail et al. also exhibited their investigation on RRAM devices with ZrO_x dielectric layers and most of them fabricated ZrO_x thin films with ALD and sputtering methods [116–120].

Similar to binary metal oxides, binary nonmental oxides like SiO_2 were also under investigation [49]. Yu et al. reported the multi-level RS performance of $Ag/SiO_2/Pt$ RRAM devices with operation voltage smaller than 1.5 V and switching ratio higher than 10^2, which was affected by the formation of multiple Ag filaments [49]. Apart from binary oxides, complex oxides with higher dielectric constants, such as

LaAlO$_3$ [10,86], SrTiO$_3$ [87,121], Pr$_{0.7}$Ca$_{0.3}$MnO$_3$ [23,122] and BiFeO$_3$ [11,88] are also explored to improve the switching performance of RRAM devices. Bailey et al. exhibited a stack-layered RRAM device with a structure of W/SrTiO$_3$ (STO)/TiN, which systematically investigated the diffusion phenomenon of ionic defects in oxides associated with various configuration states of STO layer. Consistent decay was found during the test of device retention, and then the function relationship between decay rate and conditioning voltage was confirmed [121]. The metal/Pr$_{0.7}$Ca$_{0.3}$MnO$_3$ (PCMO)/metal RRAM devices with a high-throughput electrode were studied by Tsubouchi et al. based on various metal electrodes including Mg, Ag, Al, Ti, Au, Ni, and Pt [122]. Typical RS behaviors were only observed on devices with Al electrode in the measurement process of I-V and pulsed-field resistance. In addition, the switching performance was always found near the Al/PCMO interface with the test results of the four-probe test.

Except binary and complex oxides, solid electrolyte and 2D materials are also popular in the research on RRAM-related inorganic materials. As early as in 1976, As$_2$S$_3$ solid electrolyte material has been studied as a functional layer in the RRAM device by Hirose et al. [123]. With the photodoping operation of Ag ions into the RS layer, the RRAM device with the Ag-As$_2$S$_3$ layer operated under low voltage, and CF based on the Ag element was observed by an optical microscope. After that, some solid electrolyte materials have been presented including Ag$_2$Se [24], Ge$_2$Sb$_2$Te$_5$ [25], GeTe [124], etc. At present, a series of 2D materials such as graphene [89,125,126], molybdenum disulphide (MoS$_2$) [26,127,128] and perovskite materials (CH$_3$NH$_3$SnCl$_3$ and CsPbBr$_3$) [73,90] also inspired researchers' interest due to their small size, ultra-thin thickness and excellent physical properties, which have resulted in superior performance of RRAM devices. Chen et al. proposed an electrode/oxide interface engineering technique by inserting single-layer graphene (SLG) into the TiN/HfO$_2$/Pt RRAM device [89], which enabled the RESET current to be reduced by 22 times and the programming energy consumption reduced by 47 times. Raman mapping and single-point measurement on the SLG layer noted that signals of D-band and G-band changed during the electrical cycling, which indicated that oxygen ions might drift from metal oxide layer to SLG layer. Wu et al. exhibited a flexible RRAM device based on MoS$_2$-rGO (reduced graphene oxide) hybrid layer synthesized by hydrothermal reaction between (NH$_4$)$_6$Mo$_7$O$_{24}$·4H$_2$O and (NH$_2$)$_2$CS in GO (graphene oxide) dispersion with the distilled water as dispersant [26]. As illustrated in Figure 8, their device showed not only a low SET/RESET voltage (~0.4 V) but also multi-level states through compliance current (CC) control. Excellent electrical performance could also be observed with endurance cycles over 200, retention time longer than 10^4 s and ON/OFF ratio higher than 10^3. Many parameters about RS performance of RRAM devices were compared in this review, which can be also observed in Table 1.

3.2. Thin Film Materials of Electrode

Although materials application of RS medium has played a decisive role in performance of RRAM device, the effect of electrode thin film materials on devices cannot be neglected, which act as primary transport paths for carriers. For now, a great number of thin film materials have been investigated for the selection of electrode, such as elementary substance metals, carbon-based materials, conductive oxides, complex nitrides, and silicon-based materials. Table 2 demonstrated devices with various electrodes. During the research and selection of different materials for electrode of RRAM devices, some vital factors should be under special consideration, such as the activity of materials, work function of materials and the interface between electrode and RS layer [48].

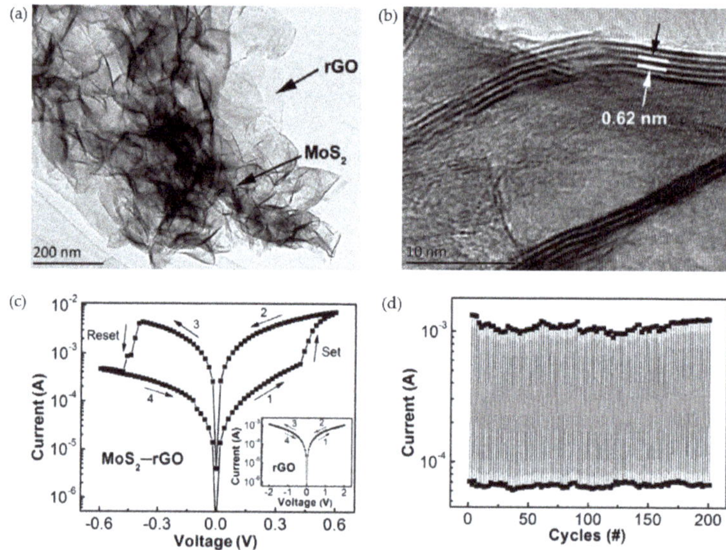

Figure 8. MoS$_2$-rGO hybrid with (**a**) TEM and (**b**) HRTEM. (**c**) Bipolar I-V curves and (**d**) endurance properties of Ti/MoS$_2$-rGO/ITO RRAM device, reproduced from [26], with permission from Elsevier, 2019.

Table 2. RRAM devices with various materials as electrodes.

Electrode Materials	Thickness (nm)	Electrode Mode	Switching Mode	V$_{SET}$ (V)	V$_{RESET}$ (V)	ON/OFF Ratio	Ref.
Hf	~40	TE	Bipolar	~4	~-4	~10^3	[129]
Al	~40	TE	Bipolar	~2	~-2	>10^2	[27]
Ti	~100	TE	Bipolar	~0.5	~-1.5	~10^3	[8]
Zr	~40	TE	Bipolar	~2	~-4	~10^3	[129]
Cr	~70	TE	Bipolar	~1.5	~-1.5	~10^4	[130]
Ni	~40	TE	Bipolar	~1.0	~-1.0	~10^3	[9]
Cu	~150	TE	Bipolar	~2.0	~-2.0	~10^3	[10]
Ag	~140	TE/BE	Bipolar	~2.0	~-2.0	~10	[11]
Pt	~200	BE	Bipolar	~6.0	~-5.0	~10^3	[12]
Au	~50	TE/BE	Bipolar	~1.0	~-2.0	~10^2	[78]
Graphene	N. A	TE	Unipolar	~1.0	~-1.0	~10^4	[131]
ITO	~50	BE	Bipolar	~1.5	~-1.0	~10^2	[20]
TiN	~20	TE	Bipolar	~0.5	~-1.0	~10^2	[85]
TaN	~60	TE	Bipolar	~1.5	~-1.0	~10^2	[20]
p-type Si	~100	BE	Bipolar	~2	~-2	>10^2	[27]

In general, the elementary substance metals are the most common and widely applied in RRAM devices, including Hf [129], Al [27], Ti [8], Zr [129], Cr [27], Ni [9], Cu [10], Ag [11], Pt [12], and Au [78]. Most of the time, RRAM devices always exhibit unipolar RS performance with noble metal electrode, such as Pt and Ru both as TE and BE [48,88]. However, the bipolar RS behavior will be always observed when one of the noble metal electrodes is replaced by an active metal electrode like Ni and TiN [8,85]. Chen et al. reported a Ta$_2$O$_5$-based RRAM device with metal Hf as TE and they covered a 50-nm-thick Pt layer onto the Hf layer to prevent the oxidation [129]. The Pt/Hf/Ta$_2$O$_5$/Pt device operated under ~4 V with ON/OFF ratio higher than 10^3, which resulted from the fact that a higher Ta atomic concentration was observed at the Hf/Ta$_2$O$_5$ interface than that at the Ta$_2$O$_5$/Pt interface. It is further indicated that considerable diffusion of oxygen ions occurred in the Hf electrode and the obvious degeneration of O/Ta ratio occurred at the Hf/Ta$_2$O$_5$ interface, which can be illustrated in Figure 9. A similar structure was also reported to replace Hf with Zr as TE in Pt/Zr/Ta$_2$O$_5$/Pt device, which showed lower operation voltage than that of the Pt/Hf/Ta$_2$O$_5$/Pt device. Shen et al.

proposed a GO-based RRAM device with Al and p-type Si as TE and BE, respectively [27]. Their RRAM device with solution-processed GO dielectric layer exhibited excellent performance with SET/RESET voltage lower than 2 V and ON/OFF ratio higher than 10^2, which was affected by the diffusion of oxygen ions between metal Al and GO thin film. The $Cr/Gd_2O_3/TiN$ RRAM device was proposed by Jana et al. [130]. The bipolar RS characteristics demonstrated that the RRAM device could operate with operation voltage lower than 1.5V and ON/OFF ratio higher than 10^4. This 9-nm-thickness device exhibited excellent repeatable RS cycles with number of endurance pulses over 10^5 and retention time longer than 3×10^4 s. In addition, some RRAM devices with noble metal as both TE and BE also demonstrated bipolar switching performance, such as Pt/sputtering-deposited Cr_2O_3/Pt reported by Chen et al. [132], Pt/sputtering-deposited TaO_x/Pt reported by Huang et al. [133], Pt/Ni/ECD (electrochemical deposition)-fabricated CuO_x/Ni/Pt, and Pt/sputtering-deposited SiO_2/Pt reported by Park et al. [134].

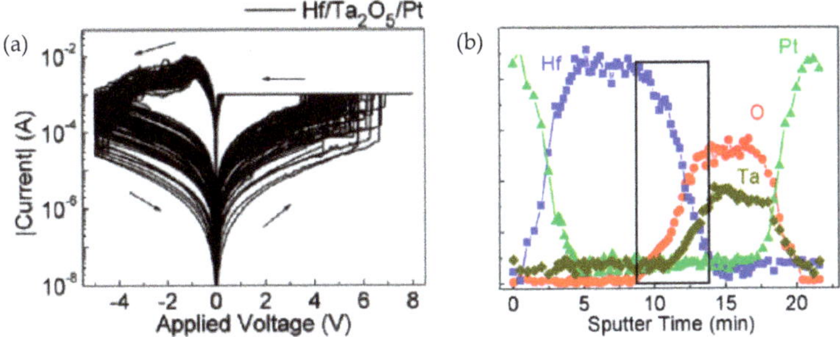

Figure 9. (**a**) Bipolar RS performance and (**b**) AES depth profiles of $Hf/Ta_2O_5/Pt$ RRAM devices, reproduced from [129], with permission from AIP Publishing, 2013.

Apart from elementary substance metals mentioned, carbon-based materials such as graphene and carbon nanotubes [48,62,131,135], conductive oxides such as ITO (indium tin oxide) [131], and complex nitrides such as TiN and TaN [20,85] are also charming as selection of RRAM electrodes. ITO electrode has been widely utilized in electronic devices as well as RRAM cells due to its excellent electrical conductivity, high light transmittance and high hardness [48]. In addition, as one of the n-type semiconductors with highly degeneration characteristic, the ITO electrode has good electrical conductivity due to internal carriers including oxygen vacancy and activated Sn^{4+} ions [48]. Zhao et al. proposed an RRAM device with multilayer graphene (MLG) and ITO as TE and BE, respectively [131]. Their transparent $MLG/Dy_2O_3/ITO$ device exhibited 80% transmittance at 550 nm wavelength due to the good light transmittance of the ITO electrode. The excellent electrical conductivity of graphene and ITO made the devices operate under low CC (<100 µA) with low voltage (<1 V) and low power consumption (<100 µW). The device also demonstrated a resistance ratio between HRS and LRS higher than 10^4 during 200 successful switching cycles and retention longer than 10^4 s.

The significance of selecting electrode materials is to optimize the performance of RRAM devices. Electrode materials like inert metals (Pt and Pd) are always investigated as carrier transportation paths, which have a slight and limited influence on RS performance. Then, electrodes of an RRAM device should have a positive impact on the formation process of CFs, which resulted from the migration of anion, which is always observed in oxygen-vacancy-based RRAM devices [8,9,32]. Finally, the selection of electrode materials for RS medium can directly affect the concentration of migrated anion and accumulated vacancy/cation; proper choice of electrode materials will have a positive influence in the formation process of CFs and also achieve stability enhancement of RRAM device.

4. Bionic Synaptic Application

With the rapid development of information technology, the neuromorphic computing applications in terms of hardware and software have stepped into the new era. For now, the neuron complexity simulation has been achieved with conventional computing technology based on animal brains of cats and rats [136,137]. However, as illustrated in Figure 10, it has been considered to be inefficient to process huge amounts of data with traditional computing architecture like Von Neumann architecture. The Von Neumann architecture was a binary-based computer concept structure proposed by John von Neumann in 1946, which indicated that a program should be considered as a part of data firstly [136,137]. Based on this architecture, the program itself and the data it processes should be stored with the same memory technology. The memory address of program instruction and the memory address of data should point to different physical locations in the same memory device [138,139]. Besides, the width of program instruction should be the same as that of processed data. As illustrated in Figure 10a, programs in matrix multiplication always process a lot of data at a slow speed, and the speed is slowed down with the increase of hidden layers. Energy power consumption also increases at the same time. Compared with CPU and GPU-based computer systems, although emerging dedicated FPGA and ASIC hardware chips have demonstrated faster operation speed and lower energy consumption, the matrix multiplication for large amounts of data is still limited by operation speed and energy consumption.

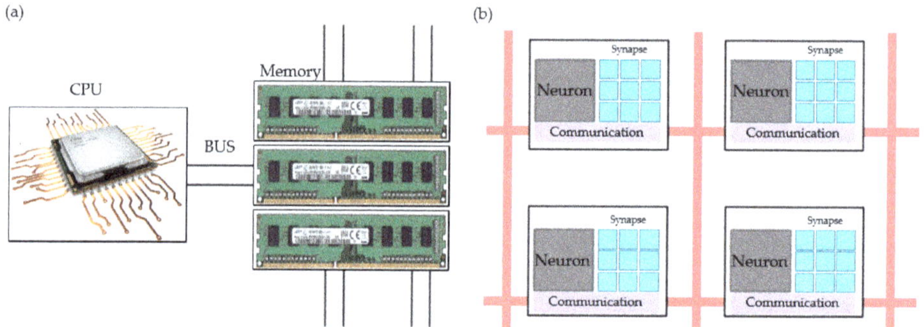

Figure 10. (**a**) Computing systems based on traditional Von Neumann architecture, the memory address of program instruction and the memory address of data point to different physical locations in the same memory device. (**b**) Computing systems based on neuromorphic architecture with integration of a single synaptic device into each unit.

Compared with a conventional computer system based on Von Neumann architecture, artificial neural networks (ANNs) have received extensive attention due to their lower power consumption and higher operation efficiency [138,139]. However, conventional synaptic devices cannot meet the requirements of ANNs because of some technological limitations like large device area, high power consumption and slow response speed. As illustrated in Figure 10b, emerging synaptic device can be integrated in to a single unit and improves the processing efficiency and processing accuracy [15,17,34–38]. The superior RS characteristics demonstrated in RRAM devices reveal its great potential in the neuromorphic application based on ANNs. An RRAM device can store multi-bit information in a non-volatile manner on a $4F^2$ device area and can also switch with the energy of ~pJ-level, which enables significant improvements in high-density integration and low power operation compared to conventional CMOS-based synapse devices [135–138].

In the analog circuit of ANNs, RRAM devices act as substitution to synapses in order to provide connection function between neurons and information storage cells [140,141]. The ultra-small size of RRAM device is to increase the synapse density of ANNs, which is expected to reach the synapse density in human brain (~10^{10} synapses/cm^2) [28,142]. As the key role during the information delivery process, a synapse is responsible for transmitting impulses from one neuron to another, which is considered to provide dynamical interconnections between two connecting neurons. Figure 11 demonstrated the basic structure of biological synapse, which mainly includes presynaptic membrane, synaptic cleft, postsynaptic membrane, and neurotransmitter [143]. Presynaptic membrane and synaptic vesicle are included by the presynaptic element, which is the dendrite/axon terminal of pre-neuron. Chemical substances in synaptic vesicle are called neurotransmitters. With the external spiking or stimulation onto the presynaptic element, synaptic vesicle filled with neurotransmitter will be released from presynaptic membrane to postsynaptic membrane due to the flux of Ca^{2+}, which has a temporary influence on synaptic transmission. The transmitting process of the neurotransmitter through a synapse represents the information delivery among neurons. The gap between presynaptic and postsynaptic membranes is the synaptic cleft, which is around 15~30-nm thick [144,145]. In general, for application in neuromorphic system and ANNs, capacitor-like synaptic devices are always used to simulate neuron bodies in the human brain; a neuron in the human brain can be compared to an electronic device with combined functions including an integrator and a device for threshold spiking. An axon acts like a connecting bridge in order to provide a transmission connection of information, which can be considered as a long wire. A dendrite works to transmit the signal input from multiple neurons to a single neuron, which can be compared to a short wire. A synapse acts as a connector to provide connection function between two next neurons [2], which have been attached to the most interests for now.

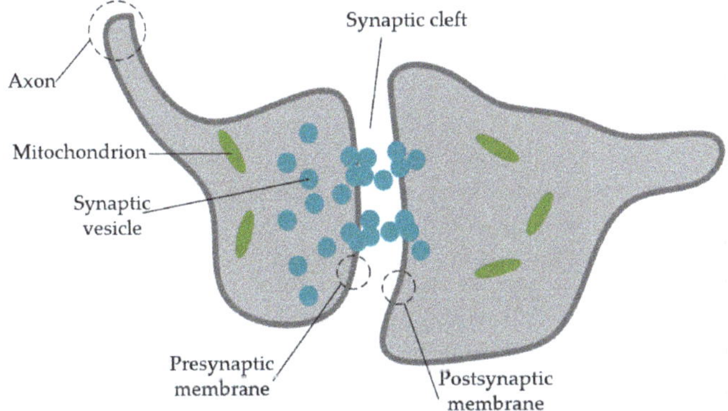

Figure 11. Structure of synapse in neural network.

Massive neurons form neural circuits through synapses in neural networks of the human brain [1–6]. These neural circuits are the foundation of advanced neural activities such as learning and memorizing, which are attributed to the plasticity of neural networks [2–4]. Basically, the plasticity of neural network is corresponding to the synaptic plasticity, which is dependent on synaptic weight [146,147]. Synaptic weight is used to demonstrate the strength or amplitude of connection between the two next neurons, which is mainly related to the amount of released or absorbed neurotransmitter. According to the theory proposed by Hebbian in 1949 [148], the connection between the two next neurons can be strengthened when they receive stimulation signals, which will result in the potentiation and depression during the neuron activities. For instance, long-term potentiation is corresponding to the synaptic potentiation behavior while long-term depression indicates depression performance of synapse. The synaptic plasticity mainly induced by the change of synaptic weight is divided into long-term plasticity (LTP) and short-term plasticity (STP) [149,150]. STP includes short-term potentiation, pair-pulse facilitation (PPF) and depression while long-term potentiation and depression are considered as representations of LTP [151]. For a biological neural system, on the one hand, STP is responsible for critical computation, on the other hand, LTP is thought of as the foundation of ability in terms of machine learning and memorizing [151]. As the molecular mechanism for learning and memorizing in the human brain, the timescale of STP can sustain from tens of milliseconds to few minutes, while the retention time of

LTP can be over a few hours, even several days. The transition from STP to LTP can be achieved through repeated stimulation during the learning process. Apart from LTP and STP, the plasticity depends on spike time (spike-time-dependent plasticity, STDP) is also under investigation, which is one of the advanced learning characteristics in the neuron system of the human brain [152,153]. As indicated by Hebb's rule, when the presynaptic membrane is stimulated earlier than the postsynaptic membrane and the postsynaptic current can be enhanced, which is called long-term potentiation. Reversely, when the spike occurs on the postsynaptic membrane earlier than that on the presynaptic membrane, the postsynaptic current will be depressed, namely long-term depression. In addition, it is noted that with the change of relative time between two stimulations (commonly pulses), the postsynaptic current will also be influenced [154]. Due to the existence of RS characteristics, the resistance states of RRAM devices can be manipulated by applied voltages and the microscopic conductive paths through the RS medium will be acquired. Therefore, resistance values of RRAM devices are the response to periodic input signals (voltage or current), which can be considered as the synaptic weights. The modification operations of resistance represent the changes of synaptic weights. When the external power supply fails to work, resistance states can also be retained [2].

4.1. Short-Term Plasticity for RRAM Devices

In general, two types of RS behaviors observed in RRAM devices, abrupt and gradual RS behaviors, were considered corresponding to digital and analog switching, respectively. The abrupt performance for resistance change was believed to be consistent with a digital signal while the gradual one with continuous conductance changes showed similar characteristics like a biological synapse [155]. Ilyas et al. demonstrated their research on a physical-vapor-deposited $Ag/SiO_x:Ag/TiO_x/p^{++}$-Si RRAM device with bilayer dielectric in terms of STP such as potentiation, depression and PPF [155]. The conductance of this bilayer device was modified by positive and negative pulses. As illustrated in Figure 12a,b, with the repeated voltage sweep, the potentiation and depression behaviors were observed and indicated the conductance states change during the processes of potentiation and depression, which proved the feasibility of conductance modification and simulated the change of synaptic weight. Figure 12c demonstrated the results of PPF for this bilayer device. PPF is always investigated to provide adjustment for conductance in order to perform short-term neural behaviors such as synaptic filtering and adaptation [156,157]. During two continuous pulse stimulations of their $Ag/SiO_x:Ag/TiO_x/p^{++}$-Si samples, the next post-synaptic response became higher than that of the previous one, which indicated that the interval time of spike was less than the recovery time. Berdan et al. also reported the STP of TiO_2-based RRAM devices through the transient conductance response [158]. As illustrated in Figure 12d, the conductance of the device increased due to the applied pulse and then decayed back to its initial state slowly. A similar performance could be observed on subsequent pulses who were dependent on the previous resistance states. Apart from RRAM devices with traditional metal oxides, short-term synaptic performance of devices fabricated with 2D materials are also under investigation [159–161]. Sun et al. reported their research on 2D-material-based devices with h-BN as a functional layer [159]. As demonstrated in Figure 12e, an obvious increase could be observed for ON-state current, which indicated the rise of synaptic strength due to the repeated pulse stimulation. The short-term facilitation with applied voltage-pulse excitation was observed on Ag/h-BN/graphene device as well. Sokolov et al. also provided their investigation results about PPF response of RRAM devices based on $TaO_x/IGZO$ (InGaZnO) bilayers [162]. Firstly, the single pulse with 0.75 V amplitude and 2 ms width was applied onto TE of RRAM device and the device exhibited a corresponding current response of ~220 mA. After that, with the RESET operation, the device transferred to HRS with initial OFF-state current. Then, two consecutive stimuli equal in amplitude and width to the first single pulse were applied onto TE, and the current response was ~350 mA, which indicated that the $TaO_x/IGZO$-based RRAM device could enhance the trans-conductance with the effect of high-frequency stimuli and further improve the PPF phenomenon.

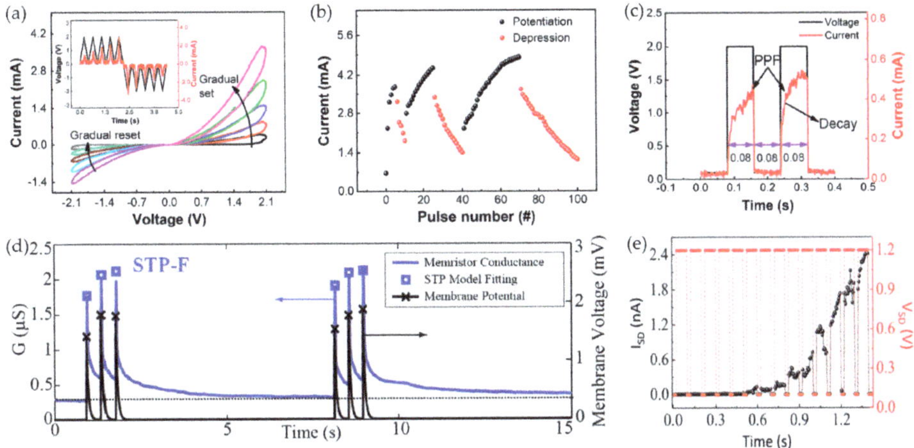

Figure 12. (**a**) Potentiation and depression response of Ag/SiO$_x$:Ag/TiO$_x$/p^{++}-Si device with repeated voltage sweeps. (**b**) Conductance modulation and (**c**) PPF of Ag/SiO$_x$:Ag/TiO$_x$/p^{++}-Si device by repeating consecutive pulses. (**d**) Repeated STP response with the model fitting of TiO$_x$-based RRAM device, reproduced from [158], with permission from Springer Nature, 2020. (**e**) Synaptic facilitation response to consecutive pulses of the device with h-BN, reproduced from [159], with permission from Elsevier, 2020.

4.2. Long-Term Plasticity for RRAM Devices

Compared with the investigation of STP, current research on LTP mainly focuses on long-term potentiation/depression and transition from STP to LTP [79,163,164]. Wang et al. proposed a flexible bipolar RRAM device with ALD-deposited Hf$_{0.5}$Zr$_{0.5}$O$_2$ (HZO) dielectric layer, as illustrated in Figure 13a–c, which worked as artificial synapses in the neuromorphic network in order to overcome the bottleneck based on traditional Von Neumann structure [79]. During the 400 continuous programming pulses, devices with gradual RS behaviors in DC sweep exhibited the feasibility of conductance modification under a sequence of consecutive programmable pulses, which indicated that the synaptic devices that emulated long-term potentiation/depression had great potential of artificial application in a neuromorphic computing system. The potentiated and depressed performance were stimulated by applying 200 0.8 V/20 ms continuous pulses and 200 0.5 V/20 ms continuous pulses in Figure 13a, respectively. This Ag/HZO/ITO/PET RRAM device showed the decayed post-synaptic current (PSC), and the PSC state turned into intermediate over time due to the forgetting effect. Figure 13b,c showed excitatory and inhibitory states of PSC with a single pre-synaptic spike, and the retention time over 1000s demonstrated the reliability of STP in Ag/HZO/ITO/PET RRAM devices. Wang et al. demonstrated their research report on a flexible RRAM device fabricated with a common polymer, poly (3,4-ethylene dioxythiophene): poly (styrene sulfonate) (PEDOT:PSS), as functional layer [163], and the PEDOT:PSS solution was spin-coated onto the ITO substrate. They chose consecutive positive and negative voltage bias carefully to apply onto TE Au of their device in order to avoid abrupt RS performance. The decreased conductance was observed when the continuous positive voltage bias swept from 0 V → 3 V → 0 V and then the gradual increased conductance was demonstrated with the consecutive negative voltage bias from 0 V → −2 V → 0 V, which indicated the potential of successful modulation of synaptic weights. The consecutive pulses with 2 V amplitude and 10 ms width were applied to TE for observing the current response, as demonstrated in Figure 13d; the potentiation and depression responses were observed with five consecutive positive pulses bias and five consecutive negative pulses bias, respectively. After that, the Au/PEDOT:PSS/ITO RRAM device was applied consecutive pulses (300 1V/10 ms positive pulses and 300 −1.5 V/10 ms negative pulses) and the long-term potentiation and depression were emulated successfully in Figure 13e. Finally, Wang et al.

repeated the pulse trains five times, as illustrated in Figure 13f, which indicated the excellent endurance property of their device.

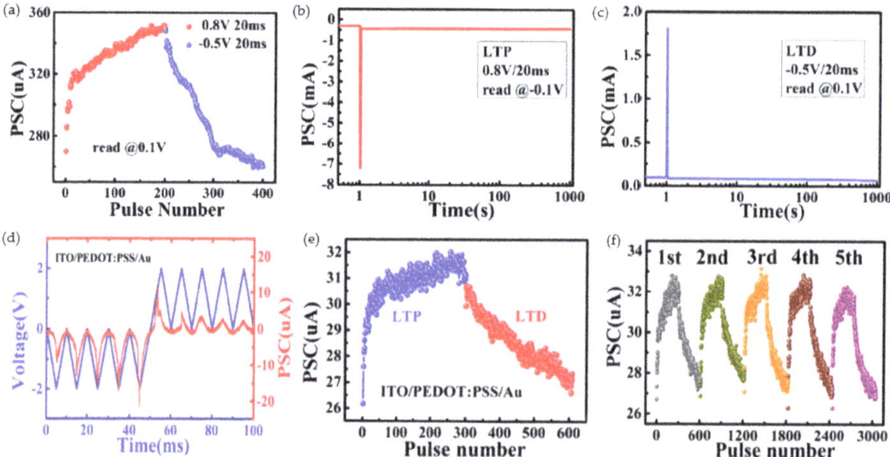

Figure 13. (**a**) Gradual modulation for conductance with long-term potentiation/depression response of Ag/HZO/ITO/PET RRAM device. Retention performance of Ag/HZO/ITO/PET RRAM device in (**b**) long-term potentiation process by consecutive positive pulses and (**c**) long-term depression process by consecutive negative pulses, reproduced from [79], with permission from Springer Nature, 2019. (**d**) Synaptic weights modulation of Au/PEDOT:PSS/ITO RRAM device by 10 consecutive pulses. (**e**) Long-term potentiation/depression under 600 consecutive pulses in one operation and (**f**) five operations of long-term potentiation/depression for Au/PEDOT:PSS/ITO RRAM device, reproduced from [163], with permission from MDPI (Basel, Switzerland), 2018.

Except for the research on long-term potentiation and depression behaviors of RRAM devices, the transition from STP to LTP also attracted significant interests [151,155,165], which is corresponding to transition from short-term memory (STM) to long-term memory (LTM). Biologically, compared with LTP, the sustainment time of STP is shorter generally, which is related to forgetting behaviors in the human brain. However, the STP can transfer into LTP with the repeated stimuli or a series rehearsal operation [155]. Zhang et al. reported the mechanism of conversion from STM to LTM in Cu/a-Si/Pt RRAM device [151]. As illustrated in Figure 14a, there were little Cu^{2+} ions drifting into the RS layer with a few stimulations, and then CF based on Cu atoms spontaneously decayed back to the initial state. However, with more repeated stimulations onto TE Cu, more Cu^{2+} ions migrated to the RS layer, which showed similar performance like training operation in the neural network. Ilyas researched the transition from STP to LTP of Ag/SiO_x:Ag/TiO_x/p++-Si device through modulating pulse strength [155]. Figure 14b,c showed the normalized current response by applying pulses with various amplitudes (1.2 V, 1.8 V, 2.0 V and 2.8 V) and each training cycle included 50 pulses. With the experimental results and the fitting results based on the equation in Figure 14b, the increased tendency of relaxation time was confirmed due to the effect of pulse strength (the red line in Figure 14c). An obvious elevation of synaptic weight was also observed when the relaxation time was around 45 s and the pulse amplitude was +2.8 V, which indicated that STP has transferred to LTP.

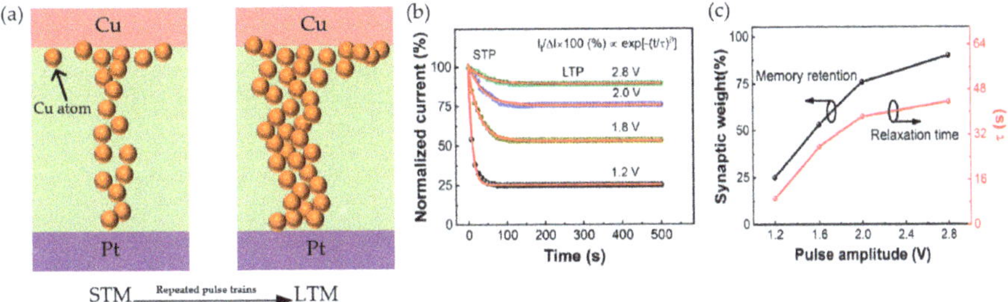

Figure 14. (**a**) Illustration of Cu atom dynamics of Cu/a-Si/Pt device during the transition from STM to LTM. (**b**) Relationship between normalized current response and retention time when the transferring process from STP to LTP occurred in Ag/SiO$_x$:Ag/TiO$_x$/p^{++}-Si device. (**c**) Synaptic weight response to changes of pulse amplitude and relaxation time τ, reproduced from [155], with permission from Springer Nature, 2020.

4.3. Spike-Time-Dependent Plasticity for RRAM Devices

Currently, many research achievements have proved that long- and short-term plasticity functions of biological neural synapse could by mimicked by RRAM devices with modulation of applied pulses (including amplitude and number modulations). On the other hand, as one of the determining factors in synapse plasticity, STDP also has received extensive attention. STPD can be defined as a function relationship between change of synaptic weight (ΔW) and time interval (Δt) resulting from activity variation of the pre- and post-neurons [155,163]. When Δt > 0, the activity of pre-synaptic neuron precedes that of post-synaptic neuron, which means that the connection strength between these two neurons will be reinforced and result in the long-term potentiation of the synapse. Conversely, the long-term depression occurs when the spike of post-synaptic neuron heads that of pre-synaptic neuron and the connection strength is weakened (Δt < 0). Obviously, for an RRAM device, top and bottom electrodes are compared to pre- and post-synaptic neurons and pulses applied onto electrode can mimic spikes of biological synapses. The polarity of ΔW is determined by the order of spikes of pre- and post-synaptic neurons. In general, the time interval and change of synaptic weight are defined as follows:

$$\Delta t = t_{post} - t_{pre} \tag{6}$$

$$\Delta W = \frac{G_t - G_i}{G_i} \times 100\% \tag{7}$$

in which t_{pre} and t_{post} are time nodes of spikes applied onto pre- and post-synaptic neurons, respectively. G_i is the conductance of the device at the initial state when t is 0, and G_t is the conductance when the time reaches node t.

Ilyas et al. emulated the STDP rule of Ag/SiO$_x$:Ag/TiO$_x$/p^{++}-Si samples, which can be observed in Figure 15a,b [155]. Through the implementation of a pair of pulses ±1.2 V/5 ms, ΔW decreased with the increase of Δt, which indicated that a more obvious conductance change could be observed when the time interval decreased. When Δt > 0, the pre-spike occurred before the post-spike, and the increased ΔW proved the enhancement of device conductance along with the decreasing Δt. Inversely (Δt > 0), the depression of device conductance was observed in ΔW when Δt increased. Mahata et al. revealed the STDP characteristic of RRAM devices with bilayer metal-oxide dielectric (TaN/HfO$_2$/Al$_2$O$_3$/ITO) [20]. As illustrated in Figure 15c,d, a series of pulses with different amplitudes were applied on to TaN and ITO electrodes. The largest values of ΔW were 97% and −84% when the |Δt| was 10 μs at both states (Δt > 0 and Δt < 0), which confirmed that their bilayer RRAM devices had STDP behaviors at various spiking timings. Wang et al. reported the STDP behaviors of the RRAM device fabricated with organic 2D materials [163]. After the test of long-term potentiation and depression with 600 consecutive programming pulses, they used a pair of pulse (±1.5 V/10 ms) to provide spikes on TE and BE of device, and the related result can be observed in Figure 15e. When the pre-synaptic neuron was spiked earlier than the post-synaptic neuron, the potentiated response of connection strength between two neighboring neurons could be demonstrated. On the contrary, the decreased synaptic weight indicated the weak connection. Wan also reported the STDP of 2D-material-based RRAM with the structure of Ag/SrTiO$_3$/RGO/FTO [166]; a similar experimental response with applied pulses (±1 V/10 ms) was observed in Figure 15f. The relationship between time interval and change of synaptic weight was similar to results proposed by others, which was more similar to their fitting results. These results proved that a more considerable change of synaptic weight could be realized with the smaller time interval of activity between two adjacent synaptic neurons.

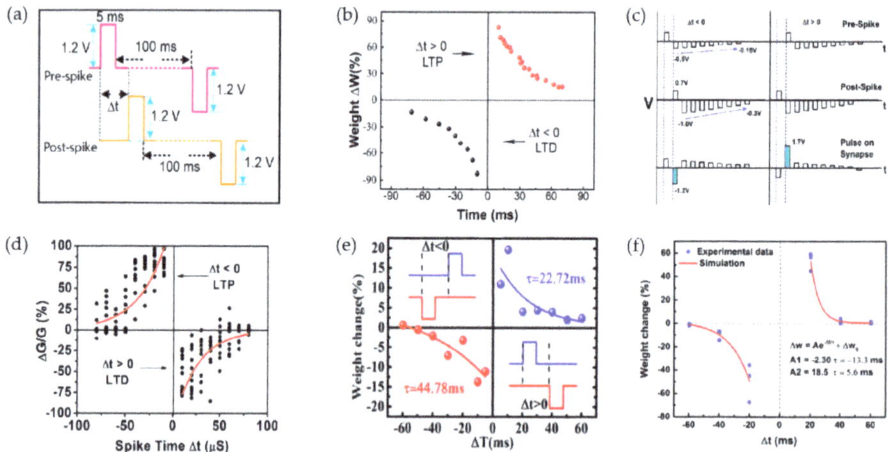

Figure 15. (**a**) Implementing programming pulses and (**b**) STDP behavior of Ag/SiO$_x$:Ag/TiO$_x$/p^{++}-Si RRAM device, reproduced from [155], with permission from Springer Nature, 2020. (**c**) Applied pre/post-spikes with sequences and (**d**) STDP characteristics of TaN/HfO$_2$/Al$_2$O$_3$/ITO RRAM device, reproduced from [20], with permission from Elsevier, 2020. (**e**) STDP results with pulse interval modulation of pre- and post-synaptic spiking for PEDOT:PSS-based RRAM device, reproduced from [163], with permission from MDPI (Basel, Switzerland), 2018. (**f**) Experimental and fitting results of STDP behaviors for Ag/SrTiO$_3$/RGO/FTO RRAM device, reproduced from [166], with permission from Elsevier, 2018.

5. Conclusions

In this work, we have provided an overview of RRAM devices with advances including various thin film materials applied in RS layer and electrode, classification of RS mechanisms and investigation on artificial synapse. Many research reports indicate that RRAM devices fabricated with inorganic materials, such as oxides, solid electrolyte and two-dimensional (2D) materials, have demonstrated relatively mature performance. There is great potential for the application of organic materials (biological and polymer materials) in RRAM devices accordingly. The performance of the devices depends largely on the RS mechanisms, which also has a strong connection with choice and processing techniques of the thin film materials. Based on the fundamental performance of RRAM devices, some outstanding enterprise or research institutions such as Samsung Electronics, Intel Corporation and Institute of Microelectronics of the Chinese Academy of Sciences (IMECAS) have put effort into promoting the development of large-scale manufacture and mature product commercialization of RRAM devices for many years. As early in 2004, Samsung Electronics reported the highly scalable TMO (binary transition-metal-oxide) RRAM devices with CMOS technology in IEDM (International Electron Devices Meeting) and NiO was used as a functional layer [167]. In 2007, the 2-MB CBRAM test chip was reported by Infineon and Samsung exhibited the 3D RRAM array with 1D1R structure [2].

From 2010 to 2013, Unity Semiconductor Corporation firstly reported their 64-MB RRAM test chip and SanDisk/Toshiba reported their 32-Gbit bilayer RRAM test chip. They realized the practical experiment in electrical circuits with devices fabricated by TaO_x and HfO_x functional layers [168]. One year later, Micron/Sony presented their 27-nm 16-Gbit CBRRAM test chip and TaO_x/HfO_x functional layers with stack structure were investigated [169]. In 2016, the four-layer 3D vertical RRAM array with self-selecting characteristic was reported by IMECAS, and RS performance of HfO_x resistive layers with the multi-level structure was verified [170]. In 2019, Intel Corporation announced they had prepared to manufacture emerging RRAM devices with 22 nm process technology, which also indicated that binary metal oxides or perovskite materials might be considered as candidates for the selection of functional layers [171]. All these developments are proving that the potential of large-scale commercialization for RRAM technology with different materials (especially TMO) is enormous and promising. Apart from the traditional large-scale commercialization process, the final objective of investigating different RRAM device performances is to provide potential assistance to artificial intelligence and neuromorphic computing systems. RRAM devices can mimic functions of biological synapse with electrical performance, which has a positive influence in hardware application of the artificial intelligence field. In addition, its human-brain-like behaviors such as STM and LTM make the development of neuromorphic computing system possible in the coming future.

Author Contributions: Conceptualization, Z.S. and C.Z. (Chun Zhao); Methodology, Z.S., C.Z. (Chun Zhao) and Y.Q.; Software, C.Z. (Chun Zhao), C.Z. (Cezhou Zhao) and L.Y.; Validation, C.Z. (Chun Zhao) and C.Z. (Cezhou Zhao); Formal Analysis, Z.S. and C.Z. (Chun Zhao); Investigation, Z.S., C.Z. (Chun Zhao), W.X. and Y.Q.; Resources, Z.S. and C.Z. (Chun Zhao); Data Curation, Z.S.; Writing-Original Draft Preparation, Z.S.; Writing-Review & Editing, Z.S., C.Z. (Chun Zhao), Y.L., L.Y., C.Z. (Cezhou Zhao) and I.Z.M.; Visualization, Z.S. and C.Z. (Chun Zhao); Supervision, C.Z. (Chun Zhao), C.Z. (Cezhou Zhao), I.Z.M., Y.L., W.X. and L.Y.; Project Administration, C.Z. (Chun Zhao) and C.Z. (Cezhou Zhao); Funding Acquisition, C.Z. (Chun Zhao), C.Z. (Cezhou Zhao) and I.Z.M. All authors have read and agreed to the published version of the manuscript.

Funding: This research was funded in part by the Natural Science Foundation of the Jiangsu Higher Education Institutions of China Program (19KJB510059), the Suzhou Science and Technology Development Planning Project: Key Industrial Technology Innovation (SYG201924), and the Key Program Special Fund in XJTLU (KSF-P-02, KSF-T-03, KSF-A-04, KSF-A-05, KSF-A-07).

Acknowledgments: The author Ivona Z. Mitrovic acknowledges the British Council UKIERI project no. IND/CONT/G/17-18/18.

Conflicts of Interest: The authors declare no conflict of interest.

References

1. Lv, Z.; Zhou, Y.; Han, S.-T.; Roy, V.A.L. From biomaterial-based data storage to bio-inspired artificial synapse. *Mater. Today* **2018**, *21*, 537–552. [CrossRef]
2. Hong, X.; Loy, D.J.; Dananjaya, P.A.; Tan, F.; Ng, C.; Lew, W. Oxide-based RRAM materials for neuromorphic computing. *J. Mater. Sci.* **2018**, *53*, 8720–8746. [CrossRef]
3. Zakaria, O.; Madi, M.; Kasugai, S. Introduction of a novel guided bone regeneration memory shape based device. *J. Biomed. Mater. Res. B Appl. Biomater.* **2020**, *108*, 460–467. [CrossRef] [PubMed]
4. Mullani, N.; Ali, I.; Dongale, T.D.; Kim, G.H.; Choi, B.J.; Basit, M.A.; Park, T.J. Improved resistive switching behavior of multiwalled carbon nanotube/TiO_2 nanorods composite film by increased oxygen vacancy reservoir. *Mater. Sci. Semicond. Process.* **2020**, *108*, 104907–104916. [CrossRef]
5. Du, X.; Tadrous, J.; Sabharwal, A. Sequential Beamforming for Multiuser MIMO With Full-Duplex Training. *IEEE Trans. Wirel. Commun.* **2016**, *15*, 8551–8564. [CrossRef]
6. Jamkhande, P.G.; Ghule, N.W.; Bamer, A.H.; Kalaskar, M.G. Metal nanoparticles synthesis: An overview on methods of preparation, advantages and disadvantages, and applications. *J. Drug Deliv. Sci. Technol.* **2019**, *53*, 101174–101184. [CrossRef]
7. Simmons, J.G.; Verderber, R.R. New conduction and reversible memory phenomena in thin insulating films. *Proc. R. Soc. Lond. Ser. A Math. Phys. Sci.* **1997**, *301*, 77–102.
8. Kang, X.; Guo, J.; Gao, Y.; Ren, S.; Chen, W.; Zhao, X. NiO-based resistive memory devices with highly improved uniformity boosted by ionic liquid pre-treatment. *Appl. Surf. Sci.* **2019**, *480*, 57–62. [CrossRef]

9. Shen, Z.; Qi, Y.; Mitrovic, I.Z.; Zhao, C.; Hall, S.; Yang, L.; Luo, T.; Huang, Y.; Zhao, C. Effect of Annealing Temperature for Ni/AlOx/Pt RRAM Devices Fabricated with Solution-Based Dielectric. *Micromachines* **2019**, *10*, 446. [CrossRef] [PubMed]

10. Wang, Y.; Liu, H.; Wang, X.; Zhao, L. Impacts of Cu-Doping on the Performance of La-Based RRAM Devices. *Nanoscale Res. Lett.* **2019**, *14*, 22401–22409. [CrossRef] [PubMed]

11. Liang, D.; Li, X.; Wang, J.; Wu, L.; Chen, P. Light-controlled resistive switching characteristics in ZnO/BiFeO3/ZnO thin film. *Solid State Electron.* **2018**, *145*, 46–48. [CrossRef]

12. Wang, G.; Hu, L.; Xia, Y.; Li, Q.; Xu, Q. Resistive switching in FeNi/Al2O3/NiO/Pt structure with various Al2O3 layer thicknesses. *J. Magn. Magn. Mater.* **2020**, *493*, 165728–165734. [CrossRef]

13. Patel, K.; Cottom, J.; Bosman, M.; Kenyon, A.J.; Shluger, A.L. An oxygen vacancy mediated Ag reduction and nucleation mechanism in SiO2 RRAM devices. *Microelectron. Reliab.* **2019**, *98*, 144–152. [CrossRef]

14. Li, L.; Chang, K.C.; Ye, C.; Lin, X.; Zhang, R.; Xu, Z.; Zhou, Y.; Xiong, W.; Kuo, T.P. An indirect way to achieve comprehensive performance improvement of resistive memory: When hafnium meets ITO in an electrode. *Nanoscale* **2020**, *12*, 3267–3272. [CrossRef] [PubMed]

15. Teherani, F.H.; Look, D.C.; Rogers, D.J.; Prezioso, M.; Merrikh-Bayat, F.; Chakrabarti, B.; Strukov, D. RRAM-based hardware implementations of artificial neural networks: Progress update and challenges ahead. Presented at the Oxide-Based Materials and Devices VII, San Francisco, CA, USA, 14–17 February 2016.

16. Cataldo, A.; Biagetti, G.; Mencarelli, D.; Micciulla, F.; Crippa, P.; Turchetti, C.; Pierantoni, L.; Bellucci, S. Modeling and Electrochemical Characterization of Electrodes Based on Epoxy Composite with Functionalized Nanocarbon Fillers at High Concentration. *Nanomaterials* **2020**, *10*, 850. [CrossRef]

17. Moon, K.; Lim, S.; Park, J.; Sung, C.; Oh, S.; Woo, J.; Lee, J.; Hwang, H. RRAM-based synapse devices for neuromorphic systems. *Faraday Discuss.* **2019**, *213*, 421–451. [CrossRef]

18. Yun, H.J.; Choi, B.J. Effects of moisture and electrode material on AlN-based resistive random access memory. *Ceram. Int.* **2019**, *45*, 16311–16316. [CrossRef]

19. Qi, Y.; Shen, Z.; Zhao, C.; Zhao, C.Z. Effect of electrode area on resistive switching behavior in translucent solution-processed AlOx based memory device. *J. Alloy. Compd.* **2020**, *822*, 153603–153621. [CrossRef]

20. Mahata, C.; Lee, C.; An, Y.; Kim, M.-H.; Bang, S.; Kim, C.S.; Ryu, J.-H.; Kim, S.; Kim, H.; Park, B.-G. Resistive switching and synaptic behaviors of an HfO2/Al2O3 stack on ITO for neuromorphic systems. *J. Alloy. Compd.* **2020**, *826*, 154434–154460. [CrossRef]

21. Cheng, C.H.; Chin, A.; Hsu, H.H. Forming-Free SiGeOx/TiOy Resistive Random Access Memories Featuring Large Current Distribution Windows. *J. Nanosci. Nanotechnol.* **2019**, *19*, 7916–7919. [CrossRef]

22. Ha, S.; Lee, H.; Lee, W.-Y.; Jang, B.; Kwon, H.-J.; Kim, K.; Jang, J. Effect of Annealing Environment on the Performance of Sol–Gel-Processed ZrO2 RRAM. *Electronics* **2019**, *8*, 947. [CrossRef]

23. Kim, I.; Siddik, M.; Shin, J.; Biju, K.P.; Jung, S.; Hwang, H. Low temperature solution-processed graphene oxide/Pr0.7Ca0.3MnO3 based resistive-memory device. *Appl. Phys. Lett.* **2011**, *99*, 042101–042104. [CrossRef]

24. Hu, Q.; Lee, T.S.; Lee, N.J.; Kang, T.S.; Park, M.R.; Yoon, T.-S.; Lee, H.H.; Kang, C.J. Resistive Switching Characteristics in MnO Nanoparticle Assembly and Ag2Se Thin Film Devices. *J. Nanosci. Nanotechnol.* **2017**, *17*, 7189–7193. [CrossRef]

25. Lin, Y.Y.; Chen, Y.C.; Rettner, C.T.; Raoux, S.; Cheng, H.Y.; Chen, S.H.; Lung, S.L.; Lam, C.; Liu, R. Fast Speed Bipolar Operation of Ge-Sb-Te Based Phase Change Bridge Devices. *IEEE Electron. Device Lett.* **2008**, *8*, 462–463.

26. Wu, L.; Guo, J.; Zhong, W.; Zhang, W.; Kang, X.; Chen, W.; Du, Y. Flexible, multilevel, and low-operating-voltage resistive memory based on MoS2–rGO hybrid. *Appl. Surf. Sci.* **2019**, *463*, 947–952. [CrossRef]

27. Shen, Z.J.; Zhao, C.; Zhao, C.Z.; Mitrovic, I.Z.; Yang, L.; Xu, W.Y.; Lim, E.G.; Luo, T.; Huang, Y.B. Al/GO/Si/Al RRAM with Solution-processed GO dielectric at Low Fabrication Temperature. In Proceedings of the 2019 Joint International EUROSOI Workshop and International Conference on Ultimate Integration on Silicon (EUROSOI-ULIS), Grenoble, France, 1–3 April 2019.

28. Tsai, T.-M.; Lin, C.-C.; Chen, W.-C.; Wu, C.-H.; Yang, C.-C.; Tan, Y.-F.; Wu, P.-Y.; Huang, H.-C.; Zhang, Y.-C.; Sun, L.-C.; et al. Utilizing compliance current level for controllability of resistive switching in nickel oxide thin films for resistive random-access memory. *J. Alloy. Compd.* **2020**, *826*, 154126–154151. [CrossRef]

29. Chen, Y. ReRAM: History, Status, and Future. *IEEE Trans. Electron. Devices* **2020**, *67*, 1420–1433. [CrossRef]

30. Das, U.; Bhattacharjee, S.; Mahato, B.; Prajapat, M.; Sarkar, P.; Roy, A. Uniform, large-scale growth of WS2 nanodomains via CVD technique for stable non-volatile RRAM application. *Mater. Sci. Semicond. Process.* **2020**, *107*, 104837–104842. [CrossRef]

31. Sun, C.; Lu, S.M.; Jin, F.; Mo, W.Q.; Song, J.L.; Dong, K.F. Multi-factors induced evolution of resistive switching properties for TiN/Gd2O3/Au RRAM devices. *J. Alloy. Compd.* **2020**, *816*, 152564–152581. [CrossRef]

32. Roy, S.; Niu, G.; Wang, Q.; Wang, Y.; Zhang, Y.; Wu, H.; Zhai, S.; Shi, P.; Song, S.; Song, Z.; et al. Toward a Reliable Synaptic Simulation Using Al-Doped HfO2 RRAM. *ACS Appl. Mater. Interfaces* **2020**, *12*, 10648–10656. [CrossRef]

33. Cui, X.; Li, X.; Zhang, M.; Cui, X. Design of the RRAM-Based Polymorphic Look-Up Table Scheme. *IEEE J. Electron. Devices Soc.* **2019**, *7*, 949–953. [CrossRef]

34. Chuang, K.-C.; Chu, C.-Y.; Zhang, H.-X.; Luo, J.-D.; Li, W.-S.; Li, Y.-S.; Cheng, H.-C. Impact of the Stacking Order of HfOx and AlOx Dielectric Films on RRAM Switching Mechanisms to Behave Digital Resistive Switching and Synaptic Characteristics. *IEEE J. Electron. Devices Soc.* **2019**, *7*, 589–595. [CrossRef]

35. Mao, J.-Y.; Zhou, L.; Ren, Y.; Yang, J.-Q.; Chang, C.-L.; Lin, H.-C.; Chou, H.-H.; Zhang, S.-R.; Zhou, Y.; Han, S.-T. A bio-inspired electronic synapse using solution processable organic small molecule. *J. Mater. Chem. C* **2019**, *7*, 1491–1501. [CrossRef]

36. Lee, D.K.; Kim, M.-H.; Kim, T.-H.; Bang, S.; Choi, Y.-J.; Kim, S.; Cho, S.; Park, B.-G. Synaptic behaviors of HfO2 ReRAM by pulse frequency modulation. *Solid State Electron.* **2019**, *154*, 31–35. [CrossRef]

37. Yang, S.; Hao, X.; Deng, B.; Wei, X.; Li, H.; Wang, J. A survey of brain-inspired artificial intelligence and its engineering. *Life Res.* **2018**, *1*, 23–29.

38. Wang, H.; Yan, X. Overview of Resistive Random Access Memory (RRAM): Materials, Filament Mechanisms, Performance Optimization, and Prospects. Phys. *Status Solidi RRL Rapid Res. Lett.* **2019**, *13*, 1900073–1900084. [CrossRef]

39. Zhang, X.; Xu, L.; Zhang, H.; Liu, J.; Tan, D.; Chen, L.; Ma, Z.; Li, W. Effect of Joule Heating on Resistive Switching Characteristic in AlOx Cells Made by Thermal Oxidation Formation. *Nanoscale Res. Lett.* **2020**, *19*, 3229-1–3229-19. [CrossRef] [PubMed]

40. Tsuruoka, T.; Hasegawa, T.; Terabe, K.; Aono, M. Conductance quantization and synaptic behavior in a Ta2O5-based atomic switch. *Nanotechnology* **2012**, *23*, 435705–435711. [CrossRef]

41. Sun, P.; Li, L.; Lu, N.; Li, Y.; Wang, M.; Xie, H.; Liu, S.; Liu, M. Physical model of dynamic Joule heating effect for reset process in conductive-bridge random access memory. *J. Comput. Electron.* **2014**, *13*, 432–438. [CrossRef]

42. Zhang, T.; Ou, X.; Zhang, W.; Yin, J.; Xia, Y.; Liu, Z. High-k-rare-earth-oxide Eu2O3 films for transparent resistive random access memory (RRAM) devices. *J. Phys. D Appl. Phys.* **2014**, *47*, 065302–065308. [CrossRef]

43. Chen, K.J.; Liu, J.; Wang, Y.; Yang, H.; Ma, Z.; Huang, X. VCM Conductive defect states based filament in MOM structure RRAM. *IEEE Electron. Device Lett.* **2018**, *9*, 1–4. [CrossRef]

44. Kwon, D.H.; Kim, K.M.; Jang, J.H.; Jeon, J.M.; Lee, M.H.; Kim, G.H.; Li, X.S.; Park, G.S.; Lee, B.; Han, S.; et al. Atomic structure of conducting nanofilaments in TiO2 resistive switching memory. *Nat. Nanotechnol* **2010**, *5*, 148–153. [CrossRef] [PubMed]

45. Xue, W.; Liu, G.; Zhong, Z.; Dai, Y.; Shang, J.; Liu, Y.; Yang, H.; Yi, X.; Tan, H.; Pan, L.; et al. A 1D Vanadium Dioxide Nanochannel Constructed via Electric-Field-Induced Ion Transport and its Superior Metal-Insulator Transition. *Adv. Mater.* **2017**, *29*, 1702162(1)–1702162(9). [CrossRef] [PubMed]

46. Lee, S.; Sohn, J.; Jiang, Z.; Chen, H.Y.; Philip Wong, H.S. Metal oxide-resistive memory using graphene-edge electrodes. *Nat. Commun.* **2015**, *6*, 8407–8413. [CrossRef] [PubMed]

47. Munjal, S.; Khare, N. Valence Change Bipolar Resistive Switching Accompanied With Magnetization Switching in CoFe2O4 Thin Film. *Sci. Rep.* **2017**, *7*, 12427–12436. [CrossRef] [PubMed]

48. Ye, C.; Wu, J.; He, G.; Zhang, J.; Deng, T.; He, P.; Wang, H. Physical Mechanism and Performance Factors of Metal Oxide Based Resistive Switching Memory: A Review. *J. Mater. Sci. Technol.* **2016**, *32*, 1–11. [CrossRef]

49. Yu, D.; Liu, L.F.; Chen, B.; Zhang, F.F.; Gao, B.; Fu, Y.H.; Liu, X.Y.; Kang, J.F.; Zhang, X. Multilevel resistive switching characteristics in Ag/SiO2/Pt RRAM devices. *IEEE Electron. Device Lett.* **2011**, *11*, 1–4.

50. Tsuruoka, T.; Terabe, K.; Hasegawa, T.; Aono, M. Forming and switching mechanisms of a cation-migration-based oxide resistive memory. *Nanotechnology* **2010**, *21*, 425205–425213. [CrossRef]

51. Guo, T.; Sun, B.; Zhou, Y.; Zhao, H.; Lei, M.; Zhao, Y. Overwhelming coexistence of negative differential resistance effect and RRAM. Phys. *Chem Chem Phys.* **2018**, *20*, 20635–20640. [CrossRef]

52. Long, S.B.; Liu, Q.; Lv, H.B.; Li, Y.T.; Wang, Y.; Zhang, S.; Lian, W.-T.; Liu, M. Resistive switching mechanism of Cu doped ZrO2-based RRAM. *IEEE Electron. Device Lett.* **2010**, *5*, 1–4.

53. Long, S.; Liu, Q.; Lv, H.; Li, Y.; Wang, Y.; Zhang, S.; Lian, W.; Zhang, K.; Wang, M.; Xie, H.; et al. Resistive switching mechanism of Ag/ZrO2:Cu/Pt memory cell. *Appl. Phys. A* **2011**, *102*, 915–919. [CrossRef]

54. Yuan, F.; Shen, S.; Zhang, Z.; Pan, L.; Xu, J. Interface-induced two-step RESET for filament-based multi-level resistive memory. *Superlattices Microstruct.* **2016**, *91*, 90–97. [CrossRef]

55. Gao, S.; Zeng, F.; Chen, C.; Tang, G.; Lin, Y.; Zheng, Z.; Song, C.; Pan, F. Conductance quantization in a Ag filament-based polymer resistive memory. *Nanotechnology* **2013**, *24*, 335201–335208. [CrossRef] [PubMed]

56. Ren, Y.; Milo, V.; Wang, Z.; Xu, H.; Ielmini, D.; Zhao, X.; Liu, Y. Analytical Modeling of Organic-Inorganic CH3NH3PbI3 Perovskite Resistive Switching and its Application for Neuromorphic Recognition. *Adv. Theory Simul.* **2018**, *1*, 1700035–1700042. [CrossRef]

57. Larentis, S.; Cagli, C.; Nardi, F.; Ielmini, D. Filament diffusion model for simulating reset and retention processes in RRAM. *Microelectron. Eng.* **2011**, *88*, 1119–1123. [CrossRef]

58. Ambrogio, S.; Balatti, S.; Gilmer, D.C.; Ielmini, D. Analytical Modeling of Oxide-Based Bipolar Resistive Memories and Complementary Resistive Switches. *IEEE Trans. Electron. Devices* **2014**, *61*, 2378–2386. [CrossRef]

59. Ielmini, D. Modeling the Universal Set/Reset Characteristics of Bipolar RRAM by Field- and Temperature-Driven Filament Growth. *IEEE Trans. Electron. Devices* **2011**, *58*, 4309–4317. [CrossRef]

60. Zhou, K.-J.; Chang, T.-C.; Lin, C.-Y.; Chen, C.-K.; Tseng, Y.-T.; Zheng, H.-X.; Chen, H.-C.; Sun, L.-C.; Lien, C.-Y.; Tan, Y.-F.; et al. Abnormal High Resistive State Current Mechanism Transformation in Ti/HfO2/TiN Resistive Random Access Memory. *IEEE Electron. Device Lett.* **2020**, *41*, 224–227. [CrossRef]

61. Sheridan, P.; Kim, K.H.; Gaba, S.; Chang, T.; Chen, L.; Lu, W. Device and SPICE modeling of RRAM devices. *Nanoscale* **2011**, *3*, 3833–3840. [CrossRef]

62. Shen, Z.; Zhao, C.; Qi, Y.; Mitrovic, I.Z.; Yang, L.; Wen, J.; Huang, Y.; Li, P.; Zhao, C. Memristive Non-Volatile Memory Based on Graphene Materials. *Micromachines* **2020**, *11*, 341. [CrossRef]

63. Xiao, W.; Song, W.; Feng, Y.P.; Gao, D.; Zhu, Y.; Ding, J. Electrode-controlled confinement of conductive filaments in a nanocolumn embedded symmetric–asymmetric RRAM structure. *J. Mater. Chem. C* **2020**, *8*, 1577–1582. [CrossRef]

64. Mao, S.; Zhou, G.; Sun, B.; Elshekh, H.; Zhang, X.; Xia, Y.; Yang, F.; Zhao, Y. Mechanism analysis of switching direction transformation in an Er2O3 based RRAM device. *Curr. Appl. Phys.* **2019**, *19*, 1421–1426. [CrossRef]

65. Wang, H.; Zhu, B.; Ma, X.; Hao, Y.; Chen, X. Physically Transient Resistive Switching Memory Based on Silk Protein. *Small* **2016**, *12*, 2715–2719. [CrossRef] [PubMed]

66. Wang, H.; Zhu, B.; Wang, H.; Ma, X.; Hao, Y.; Chen, X. Ultra-Lightweight Resistive Switching Memory Devices Based on Silk Fibroin. *Small* **2016**, *12*, 3360–3365. [CrossRef] [PubMed]

67. Chen, Y.C.; Yu, H.C.; Huang, C.Y.; Chung, W.L.; Wu, S.L.; Su, Y.K. Nonvolatile Bio-Memristor Fabricated with Egg Albumen Film. *Sci. Rep.* **2014**, *5*, 1002201–1002212. [CrossRef] [PubMed]

68. Méhes, G.; Vagin, M.; Mulla, M.Y.; Granberg, H.; Che, C.; Beni, V.; Crispin, X.; Berggren, M.; Stavrinidou, E.; Simon, D.T. Simon, Solar Heat-Enhanced Energy Conversion in Devices Based on Photosynthetic Membranes and PEDOT PSS-Nanocellulose Electrodes. *Advanced Sustain. Syst.* **2019**, *19*, 1900100.

69. Poskela, A.; Miettunen, K.; Borghei, M.; Vapaavuori, J.; Greca, L.G.; Lehtonen, J.; Solin, K.; Ago, M.; Lund, P.D.; Rojas, O.J. Nanocellulose and Nanochitin Cryogels Improve the Efficiency of Dye Solar Cells. *ACS Sustain. Chem. Eng.* **2019**, *7*, 10257–10265.

70. Jin, H.; Li, J.; Iocozzia, J.; Zeng, X.; Wei, P.C.; Yang, C.; Li, N.; Liu, Z.; He, J.H.; Zhu, T.; et al. Hybrid Organic-Inorganic Thermoelectric Materials and Devices. *Angew. Chem. Int. Ed. Engl.* **2019**, *58*, 15206–15226. [CrossRef]

71. Yarimaga, O.; Lee, S.; Ham, D.-Y.; Choi, J.-M.; Kwon, S.G.; Im, M.; Kim, S.; Kim, J.-M.; Choi, Y.-K. Thermofluorescent Conjugated Polymer Sensors for Nano- and Microscale Temperature Monitoring. *Macromol. Chem. Phys.* **2011**, *212*, 1211–1220.

72. Ling, Q.D.; Lim, S.L.; Song, Y.; Zhu, C.X.; Chan DS, H.; Kang, E.T.; Neoh, K.G. Nonvolatile Polymer Memory Device Based on Bistable Electrical Switching in a Thin Film of Poly(N-vinylcarbazole) with Covalently Bonded C60. *Langmuir* **2007**, *23*, 312–319. [CrossRef]

73. Zhu, Y.; Cheng, P.; Shi, J.; Wang, H.; Liu, Y.; Xiong, R.; Ma, H.; Ma, H. Bromine Vacancy Redistribution and Metallic-Ion-Migration-Induced Air-Stable Resistive Switching Behavior in All-Inorganic Perovskite CsPbBr$_3$ Film-Based Memory Device. *Adv. Electron. Mater.* **2019**, *6*, 1900754–1900761. [CrossRef]

74. Khurana, G.; Misra, P.; Katiyar, R.S. Forming free resistive switching in graphene oxide thin film for thermally stable nonvolatile memory applications. *J. Appl. Phys.* **2013**, *114*, 124508–124514. [CrossRef]

75. Gan, K.-J.; Chang, W.-C.; Liu, P.-T.; Sze, S.M. Investigation of resistive switching in copper/InGaZnO/Al$_2$O$_3$-based memristor. *Appl. Phys. Lett.* **2019**, *115*, 143501–143506. [CrossRef]

76. Lee, B.R.; Park, J.H.; Kim, T.G. Micro-light-emitting diode with n-GaN/NiO/Au-based resistive-switching electrode for compact driving circuitry. *J. Alloy. Compd.* **2020**, *823*, 153762–153767. [CrossRef]

77. Sung, C.; Lim, S.; Hwang, H. Experimental Determination of the Tunable Threshold Voltage Characteristics in a Ag$_x$Te$_{1-x}$/Al$_2$O$_3$/TiO$_2$-based Hybrid Memory Device. *IEEE Electron Device Lett.* **2020**, *41*, 713–716.

78. Kim, J.; Cho, S.; Kim, T.; Pak, J.J. Mimicking Synaptic Behaviors with Cross-Point Structured TiO$_x$/TiO$_y$-Based Filamentary RRAM for Neuromorphic Applications. *J. Electr. Eng. Technol.* **2019**, *14*, 869–875. [CrossRef]

79. Wang, Q.; Niu, G.; Roy, S.; Wang, Y.; Zhang, Y.; Wu, H.; Zhai, S.; Bai, W.; Shi, P.; Song, S.; et al. Interface-engineered reliable HfO$_2$-based RRAM for synaptic simulation. *J. Mater. Chem. C* **2019**, *7*, 12682–12687. [CrossRef]

80. Cai, L.; Chen, W.; Zhao, Y.; Liu, X.; Kang, J.; Zhang, X.; Huang, P. Insight into Effects of Oxygen Reservoir Layer and Operation Schemes on Data Retention of HfO$_2$-Based RRAM. *IEEE Trans. Electron. Devices* **2019**, *66*, 3822–3827. [CrossRef]

81. Gupta, C.P.; Jain, P.K.; Chand, U.; Sharma, S.K.; Birla, S.; Sancheti, S. Effect of Top Electrode Materials on Switching Characteristics and Endurance Properties of Zinc Oxide Based RRAM Device. *J. Nano Electron. Phys.* **2020**, *12*, 01007-1–01007-6. [CrossRef]

82. Kumar, D.; Chand, U.; Wen Siang, L.; Tseng, T.-Y. High-Performance TiN/Al$_2$O$_3$/ZnO/Al$_2$O$_3$/TiN Flexible RRAM Device With High Bending Condition. *IEEE Trans. Electron. Devices* **2020**, *67*, 493–498. [CrossRef]

83. Wu, P.-Y.; Zheng, H.-X.; Shih, C.-C.; Chang, T.-C.; Chen, W.-J.; Yang, C.-C.; Chen, W.-C.; Tai, M.-C.; Tan, Y.-F.; Huang, H.-C.; et al. Improvement of Resistive Switching Characteristics in Zinc Oxide-Based Resistive Random Access Memory by Ammoniation Annealing. *IEEE Electron. Device Lett.* **2020**, *41*, 357–360. [CrossRef]

84. Hussain, F.; Imran, M.; Khalil, R.M.A.; Sattar, M.A.; Niaz, N.A.; Rana, A.M.; Ismail, M.; Khera, E.A.; Rasheed, U.; Mumtaz, F.; et al. A first-principles study of Cu and Al doping in ZrO$_2$ for RRAM device applications. *Vacuum* **2019**, *168*, 108842–108847. [CrossRef]

85. Wei, X.; Huang, H.; Ye, C.; Wei, W.; Zhou, H.; Chen, Y.; Zhang, R.; Zhang, L.; Xia, Q. Exploring the role of nitrogen incorporation in ZrO$_2$ resistive switching film for enhancing the device performance. *J. Alloy. Compd.* **2019**, *775*, 1301–1306. [CrossRef]

86. Liu, K.-C.; Tzeng, W.-H.; Chang, K.-M.; Huang, J.-J.; Lee, Y.-J.; Yeh, P.-H.; Chen, P.-S.; Lee, H.-Y.; Chen, F.; Tsai, M.-J. Investigation of the effect of different oxygen partial pressure to LaAlO$_3$ thin film properties and resistive switching characteristics. *Thin Solid Film.* **2011**, *520*, 1246–1250. [CrossRef]

87. Kim, C.H.; Ahn, Y.; Son, J.Y. SrTiO$_3$-Based Resistive Switching Memory Device with Graphene Nanoribbon Electrodes. *Rapid Commun.* **2016**, *99*, 9–11. [CrossRef]

88. Chen, S.-W.; Wu, J.-M. Unipolar resistive switching behavior of BiFeO$_3$ thin films prepared by chemical solution deposition. *Thin Solid Films* **2010**, *519*, 499–504. [CrossRef]

89. Chen, H.Y.; Tian, H.; Gao, B.; Yu, S.; Liang, J.; Kang, J.; Zhang, Y.; Ren, T.-L.; Wong, H.S.P. Electrode oxide interface engineering by inserting single-layer graphene Application for HfO$_x$-based resistive random access memory. *IEEE Electron. Device Lett.* **2012**, *11*, 489–492.

90. Chen, Z.; Zhang, Y.; Yu, Y.; Che, Y.; Jin, L.; Li, Y.; Li, Q.; Li, T.; Dai, H.; Yao, J. Write once read many times resistance switching memory based on all-inorganic perovskite CsPbBr$_3$ quantum dot. *Opt. Mater.* **2019**, *90*, 123–126. [CrossRef]

91. Hickmott, T.W. Low-Frequency Negative Resistance in Thin Anodic Oxide Films. *J. Appl. Phys.* **1962**, *33*, 2669–2682. [CrossRef]

92. Samanta, S.; Han, K.; Das, S.; Gong, X. Improvement in Threshold Switching Performance Using Al$_2$O$_3$ Interfacial Layer in Ag/Al$_2$O$_3$/SiO$_x$/W Cross-Point Platform. *IEEE Electron. Device Lett.* **2020**, *41*, 924–927. [CrossRef]

93. Banerjee, W.; Rahaman, S.Z.; Maikap, S. Excellent Uniformity and Multilevel Operation in Formation-Free Low Power Resistive Switching Memory Using IrO$_x$/AlO$_x$/W Cross-Point. *Jpn. J. Appl. Phys.* **2012**, *51*, 04dd10–04dd16. [CrossRef]

94. Knorr, N.; Bamedi, A.; Karipidou, Z.; Wirtz, R.; Sarpasan, M.; Rosselli, S.; Nelles, G. Evidence of electrochemical resistive switching in the hydrated alumina layers of Cu/CuTCNQ/(native AlOx)/Al junctions. *J. Appl. Phys.* **2013**, *114*, 124510–124518. [CrossRef]

95. Sleiman, A.; Sayers, P.W.; Mabrook, M.F. Mechanism of resistive switching in Cu/AlO$_x$/W nonvolatile memory structures. *J. Appl. Phys.* **2013**, *113*, 164506–164511. [CrossRef]

96. Anwar, F.; Nogan, J.; Zarkesh-Ha, P.; Osinski, M. Multilevel resistance in Ti/Pt/AlO$_x$/HfO$_y$/Ti/Pt/Ag resistive switching devices. In Proceedings of the 2015 IEEE Nanotechnology Materials and Devices Conference (NMDC), Anchorage, AK, USA, 13–16 September 2015.

97. Varun, I.; Bharti, D.; Raghuwanshi, V.; Tiwari, S.P. Multi-temperature deposition scheme for improved resistive switching behavior of Ti/AlO$_x$/Ti MIM structure. *Solid State Ion.* **2017**, *309*, 86–91. [CrossRef]

98. Kurnia, F.; Liu, C.; Raveendra, N.; Jung, C.U.; Vasudevan, R.K.; Valanoor, N. Self-Assembled NiO Nanocrystal Arrays as Memristive Elements. *Adv. Electron. Mater.* **2020**, *6*, 1901153–1901160. [CrossRef]

99. Yoshida, C.; Kinoshita, K.; Yamasaki, T.; Sugiyama, Y. Direct observation of oxygen movement during resistance switching in NiO/Pt film. *Appl. Phys. Lett.* **2008**, *93*, 042106–042109. [CrossRef]

100. Russo, U.; Ielmini, D.; Cagli, C.; Lacaita, A.L. Filament Conduction and Reset Mechanism in NiO-Based Resistive-Switching Memory (RRAM) Devices. *IEEE Trans. Electron. Devices* **2009**, *56*, 186–192. [CrossRef]

101. Cagli, C.; Nardi, F.; Ielmini, D. Modeling of Set/Reset Operations in NiO-Based Resistive-Switching Memory Devices. *IEEE Trans. Electron. Devices* **2009**, *56*, 1712–1720. [CrossRef]

102. Ielmini, D.; Nardi, F.; Cagli, C.; Lacaita, A.L. Size-Dependent Retention Time in NiO-Based Resistive-Switching Memories. *IEEE Electron. Device Lett.* **2010**, *31*, 353–355. [CrossRef]

103. Sullaphen, J.; Bogle, K.; Cheng, X.; Gregg, J.M.; Valanoor, N. Interface mediated resistive switching in epitaxial NiO nanostructures. *Appl. Phys. Lett.* **2012**, *100*, 203115–203120. [CrossRef]

104. Sakellaropoulos, D.; Bousoulas, P.; Nikas, G.; Arvanitis, C.; Bagakis, E.; Tsoukalas, D. Enhancing the synaptic properties of low-power and forming-free HfO$_x$/TaO$_y$/HfO$_x$ resistive switching devices. *Microelectron. Eng.* **2020**, *229*, 111358–111364. [CrossRef]

105. Nauenheim, C.; Kuegeler, C.; Ruediger, A.; Waser, R. Investigation of the electroforming process in resistively switching TiO$_2$ nanocrosspoint junctions. *Appl. Phys. Lett.* **2010**, *96*, 122902–122905. [CrossRef]

106. Hermes, C.; Bruchhaus, R.; Waser, R. Forming-Free TiO$_2$-Based Resistive Switching Devices on CMOS-Compatible W-Plugs. *IEEE Electron. Device Lett.* **2011**, *32*, 1588–1590. [CrossRef]

107. Salaoru, I.; Prodromakis, T.; Khiat, A.; Toumazou, C. Resistive switching of oxygen enhanced TiO$_2$ thin-film devices. *Appl. Phys. Lett.* **2013**, *102*, 013506–013510. [CrossRef]

108. Otsuka, S.; Hamada, Y.; Shimizu, T.; Shingubara, S. Ferromagnetic nano-conductive filament formed in Ni/TiO$_2$/Pt resistive-switching memory. *Appl. Phys. A* **2014**, *118*, 613–619. [CrossRef]

109. Vishwanath, S.K.; Kim, J. Resistive switching characteristics of all-solution-based Ag/TiO$_2$/Mo-doped In$_2$O$_3$ devices for non-volatile memory applications. *J. Mater. Chem. C* **2016**, *4*, 10967–10972. [CrossRef]

110. Sharath, S.U.; Bertaud, T.; Kurian, J.; Hildebrandt, E.; Walczyk, C.; Calka, P.; Zaumseil, P.; Sowinska, M.; Walczyk, D.; Gloskovskii, A.; et al. Towards forming-free resistive switching in oxygen engineered HfO$_{2-x}$. *Appl. Phys. Lett.* **2014**, *104*, 063502–063507. [CrossRef]

111. Clima, S.; Chen, Y.Y.; Degraeve, R.; Mees, M.; Sankaran, K.; Govoreanu, B.; Jurczak, M.; De Gendt, S.; Pourtois, G. First-principles simulation of oxygen diffusion in HfO$_x$: Role in the resistive switching mechanism. *Appl. Phys. Lett.* **2012**, *100*, 133102–133106. [CrossRef]

112. Lanza, M.; Zhang, K.; Porti, M.; Nafría, M.; Shen, Z.Y.; Liu, L.F.; Kang, J.F.; Gilmer, D.; Bersuker, G. Grain boundaries as preferential sites for resistive switching in the HfO$_2$ resistive random access memory structures. *Appl. Phys. Lett.* **2012**, *100*, 123508. [CrossRef]

113. De Stefano, F.; Houssa, M.; Kittl, J.A.; Jurczak, M.; Afanas'ev, V.V.; Stesmans, A. Semiconducting-like filament formation in TiN/HfO$_2$/TiN resistive switching random access memories. *Appl. Phys. Lett.* **2012**, *100*, 142102–142105. [CrossRef]

114. Ambrogio, S.; Balatti, S.; Cubeta, A.; Calderoni, A.; Ramaswamy, N.; Ielmini, D. Statistical Fluctuations in HfO$_x$ Resistive-Switching Memory: Part I - Set/Reset Variability. *IEEE Trans. Electron. Devices* **2014**, *61*, 2912–2919. [CrossRef]

115. Abbas, Y.; Han, I.S.; Sokolov, A.S.; Jeon, Y.-R.; Choi, C. Rapid thermal annealing on the atomic layer-deposited zirconia thin film to enhance resistive switching characteristics. *J. Mater. Sci. Mater. Electron.* **2019**, *31*, 903–909. [CrossRef]

116. Verbakel, F.; Meskers, S.C.; de Leeuw, D.M.; Janssen, R.A. Resistive Switching in Organic Memories with a Spin-Coated Metal Oxide Nanoparticle Layer. *J. Phys. Chem. C Lett.* **2008**, *112*, 5254–5257. [CrossRef]

117. Awais, M.N.; Muhammad, N.M.; Navaneethan, D.; Kim, H.C.; Jo, J.; Choi, K.H. Fabrication of ZrO_2 layer through electrohydrodynamic atomization for the printed resistive switch (memristor). *Microelectron. Eng.* **2013**, *103*, 167–172. [CrossRef]

118. Kärkkänen, I.; Shkabko, A.; Heikkilä, M.; Niinistö, J.; Ritala, M.; Leskelä, M.; Hoffmann-Eifert, S.; Waser, R. Study of atomic layer deposited ZrO_2 and ZrO_2/TiO_2 films for resistive switching application. *Phys. Status Solidi (a)* **2014**, *211*, 301–309. [CrossRef]

119. Ismail, M.; Huang, C.Y.; Panda, D.; Hung, C.J.; Tsai, T.L.; Jieng, J.H.; Lin, C.-A.; Chand, U.; Rana, A.M.; Ahmed, E.; et al. Muhammad Younus Nadeem and Tseung-Yuen Tseng, Forming-free bipolar resistive switching in nonstoichiometric ceria films. *Nanoscale Res. Lett.* **2014**, *9*, 1–8. [CrossRef] [PubMed]

120. Parreira, P.; Paterson, G.W.; McVitie, S.; MacLaren, D.A. Stability, bistability and instability of amorphous ZrO_2 resistive memory devices. *J. Phys. D Appl. Phys.* **2016**, *49*, 095111–095117. [CrossRef]

121. Bailey, T.J.; Jha, R. Understanding Synaptic Mechanisms in $SrTiO_3$ RRAM Devices. *IEEE Trans. Electron. Devices* **2018**, *65*, 3514–3520. [CrossRef]

122. Tsubouchi, K.; Ohkubo, I.; Kumigashira, H.; Oshima, M.; Matsumoto, Y.; Itaka, K.; Ohnishi, T.; Lippmaa, M.; Koinuma, H. High-Throughput Characterization of Metal Electrode Performance for Electric-Field-Induced Resistance Switching in Metal $Pr_{0.7}Ca_{0.3}MnO_3$ Metal Structure. *Adv. Mater.* **2007**, *19*, 1711–1713. [CrossRef]

123. Hirose, Y.; Hirose, H. Polarity-dependent memory switching and behavior of Ag dendrite in Ag-photodoped amorphous As_2S_3 films. *J. Appl. Phys.* **1976**, *47*, 2767–2772. [CrossRef]

124. Imanishi, Y.; Kida, S.; Nakaoka, T. Direct observation of Ag filament growth and unconventional SET-RESET operation in GeTe amorphous films. *AIP Adv.* **2016**, *6*, 075003-1–075003-9. [CrossRef]

125. Bai, Y.; Wu, H.; Wang, K.; Wu, R.; Song, L.; Li, T.; Wang, J.; Yu, Z.; Qian, H. Stacked 3D RRAM Array with Graphene/CNT as Edge Electrodes. *Sci Rep.* **2015**, *5*, 13785–13793. [CrossRef] [PubMed]

126. Jeon, H.; Park, J.; Jang, W.; Kim, H.; Ahn, S.; Jeon, K.-J.; Seo, H.; Jeon, H. Detection of oxygen ion drift in $Pt/Al_2O_3/TiO_2/Pt$ RRAM using interface-free single-layer graphene electrodes. *Carbon* **2014**, *75*, 209–216. [CrossRef]

127. Xia, Y.; Sun, B.; Wang, H.; Zhou, G.; Kan, X.; Zhang, Y.; Zhao, Y. Metal ion formed conductive filaments by redox process induced nonvolatile resistive switching memories in MoS_2 film. *Appl. Surf. Sci.* **2017**, *426*, 812–816. [CrossRef]

128. Kadhim, M.S.; Yang, F.; Sun, B.; Hou, W.; Peng, H.; Hou, Y.; Jia, Y.; Yuan, L.; Yu, Y.; Zhao, Y. Existence of Resistive Switching Memory and Negative Differential Resistance State in Self-Colored MoS_2/ZnO Heterojunction Devices. *ACS Appl. Electron. Mater.* **2019**, *1*, 318–324. [CrossRef]

129. Chen, C.; Gao, S.; Zeng, F.; Tang, G.S.; Li, S.Z.; Song, C.; Fu, H.D.; Pan, F. Migration of interfacial oxygen ions modulated resistive switching in oxide-based memory devices. *J. Appl. Phys.* **2013**, *114*, 014502–014509. [CrossRef]

130. Jana, D.; Dutta, M.; Samanta, S.; Maikap, S. Subhranu Samanta and Siddheswar Maikap, RRAM characteristics using a new $Cr GdO_x TiN$ structure. *Nanoscale Res. Lett.* **2014**, *9*, 680-1–680-9. [CrossRef]

131. Hongbin, Z.; Hailing, T.; Feng, W.; Jun, D. Highly Transparent Dysprosium Oxide-Based RRAM with Multilayer Graphene Electrode for Low-Power Nonvolatile Memory Application. *IEEE Trans. Electron. Devices* **2014**, *61*, 1388–1393. [CrossRef]

132. Chen, S.-C.; Chang, T.-C.; Chen, S.-Y.; Chen, C.-W.; Chen, S.-C.; Sze, S.M.; Tsai, M.-J.; Kao, M.-J.; Yeh, F.-S. Bipolar resistive switching of chromium oxide for resistive random access memory. *Solid State Electron.* **2011**, *62*, 40–43. [CrossRef]

133. Huang, Y.; Huang, R.; Pan, Y.; Zhang, L.; Cai, Y.; Yang, G.; Wang, Y. A New Dynamic Selector Based on the Bipolar RRAM for the Crossbar Array Application. *IEEE Trans. Electron. Devices* **2012**, *59*, 2277–2280. [CrossRef]

134. Park, K.; Lee, J.S. Controlled synthesis of $Ni/CuO_x/Ni$ nanowires by electrochemical deposition with self-compliance bipolar resistive switching. *Sci. Rep.* **2016**, *6*, 23069–23074. [CrossRef] [PubMed]

135. Seo, S.; Lim, J.; Lee, S.; Alimkhanuly, B.; Kadyrov, A.; Jeon, D.; Lee, S. Graphene-Edge Electrode on a Cu-Based Chalcogenide Selector for 3D Vertical Memristor Cells. *ACS Appl Mater. Interfaces* **2019**, *11*, 43466–43472. [CrossRef] [PubMed]

136. Chang, T.; Jo, S.H.; Lu, W. Short-Term Memory to Long-Term memory transition in a nanoscale memristor. *ACS Nano* **2011**, *5*, 7669–7676. [CrossRef] [PubMed]

137. Ambrogio, S.; Balatti, S.; Milo, V.; Carboni, R.; Wang, Z.-Q.; Calderoni, A.; Ramaswamy, N.; Ielmini, D. Neuromorphic Learning and Recognition With One-Transistor-One-Resistor Synapses and Bistable Metal Oxide RRAM. *IEEE Trans.Electron. Devices* **2016**, *63*, 1508–1515. [CrossRef]

138. Wang, S.; Zhang, D.W.; Zhou, P. Two-dimensional materials for synaptic electronics and neuromorphic systems. *Sci. Bull.* **2019**, *64*, 1056–1066. [CrossRef]

139. Shi, Y.; Liang, X.; Yuan, B.; Chen, V.; Li, H.; Hui, F.; Yu, Z.; Yuan, F.; Pop, E.; Wong, H.S.P.; et al. Electronic synapses made of layered two-dimensional materials. *Nat. Electron.* **2018**, *1*, 458–465. [CrossRef]

140. Wang, Z.; Zeng, T.; Ren, Y.; Lin, Y.; Xu, H.; Zhao, X.; Liu, Y.; Ielmini, D. Toward a generalized Bienenstock-Cooper-Munro rule for spatiotemporal learning via triplet-STDP in memristive devices. *Nat. Commun.* **2020**, *11*, 1509–1510. [CrossRef]

141. Ielmini, D. Brain-inspired computing with resistive switching memory (RRAM): Devices, synapses and neural networks. *Microelectron. Eng.* **2018**, *190*, 44–53. [CrossRef]

142. Kim, B.; Choi, H.-S.; Kim, Y. A Study of Conductance Update Method for Ni/SiN$_x$/Si Analog Synaptic Device. *Solid State Electron.* **2020**, *171*, 107772–107799. [CrossRef]

143. Serb, A.; Corna, A.; George, R.; Khiat, A.; Rocchi, F.; Reato, M.; Maschietto, M.; Mayr, C.; Indiveri, G.; Vassanelli, S.; et al. Memristive synapses connect brain and silicon spiking neurons. *Sci. Rep.* **2020**, *10*, 2590–2597. [CrossRef]

144. Liu, B.; Tai, H.H.; Liang, H.; Zheng, E.-Y.; Sahoo, M.; Hsu, C.-H.; Chen, T.C.; Huang, C.-A.; Wang, J.-C.; Hou, T.-H.; et al. Dimensional Anisotropic Graphene with High Mobility and High On-Off Ratio in three-terminal RRAM device. *Mater. Chem. Front.* **2020**, *10*, 1–17.

145. Park, Y.; Lee, J.S. Artificial Synapses with Short- and Long-Term Memory for Spiking Neural Networks Based on Renewable Materials. *ACS Nano* **2017**, *11*, 8962–8969. [CrossRef] [PubMed]

146. Kim, H.; Bae, J.-H.; Lim, S.; Lee, S.-T.; Seo, Y.-T.; Kwon, D.; Park, B.-G.; Lee, J.-H. Efficient precise weight tuning protocol considering variation of the synaptic devices and target accuracy. *Neurocomputing* **2020**, *378*, 189–196. [CrossRef]

147. Halter, M.; Begon-Lours, L.; Bragaglia, V.; Sousa, M.; Offrein, B.J.; Abel, S.; Luisier, M.; Fompeyrine, J. Back-End, CMOS-Compatible Ferroelectric Field-Effect Transistor for Synaptic Weights. *ACS Appl. Mater. Interfaces* **2020**, *12*, 17725–17732. [CrossRef] [PubMed]

148. Song, S.; Miller, K.D.; Abbott, L.F. Competitive Hebbian learning through spike-timing-dependent synaptic plasticity. *Nat. Neurosci.* **2000**, *3*, 919–926. [CrossRef] [PubMed]

149. Patil, V.L.; Patil, A.A.; Patil, S.V.; Khairnar, N.A.; Tarwal, N.L.; Vanalakar, S.A.; Bulakhe, R.N.; In, I.; Patil, P.S.; Dongale, T.D. Bipolar resistive switching, synaptic plasticity and non-volatile memory effects in the solution-processed zinc oxide thin film. *Mater. Sci. Semicond. Process.* **2020**, *106*, 104769–104776. [CrossRef]

150. Jeon, D.S.; Park, J.H.; Kang, D.Y.; Dongale, T.D.; Kim, T.G. Forming-ready resistance random access memory using randomly pre-grown conducting filaments via pre-forming. *Mater. Sci. Semicond. Process.* **2020**, *110*, 104951–104956. [CrossRef]

151. Zhang, X.; Liu, S.; Zhao, X.; Wu, F.; Wu, Q.; Wang, W.; Cao, R.; Fang, Y.; Lv, H.; Long, S.; et al. Emulating Short-Term and Long-Term Plasticity of Bio-Synapse Based on Cu/a-Si/Pt Memristor. *IEEE Electron. Device Lett.* **2017**, *38*, 1208–1211. [CrossRef]

152. Lobov, S.A.; Mikhaylov, A.N.; Shamshin, M.; Makarov, V.A.; Kazantsev, V.B. Spatial Properties of STDP in a Self-Learning Spiking Neural Network Enable Controlling a Mobile Robot. *Front. Neurosci.* **2020**, *14*, 88–97. [CrossRef]

153. Kim, J.; Kim, C.-H.; Yun Woo, S.; Kang, W.-M.; Seo, Y.-T.; Lee, S.; Oh, S.; Bae, J.-H.; Park, B.-G.; Lee, J.-H. Initial synaptic weight distribution for fast learning speed and high recognition rate in STDP-based spiking neural network. *Solid State Electron.* **2020**, *165*, 107742–107747. [CrossRef]

154. Kim, K.Y.; Rios, L.C.; Le, H.; Perez, A.J.; Phan, S.; Bushong, E.A.; Deerinck, T.J.; Liu, Y.H.; Ellisman, M.A.; Lev-Ram, V.; et al. Synaptic Specializations of Melanopsin-Retinal Ganglion Cells in Multiple Brain Regions Revealed by Genetic Label for Light and Electron Microscopy. *Cell Rep.* **2019**, *29*, 628–644. [CrossRef] [PubMed]

155. Ilyas, N.; Li, D.; Li, C.; Jiang, X.; Jiang, Y.; Li, W. Analog Switching and Artificial Synaptic Behavior of Ag/SiO$_x$:Ag/TiO$_x$/p^{++}-Si Memristor Device. *Nanoscale Res. Lett* **2020**, *15*, 30–40. [CrossRef] [PubMed]

156. Pignatelli, M.; Tejeda, H.A.; Barker, D.J.; Bontempi, L.; Wu, J.; Lopez, A.; Palma Ribeiro, S.; Lucantonio, F.; Parise, E.M.; Torres-Berrio, A.; et al. Cooperative synaptic and intrinsic plasticity in a disynaptic limbic circuit drive stress-induced anhedonia and passive coping in mice. *Mol. Psychiatry* **2020**, *2*, 6868–6887. [CrossRef] [PubMed]

157. NagaJyothi, G.; Sridevi, S. High speed low area OBC DA based decimation filter for hearing aids application. *Int. J. Speech Technol.* **2019**, *23*, 111–121. [CrossRef]

158. Berdan, R.; Vasilaki, E.; Khiat, A.; Indiveri, G.; Serb, A.; Prodromakis, T. Emulating short-term synaptic dynamics with memristive devices. *Sci. Rep.* **2016**, *6*, 18639–18647.

159. Sun, L.; Hwang, G.; Choi, W.; Han, G.; Zhang, Y.; Jiang, J.; Zheng, S.; Watanabe, K.; Taniguchi, T.; Zhao, M.; et al. Ultralow switching voltage slope based on two-dimensional materials for integrated memory and neuromorphic applications. *Nano Energy* **2020**, *69*, 104472–104477. [CrossRef]

160. Chen, P.A.; Ge, R.J.; Lee, J.W.; Hsu, C.H.; Hsu, W.C.; Akinwande, D.; Chiang, M.H. An RRAM with a 2D Material Embedded Double Switching Layer for Neuromorphic Computing. In Proceedings of the 2018 IEEE 13th Nanotechnology Materials and Devices Conference (NMDC), Portland, OR, USA, 14–17 October 2018.

161. Shi, Y.; Pan, C.; Chen, V.; Raghavan, N.; Pey, K.L.; Puglisi, F.M.; Pop, E.; Wong, H.-P.; Lanza, M. Coexistence of volatile and non volatile resistive switching in 2D *h*- BN based electronic synapses. In Proceedings of the 2017 IEEE International Electron Devices Meeting (IEDM), San Francisco, CA, USA, 2–6 December 2017.

162. Sokolov, A.S.; Jeon, Y.-R.; Ku, B.; Choi, C. Ar ion plasma surface modification on the heterostructured TaO_x/InGaZnO thin films for flexible memristor synapse. *J. Alloy. Compd.* **2020**, *822*, 153625–153635. [CrossRef]

163. Wang, T.Y.; He, Z.Y.; Chen, L.; Zhu, H.; Sun, Q.Q.; Ding, S.J.; Zhou, P.; Zhang, D.W. An Organic Flexible Artificial Bio-Synapses with Long-Term Plasticity for Neuromorphic Computing. *Micromachines* **2018**, *9*, 239. [CrossRef]

164. Ohno, T.; Hasegawa, T.; Tsuruoka, T.; Terabe, K.; Gimzewski, J.K.; Aono, M. Short-term plasticity and long-term potentiation mimicked in single inorganic synapses. *Nat. Mater.* **2011**, *10*, 591–595. [CrossRef]

165. Ku, B.; Abbas, Y.; Kim, S.; Sokolov, A.S.; Jeon, Y.-R.; Choi, C. Improved resistive switching and synaptic characteristics using Ar plasma irradiation on the Ti/HfO_2 interface. *J. Alloy. Compd.* **2019**, *797*, 277–283. [CrossRef]

166. Wan, T.; Qu, B.; Du, H.; Lin, X.; Lin, Q.; Wang, D.W.; Cazorla, C.; Li, S.; Liu, S.; Chu, D. Digital to analog resistive switching transition induced by graphene buffer layer in strontium titanate based devices. *J. Colloid Interface Sci.* **2018**, *512*, 767–774. [CrossRef] [PubMed]

167. Baek, I.G.; Lee, M.S.; Seo, S.; Lee, M.J.; Seo, D.H.; Suh, D.S.; Park, J.C.; Park, S.O.; Kim, H.S.; Yoo, I.K.; et al. Moon, Highly scalable nonvolatile resistive memory using simple binary oxide driven by asymmetric unipolar voltage pulses, presented at the IEDM Technical Digest. In Proceedings of the IEEE International Electron Devices Meeting, San Francisco, CA, USA, 13–15 December 2004.

168. Zha, Y.; Wei, Z.; Li, J. Recent progress in RRAM technology: From compact models to applications. In Proceedings of the 2017 China Semiconductor Technology International Conference (CSTIC), Shanghai, China, 12–13 March 2017.

169. Karam, R.; Liu, R.; Chen, P.-Y.; Yu, S.; Bhunia, S. Security Primitive Design with Nanoscale Devices: A Case Study with Resistive RAM. In Proceedings of the 2016 International Great Lakes Symposium on VLSI (GLSVLSI), Boston, MA, USA, 18–20 May 2016.

170. Wang, W.; Li, Y.; Wang, M.; Wang, L.; Liu, Q.; Banerjee, W.; Li, L.; Liu, M. A hardware neural network for handwritten digits recognition using binary RRAM as synaptic weight element. In Proceedings of the the 2016 IEEE Silicon Nanoelectronics Workshop (SNW), Honolulu, HI, USA, 12–13 June 2016.

171. Yu, S.; Sun, X.; Peng, X.; Huang, S. Compute-in-Memory with Emerging NonvolatileMemories: Challenges and Prospects. In Proceedings of the 2020 IEEE Custom Integrated Circuits Conference (CICC), Boston, MA, USA, USA, 22–25 March 2020.

Review

Recent Advancement of Electromagnetic Interference (EMI) Shielding of Two Dimensional (2D) MXene and Graphene Aerogel Composites

Kanthasamy Raagulan [1], **Bo Mi Kim** [2,*] **and Kyu Yun Chai** [1,*]

[1] Division of Bio–Nanochemistry, College of Natural Sciences, Wonkwang University, Iksan 570-749, Korea; raagulan@live.com
[2] Department of Chemical Engineering, Wonkwang University, Iksan 570-749, Korea
*   Correspondence: 123456@wku.ac.kr (B.M.K.); geuyoon@wonkwang.ac.kr (K.Y.C.);
    Tel.: +82-108-669-0321 (B.M.K.); +82-63-850-6230 (K.Y.C.); Fax: +82-63-841-4893 (K.Y.C.)

Received: 28 February 2020; Accepted: 22 March 2020; Published: 8 April 2020

Abstract: The two Dimensional (2D) materials such as MXene and graphene, are most promising materials, as they have attractive properties and attract numerous application areas like sensors, supper capacitors, displays, wearable devices, batteries, and Electromagnetic Interference (EMI) shielding. The proliferation of wireless communication and smart electronic systems urge the world to develop light weight, flexible, cost effective EMI shielding materials. The MXene and graphene mixed with polymers, nanoparticles, carbon nanomaterial, nanowires, and ions are used to create materials with different structural features under different fabrication techniques. The aerogel based hybrid composites of MXene and graphene are critically reviewed and correlate with structure, role of size, thickness, effect of processing technique, and interfacial interaction in shielding efficiency. Further, freeze drying, pyrolysis and hydrothermal treatment is a powerful tool to create excellent EMI shielding aerogels. We present here a review of MXene and graphene with various polymers and nanomaterials and their EMI shielding performances. This will help to develop a more suitable composite for modern electronic systems.

Keywords: MXene; Graphene; EMI shielding; aerogel; composites

---

1. Introduction

The proliferation of smart electronic devices and wireless communication in the artificial intelligent age are a source of electromagnetic pollution (EMP) which is a serious universal problem. The EMP creates complexities in the natural electromagnetic environment, and is unwanted radiation which not only disturbs the general function of surrounding electronic systems but also threatens the well–being of humans [1–8]. This disturbing phenomenon is called electromagnetic interference (EMI) and it causes data theft, malfunction of the electronic devices, degradation of basic function of electronic devices, and vulnerability of the personal security in electronic components, while grounding problems in humans such as mutation, insomnia, headache, leukemia, damage the organs, thermal injuries, and cancer [1–5]. In addition, expansion of digital networks and sensitive remote-controlled systems demanding high quality densely built electronic control systems, creates un-compatible environments (UCE) or electromagnetic noise (EMN) or EMI and the current of electrodynamic and basic field are essential sources of EMI. At low frequency where electric and magnetic fields act independently, whereas high frequency waves are propagating electromagnetic radiation (EMR) which causes EMI [9]. The source of the EMI are radio and TV transmission, radar, aviation, electromagnetic missiles, warfront, Bluetooth, wireless network (WLAN), remote controls in which EMI occur by radiative coupling. The radiative coupling is due to the disturbance in all conductive parts by propagating high frequency radiation. In addition, the other types of coupling such as inductive, impedance and capacitive which are due to the current, voltage, and resistance in electric circuits [9].

In 2020, the world utilizes electromagnetic radiation as a jamming tool or weapon in battlefields or to destroy electronic components of the ships, radars, and flights, and destroy security of the countries. Hence, electromagnetic interference shielding is the inevitable choice around the globe [9–15]. So far, scientists have studied EMI shielding in various frequency range by using numerous materials which include zero−dimension (0D), one−dimension (1D), two−dimension (2D), and three−dimension (3D) materials. The combination of these materials gives rise to different structural feature influences with excellent EMI shielding behavior. The MXene, MAX phase, quantum dots (QDs) metal nanoparticle, oxide nanoparticle, carbon black, graphene (GN), graphite, single wall carbon nanotube (SWCNT), multiwall carbon nanotube (MWCNT), magnetic nanoparticle, metal oxides, core shell Nano materials, metal plates, organic substances, nonconductive polymers (NCP), conductive polymers (CP), and craft polymers are being utilized for EMI shielding application. Graft and nonconductive polymers are being used as a matrix and used to connect the (0, 1, 2, and 3)−dimensional materials (Table 1) [10–19]. CPs which act as fillers and help to create different structural feature meanwhile electric conductivity (EC), thermal conductivity (TC), tensile strength, and EMI shielding of the composites also improved substantially. In addition, the different organic substances and nanomaterials are being used to improve adhesivity, dispersity, porosity, and other physiochemical properties of the composites [1,20,21].

Table 1. Categorization of the nanomaterial with examples and properties.

Dimension	Examples	Properties	Reference
0D	CdSe/V_2O_5 QDs, CdS/CdSe QDs, ZnO QDs, C−QDs, GN QDs, CoFe QDs, ZnCo QDs, metal hybrid QDs, SiO_2 QDs, Au QDs	Improve microwave absorption, and magnetic properties, limited use because of narrow absorption bandwidth, high−density and perishable, large specific surface area, less thickness, can be mixed with other materials,	[10–15]
1D	single wall carbon nanotube (CNT) (SWCNT), multiwall CNT (MWCNT), decorated CNT (dCNT), nanowires (Ag, Cu, Si−C)	Carbonaceous materials show excellent chemical−physio−chemical stability, light weight, good complex permittivity, lack interfacial adhesion, deficiency of magnetic properties, poor dispersion, high production costs, and have impurities. Pure carbon materials show poor EMI attenuation and metal nanowire especially Ag greatly improve EMI shielding.	[10–13]
2D	2D−MXene, graphene (GN), doped GN, reduced GN, graphene oxide (GNO), nanoplates, hexagonal boron nitride (h−BN), layered double hydroxides (LDHs), transition metal dichalcogenides (TMDS), metal−organic Frameworks (MOF), layered metal Oxides (LMOs), covalent organic frameworks (COFs), metals, black phosphorus (BP) and silicene	MXene, GN and MoS_2 are commonly used for EMI shielding, structure can be easily modified, MXene is excellent EMI shielder, nanoparticle decorated GN based composites show excellent EMI SE and functionalization of GN improve dispersivity, and di−electric properties, Pyrolyzed− MOFs/nanomaterials is used to design porous−magnetic high−efficient EMW absorption material, due to the synergy effect between magnetic loss and dielectric loss	[16–18]
3D	MAX phase, expanded graphite, graphite, metal plates, the 3D structure designed by using 0D, 1D, 2D nano structure and polymers	3D materials show less shielding ability, act as precursor to synthesis other nanomaterials, 3D structure made by mixture of nanomaterials show excellent EMI shielding properties.	[10–18]

EMI shielding can be achieved by absorption (SE_A), reflection (SE_R), and multiple reflection (SE_{MR}). SE_R occurs on the surface whereas SE_A and SE_{MR} are happening within the shielding material. The wave propagating on the other side of the shielding materials is called the transmittance (T), the magnitude of which is lower than that of incident wave (I) and value of SE_A, SE_R, and SE_{MR} is smaller than that of I (Figure 1). The electrical conductivity of the composite responsible for the reflection while porous in structure, electric and magnetic bipolarity of the composite and thickness are responsible for the absorption and multiple reflection occurs due to the porous nature and multiple layer structure [1,13,15–18]. In addition, the thickness of the shielding material is higher than that of the skin depth leading to multiple reflection and finally roots the absorption and is a negligible component compared to the absorption and reflection. Though, one of the factors domineers over others which is dependent on the types of material being used [19–21]. Most of the studies show that the absorption is higher than that of the reflection and a recent study disclosed that the reflection can be eliminated from the basic mechanism, and absorption only determined the total EMI shielding. This phenomenon can be achieved by creating different internal structure with higher electric conductivity. The pure material showed higher EMI shielding, but, when a foreign material is introduced into the pure material the EMI shielding is considerably reduced by those foreign polymers. Even though, without polymers or binders, it is difficult to use EMI shielding applications in different electronic devices [1,13,15,17–20].

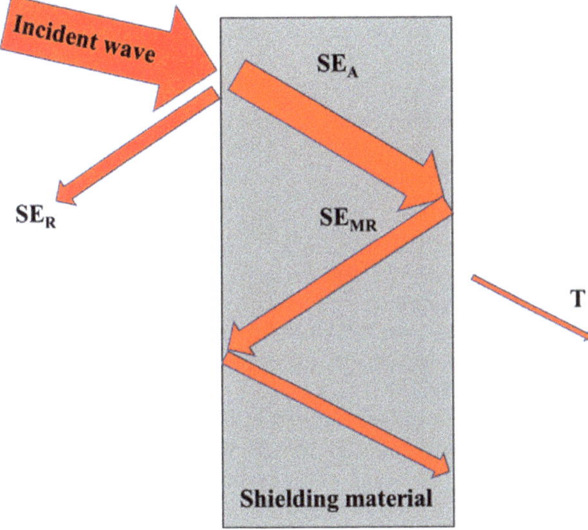

Figure 1. Transformation of the electromagnetic radiation (EMR) at the shielding material.

We focused on a different synthetic route of graphene (GN), MXene, and composites. GN and MXene structural features and synthesis are discussed in detail. In addition, different composite preparation methods are listed, and its effect is discussed in detail with proper comparison in the discussion. In general, aerogel or foam based composite are most attractive to scientists and have excellent EMI shielding with excellent absorption ability.

2. Results and Discussion

2.1. Synthesis of Graphene (GN) and Structural Features

Top–down and bottom–up approaches are being used to synthesize graphene. The top–down approach where graphite is used as a precursor and mechanical exfoliation, graphite intercalation, nanotube slicing, pyrolysis method, reduction of the graphene oxide (GNO), electrochemical exfoliation, sonication, radiation based method, and ball–milling are being used to produce GN while bottom–up follows the strategy that those are grown from metal–carbon melt, epitaxial growth on silicon carbide (SiC), dry ice method, and deposition (Table 2). Multilayer highly ordered pyrolytic graphite (HOPG) is scraped off to produce GN by mechanical exfoliation for which scotch tape, ultrasonic oscillations, and hot press techniques are being used. In the mechanical exfoliation process two types of normal force used to peel of the graphite and shear force utilize ball milling process [9,22–25]. The liquid phase exfoliation (LPE) method includes three basic processes, those are dispersion of graphite in suitable solvent, exfoliation, and purifying where van der Waals forces break down by solvents and ultrasound. The arc discharge method is utilized to synthesize the allotropes of carbon such as graphene, CNT, and fullerene. In which, the carbon precursor is used as the anode and the graphite rod as the cathode, where applied electrical current creates the plasma at higher temperature (3727–5727 °C) and finally form graphene. Various intercalants are being utilized by the intercalation technique. By carbon nanotube (CNT) slicing/unzipping micron size GN are produced for which plasma or chemical etching, intercalation and exfoliation, metal catalyst cutting, abrade on the glass surface, and the CNT tube opens into a lay-flat single layer graphene ribbon (GNR). The pyrolysis is the type of the solvothermal technique where equal molar sodium and ethanol are used to separate the graphite layers. The thermal annealing technique is where amorphous carbon is converted into single layer graphene on the nickel and cobalt surface with the aid of temperature. The reduction of the graphene oxide (GO) can be performed by using chemicals, biomass, radiation, bacteria, electrochemical methods, and heat treatment. Electrochemical exfoliation is practiced in an acidic environment by applying voltage differences between the anode (graphite) and cathode (platinum) while solvent-based high energy is used in the sonication technique. The ball milling utilized solvent or chemical assistance and magnetic assistance technique, although higher quality GN are being produced by using radiation techniques. Laser and electron beams are used in radiation techniques. The deposition method utilizes the Chemical Vapor Deposition (CVD) techniques where solid, liquid, and gaseous carbon precursors are being used, filtration with reduction, spin coating, and spray coating process for GN synthesis. The CVD produces graphene with low defects and during the process the precursors are atomized, and the graphene is formed on the metal catalyst (Cu and Ni). The epitaxial growth is performed on a silicon carbide wafer where stacks of graphene are formed. The epitaxial growth is the exothermic process during which the sublimated (1200–1600 °C) silicon leaves excessive sp^2 hybridized carbon network formed graphene. In addition, burning the dry ice by using magnesium followed by acid treatment produces GN (Table 2) [9,23–27].

The graphene (GN) is a 2D single layer crystalline material with a honeycomb (HC) structure, comprised of sp^2 hybridized carbon atoms. In early 1947, Wallance et al. reported the 2D graphite with zero activation energy by using "tight binding" approximation which was later experimentally observed in 2004 [28–31]. Due the excellent properties of GN, it is an excellent candidate for use in modern electronics. There are two major types of GN, those are zigzag and armchair [28]. Figure 2 exhibits the electrons in carbon atoms (Figure 2a), energy comparison at ground state electronic configuration (Figure 2b), shape of the orbitals and hybridized orbitals (Figure 2c), crystal lattice and unit cells (Figure 2d), and bond formation (σ and π) (Figure 2e). The carbon is a group IV element on the periodic table, with $1s^2$, $2s^2$ sp^2 ground state electronic configuration where all 2p orbitals are regenerated and $2p_z$ is an empty orbital while each $2p_x$ and $2p_y$ holding one electron, respectively. The empty $2p_z$ orbital plays a major role in creating out of the plane π bond and sp^2 hybridized carbon form in−plane σ bond which extends to hexagonal web of carboned leads to formation of mono layer graphene. The average inter atomic distance of GN is 1.42 Å, thus the graphene co−valent bond is stronger than that of C−C bond of alkanes [29]. The monolayer graphene possesses 130.5 GPa of intrinsic tensile strength and 1 TPa of Young's modulus. In addition, the monolayer GN has no band gap which is due to the free moving π electrons which offer weak van der Waals forces between the graphene layers [29]. Due to the weak Van der Waals forces the graphene can be synthesized from bulk graphite (Figures 3 and 4) [28–30].

Table 2. Comparison of different synthetic process of graphene.

Methods	Synthetic Method		Properties	Reference
Bottom–Up	Epitaxial growth of GN on SiC		SiC precursor, lacks homogeneity and quality, expensive due to energy consumption, have environmental concern because of tetraflu– oroethylene (C_2F_4)	[23–27]
	Dry ice method		Produced by complete burning of Mg ribbon inside the dry ice bowl.	
	Chemical Vapor deposition (CVD)		Is type of deposition process, gas phase precursors (CH_4, C_2H_4, C_2H_2, and C_6H_{14}) are used, elevated temperature (450–1000 °C), metallic catalyst (Cu, Ni), low defective GN, and excellent electrical and optical property.	
	Template route		Good quality and well–defined structure, can get high yield by using pyrrole under mild condition, and less desire method due to the damage during purification.	
	Total Organic synthesis		Synthesis from polycyclic aromatic hydrocarbons, high quality GN with high yield, and limited size.	
	Substrate–free gas phase synthesis		New method, gas phase precursor (isopropyl alcohol and dimethyl ether and ethanol), clean and high quality GN.	
Top–Down	Arch discharge		Conventional method, used to synthesis fullerene, CNT and GN, high temperature plasma reaction (3727–5727 °C) in inert and air condition, and affordable cost.	[23–27]
	Liquid phase exfoliation		Common synthetic method, exfoliation occurs in aqueous and no–aqueous medium	
	Graphite intercalation		Intercalation of chemical species into graphite interlayer and improve electrical conductivity.	
	Radiation based methods		Short processing time, High quality, financially not viable, and radiation source are UV and laser.	
	Pyrolysis method		Solvothermal process, can be scaled up, good yield, and speed method.	
	Un–zipping of CNT– GN nanoribbon (GNR)		Cutting the cylindrical CNT by various methods (metal–catalyzed cutting, chemical unzipping, plasma etching, intercalation and exfoliation), low yield, and expensive precursors and chemicals.	
	Mechanical exfoliation		Use normal force (roll milling) and shear force (ball milling), high production cost, large processing time (24–48 h), low yield, and undesirable for large scale production.	
	Sonication		Ultrasonic energy, need large amount of energy, difficult to remove impurities, surfactants are used for sonication, and electrical conductivity.	
	Oxidative exfoliation and reduction	Thermal or hydrothermal reduction	Reduced to rGO, high temperature, greenhouse gas effect, and high operational cost.	
		Chemical reduction	Reduced to rGO and GN, many reducing agent are used (hydrazine (N2H4), zinc/hydrochloric acid, aluminum hydride, borohydrides, nitrogen–based reagents, sulfur–based reagents, sodium borohydride, microorganisms, and caffeic acid), lengthy synthesis time, additional chemical cost, environmental pollution, and toxic.	
		Electro–chemical reduction	Cost effective, less toxic, environmentally friendly, and rapid process.	
		Other reduction methods	photothermal, laser, microwave, photocatalytic, sonochemical, and plasma treatment	

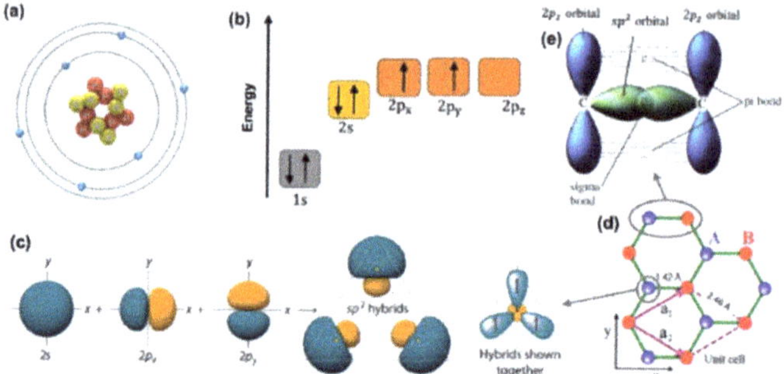

Figure 2. (a). Atomic structure of a carbon atom. (b) Energy levels of outer electrons in carbon atoms. (c) The formation of sp² hybrids. (d) The crystal lattice of graphene, where A and B are carbon atoms belonging to different sub−lattices, a1 and a2 are unit−cell vectors. (e) Sigma bond and pi bond formed by sp² hybridization [29] Copyright Science and technology of advanced materials, 2018.

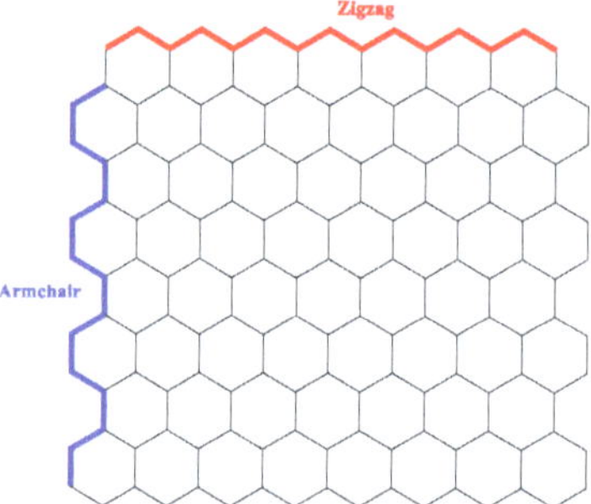

Figure 3. Zigzag–edged and armchair–edged of graphene (GNs).

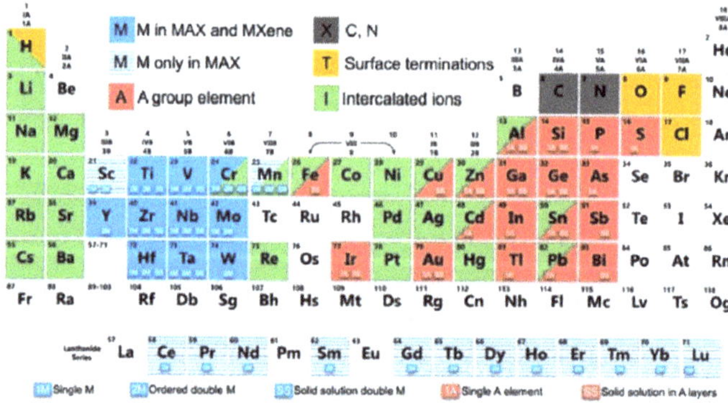

Figure 4. Elements used to create MAX phases, MXenes, and their intercalated ions. The A elements are denoted by a red background and are used to synthesize MAX phases that can possibly be utilized to make MXenes. The elements denoted by a green background, have been intercalated into MXenes (to date) and the symbols are at the bottom, 1M and 1A designate the formation of a single (pure) transition metal and A element MAX phase (and MXene). Solid solutions are indicated by an SS in transition metal atomic planes (blue) or A element planes (red); and 2M indicates the formation of an ordered double-transition metal MAX phase or MXene (in-plane or out-of-plane). The MAX phase elements denoted by blue striped background have not yet been used to synthesis MXene (Figure 5) [32] Copyright American Chemical Society, 2019.

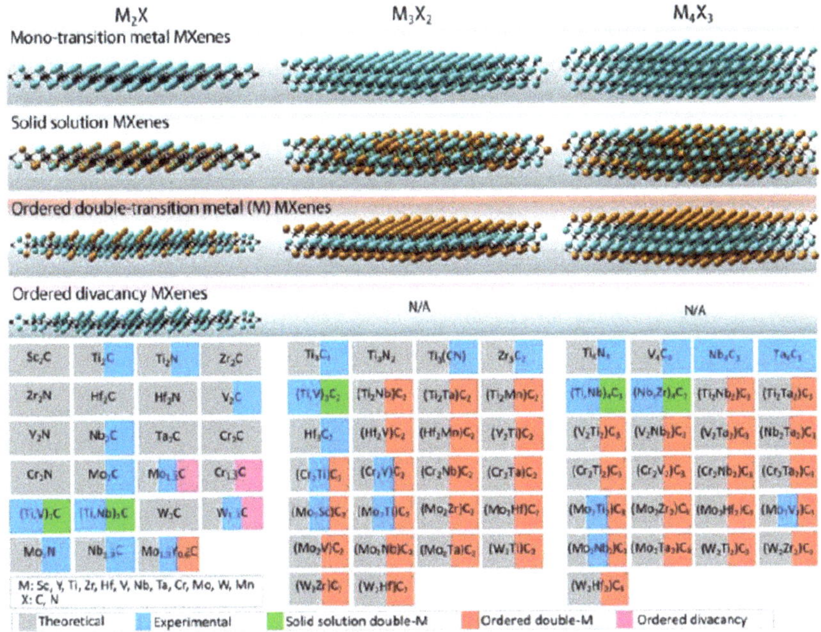

Figure 5. MXenes synthesized up to date. The top row illustrates structures of (top–down) mono–M MXenes, double-M solid solutions (SS) (marked in green), ordered double-M MXenes (marked in red), and ordered divacancy structure (only for the M_2C MXenes), respectively. This table shows the MXene reported both experimentally (blue) and theoretically (gray) so far [32] Copyright American Chemical Society, 2019.

2.2. MXene Synthesis and Structural Features.

The MXene is a fast-growing two–dimensional (2D) material, derived from its corresponding 3D MAX phase by an etching process. The general formula of MXene is $M_{n+1}X_nT_x$ where M is an early transition element, n = 1–3, X is a carbon or nitrogen and T_x is a surface functional groups –F, –OH, =O, and Cl, which are directly attached to the M. The $M_{n+1}AX_n$ is used to denote the MAX phase where generally is group A 13/14 element, but, several other elements also utilize for A layer (Figure 4 and Table 1). The different combination of elements has been used for both MAX and MXene synthesis shown below in Figures 5 and 6. The number of layers in MXene is determined by n where the n+1 layer of MXene is formed which is true for the MAX phase as well (Figure 5) [31–33].

During the development of MXene scientists practiced various strategies to produce good quality MXene. During the etching process, the A layer of the MAX phase is eradicated and surface functionalities are introduced. Fluoride based etchants are being used for the etching process, those are HF, NH_4HF_2, LiF/HCl, and FeF_3/HCl. The HF based etching process causes risk compared to the in situ etching process where fluoride salt and acid are used. According to HF protocol, 1 g of Ti_3AlC_2 is mixed with 20 mL of the HF etchant, and the concentration of HF varies based on its requirement. The 10% of HF is enough to remove Al with accordion–like morphology while 5% of HF is not enough to get accordion-like morphology, but is good enough to remove Al. The in–situ process based on the salt/acid etching process where HF is formed during the reaction is called minimally intensive layer delamination (MILD). The LiF/HCl reaction is being widely used as it produces less defective MXene compared to other studies. Recently, most of the scientists around the globe use 9M HCl/LiF as a standard. Exfoliation is a process where single the MXene layer is separated by various process such as intercalation and sonication. The sonication is the physical method where MXene is ultra-sonicated in water while intercalation used organic molecules followed by sonication. Dimethyl sulfoxide

(DMSO), tetrabutylammonium hydroxide (TBAOH), tetramethylammonium hydroxide (TMAOH), and urea are widely used in the HF based etching process, whereas the MILD method does not need the intercalation process utilized in sonication at about 15 °C under an inert environment and lithium ion induced exfoliation process (Figure 6). Raagulan et al. reported a method for the mass production of exfoliated MXene and its by-products. In which, the evaporation technique is used to concentrate the colloidal solution and filtration process is explored to separate solvent and delaminated MXene (Figure 7) [31–33].

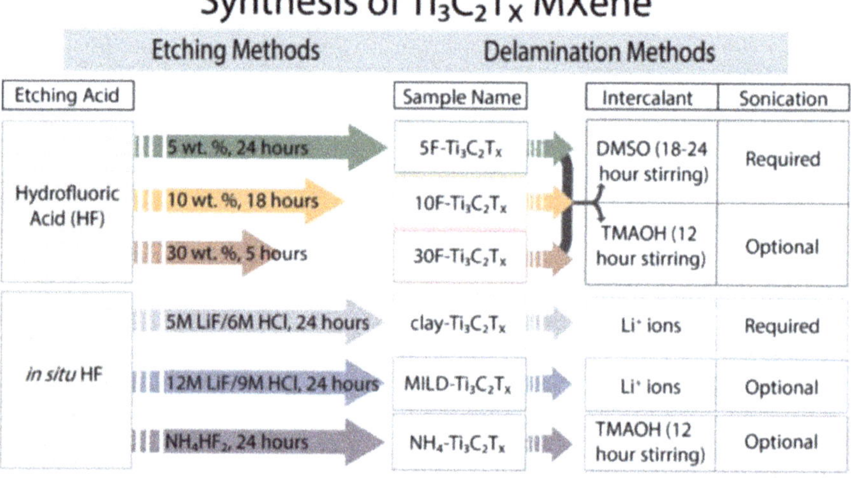

Figure 6. The general MXene synthetic route with both the HF and in−situ HF etching process and the delamination process [33] Copyright American Chemical Society, 2017.

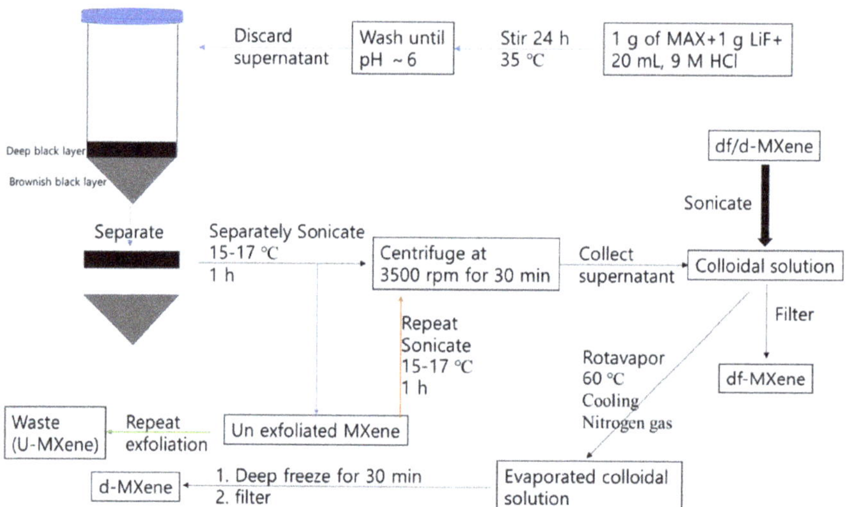

Figure 7. Flow chart of the synthetic approach of the exfoliated MXene synthesis [31] Copyright Royal Society of Chemistry, 2020.

2.3. EMI Shielding Theory and Mechanisms

The EMI shielding can be defined as how well a material quantitatively weakens the energy of the propagating electromagnetic radiation (EMR) in a certain frequency range (L, S, C, X, K, Ku, etc. bands). The EMR with specific power (P_I) hits the surface of the shielding materials undergoing different transformations such as reflection (P_R), absorption (P_A), and transmittance (P_T). The reflection occurs on the surface while absorption happens within the materials and remainder comes out as transmittance EMR [34].

The absorption (A) is depend on the type of material used. The A and absorption coefficient (A_e) can be expressed as follows, where reflection (R) and transmittance (T) are correlated (Equations (1) and (2)) [35].

$$A = 1 - R \tag{1}$$

$$A_e = \left[\frac{1 - R - T}{1 - R}\right] \tag{2}$$

The attenuation of EMR is expressed by shielding effectiveness (SE) and the corresponding unit is given in dB. According to the EMI shielding theory, the SE could be defined as the logarithmic ratio between P_i and power of transmittance (P_t) and can also be defined by using electric intensity (E), magnetic intensity (H), wavelength (λ), and slot length (l) of the EMR [36]. The total shielding effectiveness (SE_T) is expressed by using the following equations where i and t are denoted as incident and transmittance waves, respectively (Equation (3)) [37].

$$SE_T = 10\log\left(\frac{P_i}{P_t}\right) = 20\log\left(\frac{E_i}{E_t}\right) = 20\log\left(\frac{H_i}{H_t}\right) = 20\log\left(\frac{\lambda}{2l}\right) \tag{3}$$

Further, SE_T can be calculated by adding reflection (SE_R), absorption (SE_A), and multiple reflection (SE_{MR}) (Equation (4)).

$$SE_T = SE_R + SE_A + SE_{MR} \tag{4}$$

If the $SE_T > 15$ dB, the SE_M is neglected and equation can be written as follow (Equation (5)),

$$SE_T = SE_R + SE_A \tag{5}$$

Moreover, the T and R can be expressed by using electric intensity (E) and scattering parameters in which t is the transmittance wave, i is the incident wave and r is the reflection wave (Equations (6) and (7)).

$$T = \left|\frac{E_t}{E_i}\right|^2 = |S_{12}|^2 = |S_{21}|^2 \tag{6}$$

$$R = \left|\frac{E_r}{E_i}\right|^2 = |S_{11}|^2 = |S_{22}|^2 \tag{7}$$

Further, the SE_T, SE_R, and SE_{MR} can be expressed in terms of scattering parameters, wave impedance of air (Z_o), wave impedance of the material (Z_m), propagation constant (β), relative magnetic permeability (μ_r), thickness of the shielding materials (t), and imaginary unit (j) (Equations (8)–(10)).

$$SE_T = 10\log(T) = SE_R + SE_A = 10\log\left(\frac{1}{1 - |S_{12}|^2}\right) = 10\log\left(\frac{1}{|S_{21}|^2}\right) \tag{8}$$

$$SE_R = 10\log(1 - R) = \left(\frac{1}{1 - |S_{11}|^2}\right) = 20\log\left|\frac{(Z_o + Z_m)^2}{4Z_oZ_m}\right| \cong 20\log\left|\frac{Z_o}{4Z_m}\right| \tag{9}$$

$$SE_M = 20\log\left(\frac{1}{4}\sqrt{\frac{\sigma}{\omega\mu_r\varepsilon_o}}\right) = 20\log\left|1 - \left(\frac{Z_o - Z_m}{Z_o + Z_m}\right)^2 e^{-2t/\delta}e^{-2j\beta t}\right| = 20\log\left|1 - e^{-2t/\delta}\right| \tag{10}$$

In addition, the SE_A, SE_R, and SE_{MR} can differently be described by using parameters of the shielding materials such as t, skin depth (δ), μ_r, refractive index (n), relative conductivity (σ_r), and imaginary part of wave vector (ik) (Equations (11)–(13)).

$$SE_A = \frac{8.7t}{\delta} = 131.4d\,\sqrt{f\mu_r\sigma_r} = K\left(\tfrac{t}{\delta}\right) = 10\log\left[\frac{T}{(1-R)}\right] = 10\log(1-A_e) = 20\log e^{t/\delta} = 20lm(k)d\,\log e \tag{11}$$

$$SE_R = 108 + \log\left(\frac{\sigma}{f\mu}\right) = 39.5 + 10\log\left(\frac{\sigma}{2\pi f\mu}\right) = 20\log\left|\frac{1+n^2}{4n}\right| \tag{12}$$

$$SE_M = 20\log\left|1 - 10^{\frac{SE_A}{10}}\right| = 168 + 10\log\left(\frac{\sigma_r}{\mu f}\right) = 20\log\left|\frac{1-\left(1-n^2\right)}{(1+n)^2}\exp(2ikd)\right| \tag{13}$$

Skin depth is inversely proportional to square root of $\pi f\sigma\mu$ of the composition where f is frequency of EMR, μ is magnetic permeability, and σ is electric conductivity (Equation (14)) [36].

$$\delta = \frac{1}{\sqrt{\pi f\sigma\mu}} \tag{14}$$

When the electromagnetic radiation propagates, it undergoes the changes from near field to far field which is depend on the distance. The $r < \lambda/2\pi$ is considered as near field and $r > \lambda/2\pi$ is denoted as far field. Thus, most of the EMRs are far field and are regarded as planar waves. The impedance of the wave (intrinsic impedance) Z can be articulated that the amplitude ratio between electric fielding (E) and magnetic field (H) waves, which are perpendicular to each other ($E \perp H$). Furthermore, the Z is influenced by σ, μ, angular frequency ($\omega = 2\pi f$), j, and electric permeability (ε). Z of air is symbolized as Z_o, and has a value of 377 Ω and at this stage j and ω is considered as one and σ is zero (Equations (15)–(17)).

$$Z = \frac{|E|}{|H|} \tag{15}$$

$$Z = \sqrt{\frac{j\omega\mu}{\sigma - j\omega\varepsilon}} \tag{16}$$

$$Z_o = \sqrt{\frac{\mu_o}{\varepsilon_o}} \tag{17}$$

The EMI shielding of the composites are complicated and the physiochemical properties of constitutional composition of the composites which are significantly different from the homogeneous shielding materials. The most imperative parameter for the theoretical calculation of the EMI shielding is the effective relative permittivity ε_{eff} of the composite that can be calculated by using the Maxwell Garnett formula. The ε_{eff} is determined by the relative permittivity of the matrix (ε_e), relative permittivity of the fillers (ε_i), and f is the volume fraction of the filler. The ε_i is calculated by using the imaginary part of the complex relative permittivity (ε' and ε''), imaginary unit (j), σ, ω, and ε_o (Equations (18) and (19)).

$$\varepsilon_{eff} = \varepsilon_e + 3f\varepsilon_e\frac{\varepsilon_i - \varepsilon_e}{\varepsilon_i + 2\varepsilon_e - f(\varepsilon_i - \varepsilon_e)} \tag{18}$$

$$\varepsilon_i = \varepsilon' - j\varepsilon'' = \varepsilon' - j\frac{\sigma}{\omega\varepsilon_o} \tag{19}$$

On the other hand, the EMI shielding can be expressed as how far a composite has transmitted the EMR, which can be explained by using the transmission coefficient (T). The T is depending on the transmission coefficient at the $0-t$ boundary (T_1 and T_2), reflection coefficient at the $0-t$ boundary (R_1

and R_2), where 0 is considered as 1 and t as 2, and complex propagation constant (γ_m). The μ, ε, j, and ω affect the value of γ_m of the composite (Equations (20) and (21)).

$$T = \frac{T_1 T_2 e^{-\gamma_m D}}{1 + R_1 R_2 e^{-2\gamma_m D}} \tag{20}$$

$$\gamma_m = j\omega \sqrt{\varepsilon_0 \mu_0 (\varepsilon'_{eff} - j\varepsilon''_{eff})} \tag{21}$$

Z_o and Z_m determines the magnitude of T_1 and R. Moreover, Z_o, μ_r, and ε_{eff} have the impact on the value of Z_m (Equations (22)–(26) and Scheme 1).

$$T_1 = \frac{2Z_m}{Z_m + Z_o} \tag{22}$$

$$T_2 = \frac{2Z_o}{Z_m + Z_o} \tag{23}$$

$$R_1 = \frac{Z_m - Z_o}{Z_m + Z_o} \tag{24}$$

$$R_2 = \frac{Z_o - Z_m}{Z_m + Z_o} \tag{25}$$

$$Z_m = Z_o \sqrt{\frac{\mu_r}{\varepsilon_{eff}}} \tag{26}$$

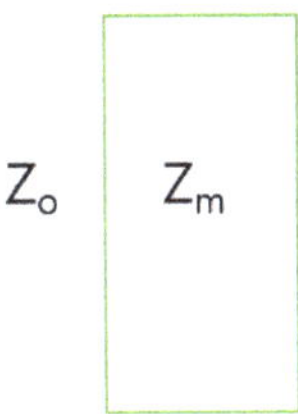

Scheme 1. Indication of the Z_o and Z_m in a composite.

Hence, the SE can be calculated in terms of T and the shielding efficiency of the materials can be calculated based on the SE_T of the composite (Equations (27) and (28)) [9,38].

$$SE = -20 \log(|T|) \tag{27}$$

$$\text{Shielding efficiency } (\%) = 100 - \left(\frac{1}{10^{SE/10}}\right) \times 100 \tag{28}$$

2.4. The MXene Composite Preparation Techniques

The composite of MXene and graphene are prepared by various processes such as vacuum assistant filtration (VAF) [38], dipped coating [39], spray coating [40], solvent casting techniques [41], freeze drying [42], and spin coating [43,44]. Practicing the processes varies based on the composition, purpose, and type of material used. The VAF technique is where a homogenized mixture is filtered through the filter paper and dried. The homogenization is carried out by sonication or a stirring process and various types of filters are being used such as paper, nylon, polypropylene, and Nuclepore track-etched polycarbonate (PC). Further, hybrid films are prepared by alternative filtering of the homogenized mixture (Figure 8a) [38]. In the dipped coating process, the matrix materials are immersed in the suitable solvent for a particular time period and dried, which is repeated several times based on its requirement (Figure 8b). Furthermore, the spray coating has a similar process (Figure 8c) [39]. In the spray coating process, spraying speed, time, solvents, pressure, and particle size of the materials in the solvent are controlled based on the quality of the products' needs (Figure 8c) [40]. The solvent casting is done by evaporating solvent of the homogenized mixture in an air or vacuum oven and the evaporation time depends on the type of solvent used. Then, the film is separated from the casting plate (Figure 8d) [41]. In the freeze drying technique, the homogenized water mixture is frozen under liquid nitrogen, and then the composite dried at the same temperature. During this process various structures such as honeycomb, porous, and other types of foam composites are formed, which is dependent on the type of constitutional elements present in the homogenized mixture. The resultant product is mixed with proper binder to prepare EMI shielding composite and this new technique is widely used to create highly efficient EMI shielding material (Figure 8e) [42]. Spin coating is a general technique used to prepare various thin electronic devices for which high rpm, different substrate, and evaporation techniques are used (Figure 8f) [43,44].

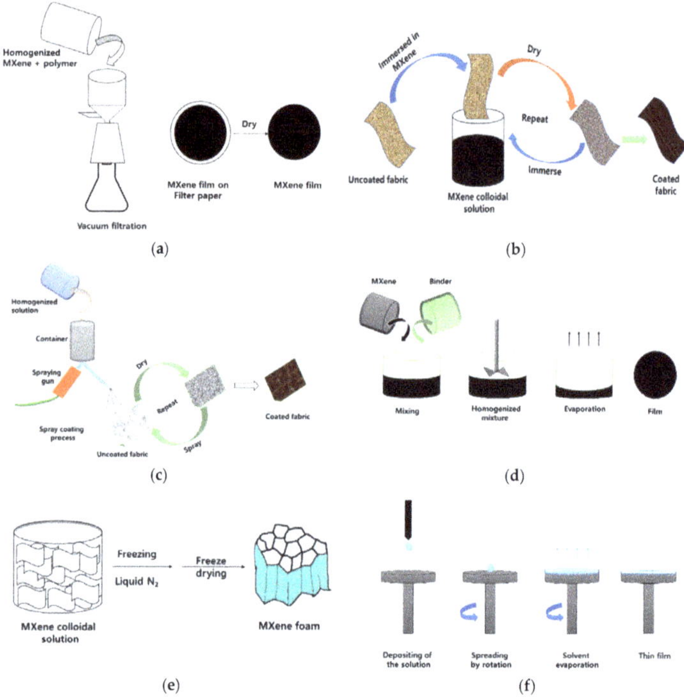

Figure 8. Schematic representation of: (**a**) vacuum assistant filtration (VAF), (**b**) dipped coating, (**c**) spray coating (**d**), solvent casting, (**e**) freeze drying, and (**f**) spin coating.

2.5. EMI Shielding of MXene Composite and Synthesis

The EMI shielding of the MXene (MX) composite varies based on the components that are integrated with MXene and the structural features of the composites. The MXene/aodium alginate (SA) shows 92 dB of EMI shielding effect (EMI SE) with 45 μm of thickness (t) and 4600 S·cm^{-1} of electric conductivity (EC) which is the highest EMI shielding reported for MXene composition in 2016. Further, the EMI shielding of the MX/SA composites diminish with reduced thickness of the composite prepared by the vacuum assistance filtration (VAF) technique, giving rise to the nacre–like structure and display good EMI SE. The spray coating of pure MXene (10 mg·mL^{-1}) on the PET surface displays EMI shielding of about 50 dB with 4 μm, which is almost similar to that of VAF MX/SA composition with 8-μm thickness. It is obvious that spraying on the polar surface increases the EMI SE while integration of the foreign element with MXene significantly reduces the EMI SE. Hence, the synthetic route influences the EMI shielding of the composite and the SA helps to arrange the MXene in a nacre array which significantly improves EMI SE of the shielding material (Figure 8 and Table 3) [45]. Hu et al. stated that the cellulose/MXene (M−filter) nanocomposite paper displays the 43 dB of EMI SE (seven cycle dipped coating) in both X and Ku band with 27.56 S·cm^{-1} of EC and thickness of 0.2 mm which is differ from the Coa et al. MX/cellulose composite performance. The MXene/cellulose composite exhibited EMI SE of 24 dB with SSE of 12 dB·cm^{3}·g^{-1}, 2647 dB.cm^{2}·g^{-1}, and 0.047 mm of thickness (Table 3) [39,46,47]. Hu et al. coated MXene with commercial filter paper (density of 0.49 g·cm^{-3}) and then, utilized the polydimethylsiloxane (PDMS) for finishing purposes, denoted as PDMS−M−filter composite while Coa et al. used MXene/cellulose nanofiber (CNFs) derived from garlic husk and both practiced dipped coating and VAF, respectively. The nacre-inspired structure of MXene/CNFs, which is absent in dipped coating cellulose composite and present in VAF composite, is an inspired structural feature in the MXene/cellulose based composites. Further, the filler loading determines the EMI, EC, and tensile strength of the composite [39].

In addition, CNTs/MXene/cellulose nanofibrils composite paper prepared by facile alternating vacuum assisted filtration process give rise the 38.4 dB of EMI SE and corresponding EC and thickness are 25.066 S·cm^{-1} and 0.038 mm, respectively. The EMI SE of CNTs/MXene/cellulose is higher than that of MX/cellulose composite reported above. Thus, the introduction of the CNT enhances the EMI SE. A similar study reported by Raagulan et al. used MXene−carbon nanotube nanocomposites (MXCS) where carbon fabric is used instead of cellulose displayed 99.999% shielding ability. The EC of CNTs/MXene/cellulose nanofibrils composite is 2.12-times higher than that of MXCS. The difference of EMI shielding in both cases is due to the thickness, EC, and curved MXene multilayer structure [47,48]. Xin et al. described that intercalation of silver nanoparticle with MXene/cellulose composition enhanced EMI shielding and exhibited EMI SE of 50.7 dB with 46 µm thickness and 5.882 S·cm^{-1} of EC. In this case, the introduction of silver nanoparticle in cellulose and MXene composition formed a similar structure reported by Cao et al. and the silver ion causes self-reduction of MXene, which improved the EC, multilayer formation, dielectric constant, and conduction loss [49]. The fiber matrix helps to diffuse EMR and fillers and fibers attenuate the EMR. The study of Liu et al. showed that the aluminum ion reinforced MXene film exhibits the excellent EMI shielding of 80 dB with 2656 S·cm^{-1} of EC and 0.005 mm of thickness where aluminum ion plays a major role in EMI SE and tensile strength which is due to the cross link formation between MXene and aluminum ion. Thus, aluminum ion has a greater tendency to induce the EMI SE than sliver ion in the matrix of the composite [49,50]. Furthermore, conductivity of the aluminum reinforced MXene is lower than that of MX/SA composite which is due to the interlayer space caused by the aluminum ion and higher than that of MXene/poly(3,4−ethylenedioxythiophene) polystyrene sulfonate (MX/PEDOT: PSS), MXCS, and d−Ti$_3$C$_2$T$_x$/CNFs reported (Table 3) [45,48–50].

The poly (vinyl alcohol)/MXene (PVA/MXene) multilayered composite reported by Jin et al. exhibits 44.4 dB (wt.%—19.5%) of EMI SE (SE$_R$—8.3 dB) with 0.027 mm of thickness and decrease with decreasing MXene loading in the PVA matrix. MXene in a PVA matrix has a similar EC trend and diminishes with less loading of MXene filler (Table 3) [51–53]. Further, the Xu et al. study shows that the PVA/MXene foam considerably minimized the EMI SE and exhibits 28 dB of EMI SE with a lower SE$_R$ (2 dB), which is due to the differences in EC and its porous nature. The SE$_R$ above 3 dB is induced by charge flow in the matrix. It is obvious that the EMI SE of the PVA/MXene composites are contingent not only structural features, but also filler loading of the composite [53,54]. The honeycomb (HC) structure can be manufactured by using reduced graphene oxide (rGO)−MXene/epoxy composition for which the Al$_2$O$_3$ HC template is used. Initially, the rGO is adsorbed on the surface of the template and then it is dissolved by using hydrochloric acid. Consequently, the HC graphene oxide is immersed into the MXene/CTAB solution and freeze dried. The yielded HC−rGO/MXene is strengthened by epoxy polymer. The HC−rGO−MXene/epoxy composition displays EMI SE of 55 dB with 3.871 S·cm^{-1} of EC and 0.5 mm of thickness [55,56]. A similar study is performed by Bian et al. who prepared MXene aerogel without a template and its corresponding EMI SE of 75 dB and SE$_R$ is about 1 dB [57].

The EC of HC–GO–MXene/epoxy composite is lower than that of MXene aerogel which is due to the interconnection between MXene and epoxy polymer comparatively lessened the EMI SE and EC. The epoxy polymer not only minimized the electron flow among MXene and graphene but also improved the tensile strength of the composite and flexibility [56,57]. Zhou et al. described that the MXene/calcium alginate aerogel ((MX/CA (t = 26 µm)) exhibits 54.43 dB of EMI SE which is higher than that of the MX/SA composite synthesized in the same condition (t = 14 µm) (Table 3) [58]. Although, the 8 µm MX/SA reported by Shahzad et al. exhibited 57 dB of EMI SE which is assumed as the quality of the exfoliated MXene synthesized and fabrication condition used (Table 3) [46,58]. Further, the thickness or inter space of the composite is increased by types of interacting ion or organic substances used which also affect the EMI SE parameters. The thickness of the MX/CA is higher than the MX/SA which is caused by calcium alginate and aerogel structure of MX/CA diminished the EC and increased the corresponding SSE/t of the composite (Table 3) [58].

The interposing of the rGO into MXene formed, MX/rGO aerogel shows 56.4 dB of EMI SE which is almost similar to the HC–MX–rGO/epoxy composite, thus, the HC structure is an effective feature with the lowest thickness and filling load [56,59]. Scientists recently focusses on the 3D aerogel which seems to be a more effective EMI absorbent than the planar composite. The 3D $Ti_3C_2T_x$/SA (95%) hybrid aerogel coated by electrically conductive polydimethylsiloxane–coated (PDMS) displays EMI SE of 70.5 dB and corresponding EC is 22.11 S·cm^{-1}. In addition, the higher amount of SA reduces the EMI SE and EC. Further, SE_A and SE_T are almost similar where SE_R seems null is an evidence that the aerogel structure greatly improved absorption and the direct interconnection of MXene is a crucial parameter for EMI SE and SE_A [42]. Further, introduction of carbon into the MXene framework (MXene/carbon foam (MCF)) showed less EMI SE of 8 dB and addition of epoxy polymer followed by annealing of MCF reached up to EMI SE of 46 dB with 1.84 S·cm^{-1} of EC (2 cm of thickness). In this study, the resorcinol–formaldehyde sol–gel mixture is used as a precursor for carbon form, created by an annealing process, nonetheless, Song et al.'s epoxy composite consisted of rGO as a carbon form gave rise to higher EMI SE (55 dB) [56,60]. Hence, the filler with good electric conductivity and proper geometry improved EMI SE [60]. Furthermore, Wang, et al. reported that the annealed MXene/Epoxy Nano composites (wt.% 15) with 41 dB of EMI SE, 2 cm of thickness and 1.05 S·cm^{-1}. From these studies, the annealing of composite internally created a carbon form from polymer that significantly changed EC and EMI SE. The direct annealing of pure MXene reduce the EMI shielding and due to the curing ability epoxy in MXene/Epoxy composite matrix enhances EMI SE [42,60].

Mixing of silver nanowire with MXene with cellulose pressured-extrusion method blocked 99.99% of incoming EMR, which is lower than that of pure MXene, and silver nanowire improved the electron flow path between 1D and 2D filler (Table 3) [45,61]. The corresponding composite assembled like brick-and-mortar like arrangement with internal pores which provide reflection and scattering interfaces that improve EMI SE [61]. Further, the heat treated monolayer MXene with 4.14×10^{-5} mm of thickness shows EMI SE of 17.13 dB and corresponding SSE and SSE/t are 7.17 dB·cm^3·g^{-1} and 1.73×106 dB·cm^2·g^{-1}, respectively. The corresponding composite (without heat process) with the same thickness displays EMI SE, SSE, and SSE/t of 13.56 dB, 5.67 dB·cm^3·g^{-1}, and 1.37×10^6 dB·cm^3·g^{-1}, respectively. It is apparent that the annealing at 600 °C improve EMI SE and other parameters significantly [43]. Wan et al. produced MXene/PEDOT: PSS composite with 6 µm of thickness gave rise to 40.5 dB of EMI SE while the same composition reported by Liu et al. showed 42.1 dB of EMI SE with 11 µm of thickness from which the preparation method plays a major role in determining the EMI SE. The removal of PSS from the matrix increase EMI shielding with 6 µm of thickness which is the main reason for Wan et al.'s results [38,62,63]. The SE_A mostly depends on the dielectric properties of the EMI shielding composite. Han et al. alter the surface of the MXene by annealing at 800 °C under the inert environment and prepare the MXene/wax composite (1 mm) display show EMI SE of 76.1 dB with 67.3 dB of SE_A and corresponding composite exhibit −48.4 dB of minimum reflection coefficient which is due to the formation of the titanium oxide on the surface of the MXene [64,65].

Table 3. Electromagnetic interference (EMI) shielding comparison of the two–dimension (2D) MXene $(MX-Ti_3C_2T_x)$ and graphene (GN) composite.

No	Composite	Filler (wt.%)	t (mm)	SE (dB)	SEE (dB cm^3·g^{-1})	SSE/t (dB·cm^2·g^{-1})	Density (g·cm^{-3})	σ (S·cm^{-1})	Ref.
1	MX/SA	90	0.008	57	24.6	30,830	2.31	4600	[45]
2	MX/SA	bulk	0.011	68	28.4	25,863	2.39	–	[45]
3	d–Ti$_3$C$_2$T$_x$/CNFs	90	0.047	24	12	2647	1.91	73.94	[46]
4	MXCS	bulk	0.386	50.5	324.15	8397.78	0.153	11.8	[48]
5	MX/PEDOT:PSS	87.5	0.011	42.1	2144.76	19,497.8	0.0196	340.5	[38]
6	MX/PEDOT:PSS	bulk	0.007	42.5	1577.08	22,529.7	0.0269	1000	[38]
7	MX/aramid nanofiber	90	0.015	32.84	20.05	13,366.67	1.638	628.272	[51]
8	MX/aramid nanofiber	80	0.02	30	20.65	10,325	1.453	173.36	[51]
9	MX/aramid nanofiber	40	0.022	19.43	16.36	7436.36	1.188	24.826	[51]
10	MX/GN	bulk	3	50.7	11021	36,736.67	0.0046	1000	[52]
11	PVA/MX	19.5	0.027	44.4	25.23	9343	1.744	7.16	[53]
12	PVA/MX	13.9	0.025	37.1	22.08	8833	1.68	3.79	[53]
13	Ti2CTx/PVA	0.15 (Vol.%)	5	28	2586	5136	0.0109	–	[54]
14	CNF@MX	bulk	0.035	39.6	24.6	7029	1.16	1.43	[55]
14	MX aerogel	bulk	2	75	9904		0.01	22	[57]
15	MX/CA aerogel	90	0.026	54.43	40.32	17,586	1.35	338.32	[58]
16	MX film	bulk	0.013	46.2	16.62	13,195	2.78	1354.29	[58]
17	MX/SA film	90	0.014	43.9	17.56	14,830	2.50	795.51	[58]
18	MX/AgNW film/Nanocellulose	86	0.017	42.74	28.49	16724	1.5	300	[61]
19.	MX foam	bulk	0.06	70	318	53,030	0.22	580	[65]
20	MXPATPA	bulk	0.62	45.18	33.26	236.45	1.217	1.241	[31]
21	TG–CN/PMMA foam	10	2.5	30.4	43.4	173.6	0.701	0.0292	[66]
22	RG–CN/PMMA foam	10	2.5	18.1	26.2	104.8	0.691	0.0015	[66]
23	GN–CN/PMMA foam	10	2.5	25.2	47.5	190	0.531	0.013	[66]
24	Fe$_3$O$_4$/GN/PDMS	bulk	1	32.4	249.23	2492.31	0.13	2.5	[67]
25	Gr–PANI10:1@PI	40	0.04	21.3	16.38	4096.2	1.299	490.3	[68]

2.6. EMI Shielding of Graphene Composite

Graphene is a 2D material and similar to MXene, whereas EMI shielding of GN depend on number of layers in GN and increasing layers enhance EMI SE of GN (EMI SE of single layer GN is 2.27 dB) [18]. The nacre-mimetic structure is an inspired structure for 2D materials such as MXene and GN which exert excellent EMI shielding (Figure 9). The graphene/PDMS polymer aerogel composite shows EMI SE of about 65 dB with the low GN loading (0.42 wt.%) and corresponding SSE and density are 100 dB·cm^3·g^{-1}, 0.0042 g·cm^{-3}, respectively [8]. The physical or chemical process of the constitutional element of the composite determine the EMI shielding and other parameter of the composite. Further, the (thermally reduced graphene oxide–carbon nanotubes (TG–CN)–poly (methyl methacrylate) (PMMA) composite displays EMI SE of 30.4 dB while chemically reduce–GO–carbon nanotube/PMMA composite exhibits 1.68-times lesser EMI SE. For the reduction purpose, hydrazine and 900 °C were used and this process has the impact on structure of the foam, polarization loss, and multiple reflection of the composite. Additionally, chemical reduction process greatly reduces the EC of graphene–carbon nanotube composition whereas thermal process considerably improves free electron path in the composite [46,66–68]. Yang et al. handled another strategy to create the composite aerogel that is epoxy copper nanowires/thermally annealed graphene aerogel (6.0–1.2 wt.%) and corresponding EMI SE is 47 dB with EC of 1.208 S·cm^{-1}, and 2 mm of thickness. Introduction of the copper significantly improved EMI SE and the EMI SE of the pure epoxy resin by 2 dB [69]. According to the Liang et al. report 32.5 dB of EMI SE was achieved for graphene/SiC–nanowires/poly(vinylidene fluoride) composites with a thickness of 1.2 mm and 0.015 S·cm^{-1} of EC. Thus, copper nanowire is the best choice compared to SiC–nanowire, though, SiC nanowire possesses good dielectric properties and enhances the absorption of the composite. Henceforth, the penetrating EMR is converted into heat energy in the 1D–2D network which is due to the ohmic loss greatly improved SE$_A$ [1]. The addition of the inorganic component into the composite influenced the EMI SE, EC, magnetic property, and reflection loss. The cobalt-rich glass-coated microwires with the formula of $Co_{60}Fe_{15}Si_{10}B_{15}$ graphene/silicone rubber composite prepared blocked 98.4% of incident radiation and the corresponding filler loading is 0.059 wt.%. Furthermore, the composite consists of magnetic property due to the presence of the cobalt–ion containing microwire. The microwire (M) and graphene (G) arrange with the most prominent shielding structures, such as MMMGGG and MGMGMG. The microwire plays a dominant role in EMI SE, which is inferred that higher loading of M with proper geometrical array (MMMGGG) shows greater EMI SE than that of GN. This is due to the magnetic behavior of the M, polarization relaxation, and impedance matching. In addition, the dispersion array of the M and GN lower the EMI SE. According to the Xu et al., the multiple array of one type fillers followed another type of filler significantly enhanced the EMI SE [70].

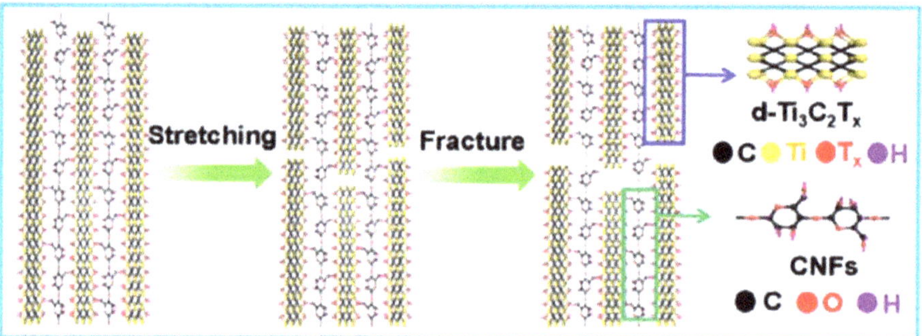

Figure 9. Interaction of the cellulose fiber with MXene and nacre–like structure formation [46] Copyright American Chemical Society, 2018.

The nickel foam can be created by a solution combustion (SC) technique and the resultant foam made up of Ni and NiO is new type of metal foam. By immersing Ni foam into the rGO solution (10 mgÅmL^{-1}) for 2–3 h, the Ni-based EMI shielding composition can be fabricated. During the immersing period, the rGO nanoplates penetrate into the skeleton of Ni–foam which gives rise to a remarkable structure that enhances the EMI SE. The maximum reflection loss (RL) of Ni–GN foam is −53.11 dB and the corresponding thickness, porosity, dielectric loss range, complex permeability range, and density are 4.5 mm, 96.74%, 0.44–0.55, 1.05–0.97, and 38.54 mg·mL^{-1}, respectively. The rGO majorly contributes dielectric loss to the Ni/NiO framework and Ni/rGO foam showed relatively higher μr than individual constitutional elements. Hence, nickel/rGO foam is a better choice than Cu nanowire/graphene aerogel [69,71]. Zhu et al. reported another type of composition that is Fe_3O_4/graphene coated Ni foam/poly dimethylsiloxane composite reached EMI SE of 32.4 dB with 2.5 S·cm^{-1} of EC and 1 mm of thickness. Introduction of the Fe_3O_4 and coating of the graphene on the Ni foam considerably affected EMI SE. The composite preparation method and type of binder used affects the shielding efficiency of the composite (Table 3) [67,69,71]. Li et al. made polyurethane based GN composite that is polyurethane/graphene (PUG20) composite with twenty percentage of weight ratio (20 wt.%) blocks 98.7% of incident radiation while epoxy encapsulated pyrolyzed PUG20 doubled the shielding ability (shielding power—99.99992%). The pyrolysis process not only improved GN content but also enhanced the porous nature, electrical conductivity, impedance mismatching, and SE$_A$, and a higher filling load was attenuated at 99.99999% of the incident wave (Table 3) [4,43,45]. Graphene with the loading rate of 8.9 vol.% is dispersed in thermoplastic polyether block amide (Trade name—PEBAX) by heat process (175 °C) exhibits 30.7 dB of EMI SE and half volume loading of the graphene give rise to 16.6 dB of EMI SE.

In addition, the SE$_A$ of GN/PEBAX is dominant factor like another composite reported above [72]. Thus, pyrolysis of the GN based polymer composite greatly improve EMI shielding and other parameters such as EC, dielectric property, porous structure, permeability, etc. The CNTs co−decorated porous carbon/graphene/PDMS foam generate 48 dB of shielding ability and corresponding SSE is 347.8 dB·g^{-1}·cm^3. The composite consists multilayer of components those design is better than that of the Zhu et al. study. The Ni foam is coated by CNT (GF) by the CVD method and Fe−Zn co−precipitated on hydrothermal method (FCC−GF) which is dispersed in the PDMS matrix [G11]. The Zhu et al. study just used an electrostatic assemble of nanomaterials on the porous structure that has less effect on EMI SE and CVD, or a hydro thermal process improved porous geometry and minimized the defect of fillers [67,73].

The many studies use PDMS as a binder to make GN-based EMI shielding materials. The 3D graphene/carbon nanotubes/polydimethylsiloxane composites (GC) was prepared by using freeze drying followed by annealing at a high temperature (2800 °C) and this is similar to the Song et al. study where they used the HC template to create the aerogel [58,74]. GC with 95.8% of porosity, 0.013 g·cm^{-3} of density, 2.4 mm of thickness, and 92.78 (vol.%) loading rate annealed at 1400 °C gave rise to 54.43 dB of highest EMI SE and the corresponding EC is 0.2012 S·cm^{-1}. The higher annealing temperature (2800 °C) destroys EMI shielding of the composite, whereas the higher temperature repairs the defects by removing functional groups, reduced polarization loss, lessens the amorphous nature (higher amorphous nature at 1400 °C) and enhances the ohmic loss. Further, the introduction of the CNT prevents the agglomeration of GN and enhance the electric channels in the 3D network [58,74]. Cheng et al. prepared graphene−polyaniline (PANI)@polyimide composite with a ratio of 10:1 (GN: PANI) shows 21.3 dB (t − 0.04 mm) with 490.3 S·cm^{-1} of EC while GN/PANI, Ni decorated GN/PANI and Ag decorated GN/PANI (5 wt.%) show EMI SE of 24.85, 29.33, and 24.93 dB, respectively. Therefore, metal nanoparticle decoration with GN in PANI matrix improved EMI SE compared to the GN/PANI in polyimide matrix [68,75,76].

The electrical conductivity of the pure materials is high compared to the composite made from corresponding pure materials. Further, MXene based composition showed excellent EC compared to the GN. This is because of the functional group present in the MXene gave rise to a better arrangement in the polymer matrix which promoted a good electron flow path. Due to this, MXene is one of the most attractive materials in the modern electronic world. The SSE and SSE/t is depend on the density and thickness of the composite. Further, proper geometrical arrangement of the fillers is a crucial factor for better EMI shielding.

3. Conclusions

In this review, we have highlighted recent progress in EMI shielding of MXene and graphene composite with different fabrication technique. The global requirement is a lightweight, flexible, cost−effective, and thinner EMI shielding material and EMI shielding over 20 dB, which is the basic requirement for the EMI shielding application in electronic devices. The graphene is synthesized by various methods in which radiative and chemical vapor deposition gives rise to good quality graphene and utilization of other methods is dependent on the types of applications. MXene synthesis also adapts various techniques and the lithium fluoride/hydrochloric acid etching is the most desirable etching method. The 2D MXene and graphene play a major role in creating different structures, especially aerogel. The freezing and freeze-drying process is used to prepare different geometric composite with excellent EMI shielding is a new approach. Further, the pyrolysis and hydrothermal process are important techniques used to achieve higher EMI shielding of the aerogel hybrid structure. The pyrolysis process of MXene created with surface titanium oxide, induces the EMI shielding while annealing of the MXene/polymer composite forms an internal conductive carbon network is another approach for excellent EMI shielding. The high temperature pyrolysis process of MXene or graphene polymer composite destroyed the EMI shielding function. The EMI shielding of graphene can be improved further by doping, decoration, oxidation, reduction, encapsulation, type of polymer, and processing technique. Thus, the proper geometrical arrangement of the fillers is a crucial factor for better EMI shielding and can be tuned by changing thickness, porosity, dielectric loss range, and complex permeability. The addition of inorganic nano composition like nanowire, ions, and nanoparticle further enhances EMI shielding and are being used to make the most powerful EMI shielding composites. The polymer/filler composites are explored to fulfil recent demand in EMI shielding in electronic devices. We hope, we have provided the fundamental understanding of MXene, graphene, and its corresponding composites to achieve a modern electronic goal in the near future.

Author Contributions: K.Y.C. designed the project; B.M.K. and K.R. wrote the manuscript. All authors have read and agreed to the published version of the manuscript.

Funding: This research received no external funding.

Acknowledgments: This study was supported by Wonkwang University 2019.

Conflicts of Interest: The authors declare no conflict of interest.

References

1. Liang, C.; Hamidinejad, M.; Ma, L.; Wang, Z.; Park, C.B. Lightweight and flexible graphene/SiC-nanowires/poly (vinylidene fluoride) composites for electromagnetic interference shielding and thermal management. *Carbon* **2020**, *156*, 58–66. [CrossRef]
2. Xing, D.; Lu, L.; Xie, Y.; Tang, Y.; Teh, K.S. Highly flexible and ultra-thin carbon-fabric/Ag/waterborne polyurethane film for ultra-efficient EMI shielding. *Mater. Des.* **2020**, *185*, 108227. [CrossRef]
3. Sheng, A.; Ren, W.; Yang, Y.; Yan, D.X.; Duan, H.; Zhao, G.; Liu, Y.; Li, Z.M. Multilayer WPU conductive composites with controllable electro-magnetic gradient for absorption-dominated electromagnetic interference shielding. *Compos. Part A: Appl. Sci. Manuf.* **2020**, *129*, 105692. [CrossRef]
4. Li, Y.; Liu, J.; Wang, S.; Zhang, L.; Shen, B. Self-templating graphene network composites by flame carbonization for excellent electromagnetic interference shielding. *Compos. Part B: Eng.* **2020**, *182*, 107615. [CrossRef]
5. Zhao, B.; Zeng, S.; Li, X.; Guo, X.; Bai, Z.; Fan, B.; Zhang, R. Flexible PVDF/carbon materials/Ni composite films maintaining strong electromagnetic wave shielding under cyclic microwave irradiation. *J. Mater. Chem. C* **2020**, *8*, 500–509.
6. Shukla, V. Role of spin disorder in magnetic and EMI shielding properties of Fe_3O_4/C/PPy core/shell composites. *J. Mater. Sci.* **2020**, *55*, 2826–2835. [CrossRef]
7. Li, T.T.; Wang, Y.; Peng, H.K.; Zhang, X.; Shiu, B.C.; Lin, J.H.; Lou, C.W. Lightweight, flexible and superhydrophobic composite nanofiber films inspired by nacre for highly electromagnetic interference shielding. *Compos. Part A: Appl. Sci. Manuf.* **2020**, *128*, 105685. [CrossRef]
8. Gao, W.; Zhao, N.; Yu, T.; Xi, J.; Mao, A.; Yuan, M.; Bai, H.; Gao, C. High-efficiency electromagnetic interference shielding realized in nacre-mimetic graphene/polymer composite with extremely low graphene loading. *Carbon* **2020**, *157*, 570–577. [CrossRef]
9. Langguth, W. Earthing & EMC Fundamentals of Electromagnetic Compatibility (EMC). *Power Qual. Appl. Guide Copp. Dev. Assoc.* **2006**, *6*, 1–16.
10. Wang, Y.; Wang, H.; Ye, J.; Shi, L.; Feng, X. Magnetic CoFe alloy@ C nanocomposites derived from ZnCo-MOF for electromagnetic wave absorption. *Chem. Eng. J.* **2020**, *383*, 123096. [CrossRef]
11. Singh, A.K.; Yadav, A.N.; Srivastava, A.; Haldar, K.K.; Tomar, M.; Alaferdov, A.V.; Moshkalev, S.A.; Gupta, V.; Singh, K. CdSe/V2O5 core/shell quantum dots decorated reduced graphene oxide nanocomposite for high-performance electromagnetic interference shielding application. *Nanotechnology* **2019**, *30*, 505704. [CrossRef] [PubMed]
12. Hawkins, S.A.; Yao, H.; Wang, H.; Sue, H.J. Tensile properties and electrical conductivity of epoxy composite thin films containing zinc oxide quantum dots and multi-walled carbon nanotubes. *Carbon* **2017**, *115*, 18–27. [CrossRef]
13. El-Shamy, A.G. Novel conducting PVA/Carbon quantum dots (CQDs) nanocomposite for high anti-electromagnetic wave performance. *J. Alloys Compd.* **2019**, *810*, 151940. [CrossRef]
14. Ge, C.; Zou, J.; Yan, M.; Bi, H. C-dots induced microwave absorption enhancement of PANI/ferrocene/C-dots. *Mater. Lett.* **2014**, *137*, 41–44. [CrossRef]
15. Lakshmi, N.V.; Tambe, P. EMI shielding effectiveness of graphene decorated with graphene quantum dots and silver nanoparticles reinforced PVDF nanocomposites. *Compos. Interfaces* **2017**, *24*, 861–882. [CrossRef]

16. Ji, B.; Giovanelli, E.; Habert, B.; Spinicelli, P.; Nasilowski, M.; Xu, X.; Lequeux, N.; Hugonin, J.P.; Marquier, F.; Greffet, J.J.; et al. Non-blinking quantum dot with a plasmonic nanoshell resonator. *Nat. Nanotechnol.* **2015**, *10*, 170. [CrossRef] [PubMed]

17. Wang, Y.; Wang, W.; Qi, Q.; Xu, N.; Yu, D. Layer-by-layer assembly of PDMS-coated nickel ferrite/multiwalled carbon nanotubes/cotton fabrics for robust and durable electromagnetic interference shielding. *Cellulose* **2020**, *27*, 2829–2845. [CrossRef]

18. Kumar, P. Ultrathin 2D Nanomaterials for Electromagnetic Interference Shielding. *Adv. Mater. Interfaces* **2019**, *6*, 1901454. [CrossRef]

19. Sangwan, V.K.; Hersam, M.C. Electronic transport in two-dimensional materials. *Annu. Rev. Phys. Chem.* **2018**, *69*, 299–325. [CrossRef]

20. Le, M.Q.; Nguyen, D.T. The role of defects in the tensile properties of silicene. *Appl. Phys. A* **2015**, *118*, 1437–1445. [CrossRef]

21. Fan, X.; Ma, Y.; Dang, X.; Cai, Y. Synthesis and Electromagnetic Interference Shielding Performance of Ti3SiC2-Based Ceramics Fabricated by Liquid Silicon Infiltration. *Materials* **2020**, *13*, 328. [CrossRef] [PubMed]

22. Guan, H.; Chung, D.D.L. Radio-wave electrical conductivity and absorption-dominant interaction with radio wave of exfoliated-graphite-based flexible graphite, with relevance to electromagnetic shielding and antennas. *Carbon* **2020**, *157*, 549–562. [CrossRef]

23. Shams, S.S.; Zhang, R.; Zhu, J. Graphene synthesis: A Review. *Mater. Sci.-Pol.* **2015**, *33*, 566–578. [CrossRef]

24. Lee, H.C.; Liu, W.W.; Chai, S.P.; Mohamed, A.R.; Aziz, A.; Khe, C.S.; Hidayah, N.M.; Hashim, U. Review of the synthesis, transfer, characterization and growth mechanisms of single and multilayer graphene. *RSC Adv.* **2017**, *7*, 15644–15693. [CrossRef]

25. Persichetti, L.; De Seta, M.; Scaparro, A.M.; Miseikis, V.; Notargiacomo, A.; Ruocco, A.; Sgarlata, A.; Fanfoni, M.; Fabbri, F.; Coletti, C.; et al. Driving with temperature the synthesis of graphene on Ge (110). *Appl. Surf. Sci.* **2020**, *499*, 143923. [CrossRef]

26. Kumar, K.T.; Reddy, M.J.K.; Sundari, G.S.; Raghu, S.; Kalaivani, R.A.; Ryu, S.H.; Shanmugharaj, A.M. Synthesis of graphene-siloxene nanosheet based layered composite materials by tuning its interface chemistry: An efficient anode with overwhelming electrochemical performances for lithium-ion batteries. *J. Power Sources* **2020**, *450*, 227618. [CrossRef]

27. Yoon, K.Y.; Dong, G. Liquid-phase bottom-up synthesis of graphene nanoribbons. *Mater. Chem. Front.* **2020**, *4*, 29–45. [CrossRef]

28. Wong, K.L.; Chuan, M.W.; Hamzah, A.; Rusli, S.; Alias, N.E.; Sultan, S.M.; Lim, C.S.; Tan, M.L.P. Electronic properties of graphene nanoribbons with line-edge roughness doped with nitrogen and boron. *Phys. E: Low-Syst. Nanostruct.* **2020**, *117*, 113841. [CrossRef]

29. Yang, G.; Li, L.; Lee, W.B.; Ng, M.C. Structure of graphene and its disorders: A review. *Sci. Technol. Adv. Mater.* **2018**, *19*, 613–648. [CrossRef]

30. Wallace, P.R. The band theory of graphite. *Phys. Rev.* **1947**, *71*, 622. [CrossRef]

31. Raagulan, K.; Braveenth, R.; Kim, B.M.; Lim, K.J.; Lee, S.B.; Kim, M.; Chai, K.Y. An effective utilization of MXene and its effect on electromagnetic interference shielding: Flexible, free-standing and thermally conductive composite from MXene-AT-poly (p-aminophenol)–polyaniline co-polymer. *RSC Adv.* **2020**, *10*, 1613–1633. [CrossRef]

32. Yury, G.; Anasori, B. The Rise of MXenes. *ACS Nano* **2019**, *13*, 8491–8494.

33. Alhabeb, M.; Maleski, K.; Anasori, B.; Lelyukh, P.; Clark, L.; Sin, S.; Gogotsi, Y. Guidelines for synthesis and processing of two-dimensional titanium carbide (Ti3C2T × MXene). *Chem. Mater.* **2017**, *29*, 7633–7644. [CrossRef]

34. Shukla, V. Review of electromagnetic interference shielding materials fabricated by iron ingredients. *Nanoscale Adv.* **2019**, *1*, 1640–1671. [CrossRef]

35. Nasouri, K.; Shoushtari, A.M.; Mojtahedi, M.R.M. Theoretical and experimental studies on EMI shielding mechanisms of multi-walled carbon nanotubes reinforced high performance composite nanofibers. *J. Polym. Res.* **2016**, *23*, 71. [CrossRef]

36. Bi, S.; Zhang, L.; Mu, C.; Liu, M.; Hu, X. Electromagnetic interference shielding properties and mechanisms of chemically reduced graphene aerogels. *Appl. Surf. Sci.* **2017**, *412*, 529–536. [CrossRef]

37. Drakakis, E.; Kymakis, E.; Tzagkarakis, G.; Louloudakis, D.; Katharakis, M.; Kenanakis, G.; Suchea, M.; Tudose, V.; Koudoumas, E. A study of the electromagnetic shielding mechanisms in the GHz frequency range of graphene based composite layers. *Appl. Surf. Sci.* **2017**, *398*, 15–18. [CrossRef]

38. Liu, R.; Miao, M.; Li, Y.; Zhang, J.; Cao, S.; Feng, X. Ultrathin Biomimetic Polymeric Ti3C2T × MXene Composite Films for Electromagnetic Interference Shielding. *ACS Appl. Mater. Interfaces* **2018**, *10*, 44787–44795. [CrossRef]

39. Hu, D.; Huang, X.; Li, S.; Jiang, P. Flexible and durable cellulose/MXene nanocomposite paper for efficient electromagnetic interference shielding. *Compos. Sci. Technol.* **2020**, *188*, 107995. [CrossRef]

40. Hantanasirisakul, K.; Zhao, M.Q.; Urbankowski, P.; Halim, J.; Anasori, B.; Kota, S.; Ren, C.E.; Barsoum, M.W.; Gogotsi, Y. Fabrication of Ti_3C_2Tx MXene transparent thin films with tunable optoelectronic properties. *Adv. Electron. Mater.* **2016**, *2*, 1600050. [CrossRef]

41. Naguib, M.; Saito, T.; Lai, S.; Rager, M.S.; Aytug, T.; Paranthaman, M.P.; Zhao, M.Q.; Gogotsi, Y. Ti_3C_2Tx (MXene)–polyacrylamide nanocomposite films. *RSC Adv.* **2016**, *6*, 72069–72073. [CrossRef]

42. Wu, X.; Han, B.; Zhang, H.B.; Xie, X.; Tu, T.; Zhang, Y.; Dai, Y.; Yang, R.; Yu, Z.Z. Compressible, durable and conductive polydimethylsiloxane-coated MXene foams for high-performance electromagnetic interference shielding. *Chem. Eng. J.* **2020**, *381*, 122622. [CrossRef]

43. Yun, T.; Kim, H.; Iqbal, A.; Cho, Y.S.; Lee, G.S.; Kim, M.K.; Kim, S.J.; Kim, D.; Gogotsi, Y.; Kim, S.O.; et al. Electromagnetic Shielding of Monolayer MXene Assemblies. *Adv. Mater.* **2020**, *32*, 1906769. [CrossRef] [PubMed]

44. Amokrane, G.; Falentin-Daudré, C.; Ramtani, S.; Migonney, V. A simple method to functionalize PCL surface by grafting bioactive polymers using UV irradiation. *Irbm* **2018**, *39*, 268–278. [CrossRef]

45. Shahzad, F.; Alhabeb, M.; Hatter, C.B.; Anasori, B.; Hong, S.M.; Koo, C.M.; Gogotsi, Y. Electromagnetic interference shielding with 2D transition metal carbides (MXenes). *Science* **2016**, *353*, 1137–1140. [CrossRef]

46. Cao, W.T.; Chen, F.F.; Zhu, Y.J.; Zhang, Y.G.; Jiang, Y.Y.; Ma, M.G.; Chen, F. Binary strengthening and toughening of MXene/cellulose nanofiber composite paper with nacre-inspired structure and superior electromagnetic interference shielding properties. *ACS Nano* **2018**, *12*, 4583–4593. [CrossRef]

47. Cao, W.; Ma, C.; Tan, S.; Ma, M.; Wan, P.; Chen, F. Ultrathin and flexible CNTs/MXene/cellulose nanofibrils composite paper for electromagnetic interference shielding. *Nano-Micro Lett.* **2019**, *11*, 72. [CrossRef]

48. Raagulan, K.; Braveenth, R.; Lee, L.R.; Lee, J.; Kim, B.M.; Moon, J.J.; Lee, S.B.; Chai, K.Y. Fabrication of Flexible, Lightweight, Magnetic Mushroom Gills and Coral-Like MXene–Carbon Nanotube Nanocomposites for EMI Shielding Application. *Nanomaterials* **2019**, *9*, 519. [CrossRef]

49. Xin, W.; Xi, G.Q.; Cao, W.T.; Ma, C.; Liu, T.; Ma, M.G.; Bian, J. Lightweight and flexible MXene/CNF/silver composite membranes with a brick-like structure and high-performance electromagnetic-interference shielding. *RSC Adv.* **2019**, *9*, 29636–29644. [CrossRef]

50. Liu, Z.; Zhang, Y.; Zhang, H.B.; Dai, Y.; Liu, J.; Li, X.; Yu, Z.Z. Electrically conductive aluminum ion-reinforced MXene films for efficient electromagnetic interference shielding. *J. Mater. Chem. C* **2020**. [CrossRef]

51. Xie, F.; Jia, F.; Zhuo, L.; Lu, Z.; Si, L.; Huang, J.; Zhang, M.; Ma, Q. Ultrathin MXene/aramid nanofiber composite paper with excellent mechanical properties for efficient electromagnetic interference shielding. *Nanoscale* **2019**, *11*, 23382–23391. [CrossRef] [PubMed]

52. Fan, Z.; Wang, D.; Yuan, Y.; Wang, Y.; Cheng, Z.; Liu, Y.; Xie, Z. A lightweight and conductive MXene/graphene hybrid foam for superior electromagnetic interference shielding. *Chem. Eng. J.* **2020**, *381*, 122696. [CrossRef]

53. Jin, X.; Wang, J.; Dai, L.; Liu, X.; Li, L.; Yang, Y.; Cao, Y.; Wang, W.; Wu, H.; Guo, S. Flame-retardant poly (vinyl alcohol)/MXene multilayered films with outstanding electromagnetic interference shielding and thermal conductive performances. *Chem. Eng. J.* **2020**, *380*, 122475. [CrossRef]

54. Xu, H.; Yin, X.; Li, X.; Li, M.; Liang, S.; Zhang, L.; Cheng, L. Lightweight Ti_2CT × MXene/Poly (vinyl alcohol) Composite Foams for Electromagnetic Wave Shielding with Absorption-Dominated Feature. *ACS Appl. Mater. Interfaces* **2019**, *11*, 10198–10207. [CrossRef] [PubMed]

55. Zhou, B.; Zhang, Z.; Li, Y.; Han, G.; Feng, Y.; Wang, B.; Zhang, D.; Ma, J.; Liu, C. Flexible, Robust and Multifunctional Electromagnetic Interference Shielding Film with Alternating Cellulose Nanofiber and MXene Layers. *ACS Appl. Mater. Interfaces* **2020**, *12*, 4895–4905. [CrossRef] [PubMed]

56. Song, P.; Qiu, H.; Wang, L.; Liu, X.; Zhang, Y.; Zhang, J.; Kong, J.; Gu, J. Honeycomb structural rGO-MXene/epoxy nanocomposites for superior electromagnetic interference shielding performance. *Sustain. Mater. Technol.* **2020**, e00153. [CrossRef]

57. Bian, R.; He, G.; Zhi, W.; Xiang, S.; Wang, T.; Cai, D. Ultralight MXene-based aerogels with high electromagnetic interference shielding performance. *J. Mater. Chem. C* **2019**, *7*, 474–478. [CrossRef]

58. Zhou, Z.; Liu, J.; Zhang, X.; Tian, D.; Zhan, Z.; Lu, C. Ultrathin MXene/Calcium Alginate Aerogel Film for High-Performance Electromagnetic Interference Shielding. *Adv. Mater. Interfaces* **2019**, *6*, 1802040. [CrossRef]

59. Zhao, S.; Zhang, H.B.; Luo, J.Q.; Wang, Q.W.; Xu, B.; Hong, S.; Yu, Z.Z. Highly electrically conductive three-dimensional Ti3C2Tx MXene/reduced graphene oxide hybrid aerogels with excellent electromagnetic interference shielding performances. *ACS Nano* **2020**, *12*, 11193–11202. [CrossRef]

60. Wang, L.; Qiu, H.; Song, P.; Zhang, Y.; Lu, Y.; Liang, C.; Kong, J.; Chen, L.; Gu, J. 3D Ti3C2Tx MXene/C hybrid foam/epoxy nanocomposites with superior electromagnetic interference shielding performances and robust mechanical properties. *Compos. Part A: Appl. Sci. Manuf.* **2019**, *123*, 293–300. [CrossRef]

61. Miao, M.; Liu, R.; Thaiboonrod, S.; Shi, L.Y.; Cao, S.; Zhang, J.; Fang, J.; Feng, X. Silver nanowires intercalating Ti3C2Tx MXene composite films with excellent flexibility for electromagnetic interference shielding. *J. Mater. Chem. C* **2020**, *8*, 3120–3126. [CrossRef]

62. He, P.; Cao, M.S.; Cai, Y.Z.; Shu, J.C.; Cao, W.Q.; Yuan, J. Self-assembling flexible 2D carbide MXene film with tunable integrated electron migration and group relaxation toward energy storage and green EMI shielding. *Carbon* **2020**, *157*, 80–89. [CrossRef]

63. Wan, Y.J.; Li, X.M.; Zhu, P.L.; Sun, R.; Wong, C.P.; Liao, W.H. Lightweight, flexible MXene/polymer film with simultaneously excellent mechanical property and high-performance electromagnetic interference shielding. *Compos. Part A: Appl. Sci. Manuf.* **2020**, *130*, 105764. [CrossRef]

64. Han, M.; Yin, X.; Wu, H.; Hou, Z.; Song, C.; Li, X.; Zhang, L.; Cheng, L. Ti3C2 MXenes with modified surface for high-performance electromagnetic absorption and shielding in the X-band. *ACS Appl. Mater. Interfaces* **2016**, *8*, 21011–21019. [CrossRef]

65. Liu, J.; Zhang, H.B.; Sun, R.; Liu, Y.; Liu, Z.; Zhou, A.; Yu, Z.Z. Hydrophobic, flexible, and lightweight MXene foams for high-performance electromagnetic-interference shielding. *Adv. Mater.* **2017**, *29*, 1702367. [CrossRef]

66. Zhang, H.; Zhang, G.; Gao, Q.; Zong, M.; Wang, M.; Qin, J. Electrically Electromagnetic Interference Shielding Microcellular Composite Foams with 3D Hierarchical Graphene-Carbon nanotube Hybrids. *Compos. Part A: Appl. Sci. Manuf.* **2020**, *130*, 105773. [CrossRef]

67. Zhu, S.; Cheng, Q.; Yu, C.; Pan, X.; Zuo, X.; Liu, J.; Chen, M.; Li, W.; Li, Q.; Liu, L. Flexible Fe3O4/graphene foam/poly dimethylsiloxane composite for high-performance electromagnetic interference shielding. *Compos. Sci. Technol.* **2020**, *189*, 108012. [CrossRef]

68. Cheng, K.; Li, H.; Zhu, M.; Qin, H.; Yang, J. In situ polymerization of graphene-polyaniline@ polyimide composite films with high EMI shielding and electrical properties. *RSC Adv.* **2020**, *10*, 2368–2377. [CrossRef]

69. Yang, X.; Fan, S.; Li, Y.; Guo, Y.; Li, Y.; Ruan, K.; Zhang, S.; Zhang, J.; Kong, J.; Gu, J. Synchronously improved electromagnetic interference shielding and thermal conductivity for epoxy nanocomposites by constructing 3D copper nanowires/thermally annealed graphene aerogel framework. *Compos. Part A: Appl. Sci. Manuf.* **2020**, *128*, 105670. [CrossRef]

70. Xu, Y.L.; Uddin, A.; Estevez, D.; Luo, Y.; Peng, H.X.; Qin, F.X. Lightweight microwire/graphene/silicone rubber composites for efficient electromagnetic interference shielding and low microwave reflectivity. *Compos. Sci. Technol.* **2020**, *189*, 108022. [CrossRef]

71. Liu, Q.; He, X.; Yi, C.; Sun, D.; Chen, J.; Wang, D.; Liu, K.; Li, M. Fabrication of ultra-light nickel/graphene composite foam with 3D interpenetrating network for high-performance electromagnetic interference shielding. *Compos. Part B: Eng.* **2020**, *182*, 107614. [CrossRef]

72. Zhao, B.; Zhang, X.; Deng, J.; Zhang, C.; Li, Y.; Guo, X.; Zhang, R. Flexible PEBAX/graphene electromagnetic shielding composite films with a negative pressure effect of resistance for pressure sensors applications. *RSC Adv.* **2020**, *10*, 1535–1543. [CrossRef]

73. Yu, C.; Zhu, S.; Xing, C.; Pan, X.; Zuo, X.; Liu, J.; Chen, M.; Liu, L.; Tao, G.; Li, Q. Fe nanoparticles and CNTs co-decorated porous carbon/graphene foam composite for excellent electromagnetic interference shielding performance. *J. Alloys Compd.* **2020**, *820*, 153108. [CrossRef]

74. Jia, H.; Kong, Q.Q.; Liu, Z.; Wei, X.X.; Li, X.M.; Chen, J.P.; Li, F.; Yang, X.; Sun, G.H.; Chen, C.M. 3D graphene/carbon nanotubes/polydimethylsiloxane composites as high-performance electromagnetic shielding material in X-band. *Compos. Part A: Appl. Sci. Manuf.* **2020**, *129*, 105712. [CrossRef]

75. Chen, Y.; Li, Y.; Yip, M.; Tai, N. Electromagnetic interference shielding efficiency of polyaniline composites filled with graphene decorated with metallic nanoparticles. *Compos. Sci. Technol.* **2013**, *80*, 80–86. [CrossRef]
76. Zhang, W.; Wei, L.; Ma, J.; Bai, S.L. Exfoliation and defect control of graphene oxide for waterborne electromagnetic interference shielding coatings. *Compos. Part A: Appl. Sci. Manuf.* **2020**, *132*, 105838. [CrossRef]

MDPI

St. Alban-Anlage 66

4052 Basel

Switzerland

Tel. +41 61 683 77 34

Fax +41 61 302 89 18

www.mdpi.com

Nanomaterials Editorial Office

E-mail: nanomaterials@mdpi.com

www.mdpi.com/journal/nanomaterials